Cambridge Imperial and Post-Colonial Studies Series

General Editors: **Megan Vaughan**, King's College, Cambridge, and **Richard Drayton**, King's College, London

This informative series covers the broad span of modern imperial history while also exploring the recent developments in former colonial states where residues of empire can still be found. The books provide in-depth examinations of empires as competing and complementary power structures encouraging the reader to reconsider their understanding of international and world history during recent centuries.

Titles include:

Miguel Bandeira Jerónimo
THE 'CIVILISING MISSION' OF PORTUGUESE COLONIALISM, 1870–1930

Miguel Bandeira Jerónimo and António Costa Pinto
THE ENDS OF EUROPEAN COLONIAL EMPIRES
Cases and Comparisons

Gregory A. Barton
INFORMAL EMPIRE AND THE RISE OF ONE WORLD CULTURE

Rachel Berger
AYURVEDA MADE MODERN
Political Histories of Indigenous Medicine in North India, 1900–1955

Ulbe Bosma and Anthony Webster
COMMODITIES, PORTS AND ASIAN MARITIME TRADE SINCE 1750

Rachel Bright
CHINESE LABOUR IN SOUTH AFRICA, 1902–10
Race, Violence, and Global Spectacle

Larry Butler and Sarah Stockwell
THE WIND OF CHANGE
Harold Macmillan and British Decolonization

Esme Cleall
MISSIONARY DISCOURSE
Negotiating Difference in the British Empire, c.1840–1895

T. J. Cribb *(editor)*
IMAGINED COMMONWEALTH
Cambridge Essays on Commonwealth and International Literature in English

Bronwen Everill
ABOLITION AND EMPIRE IN SIERRA LEONE AND LIBERIA

Anna Greenwood and Harshad Topiwala
INDIAN DOCTORS IN KENYA, 1890–1940

Róisín Healy and Enrico Dal Lago *(editors)*
THE SHADOW OF COLONIALISM IN EUROPE'S MODERN PAST

Leslie James
GEORGE PADMORE AND DECOLONIZATION FROM BELOW
Pan-Africanism, the Cold War, and the End of Empire

Robin Jeffrey
POLITICS, WOMEN AND WELL-BEING
How Kerala Became a 'Model'

Gerold Krozewski
MONEY AND THE END OF EMPIRE
British International Economic Policy and the Colonies, 1947–58

Zoë Laidlaw and Alan Lester (*editors*)
INDIGENOUS COMMUNITIES AND SETTLER COLONIALISM
Land Holding, Loss and Survival in an Interconnected World

Sophus Reinert and Pernille Røge
THE POLITICAL ECONOMY OF EMPIRE IN THE EARLY MODERN WORLD

Jonathan Saha
LAW, DISORDER AND THE COLONIAL STATE
Corruption in Burma c.1900

John Singleton and Paul Robertson
ECONOMIC RELATIONS BETWEEN BRITAIN AND AUSTRALASIA 1945–1970

Leonard Smith
INSANITY, RACE AND COLONIALISM
Managing Mental Disorder in the Post-Emancipation British Caribbean, 1838–1914

Alex Sutton
THE POLITICAL ECONOMY OF IMPERIAL RELATIONS
Britain, the Sterling Area, and Malaya 1945–1960

Miguel Suárez Bosa
ATLANTIC PORTS AND THE FIRST GLOBALISATION c.1850–1930

Jerome Teelucksingh
LABOUR AND THE DECOLONIZATION STRUGGLE IN TRINIDAD AND TOBAGO

Julia Tischler
LIGHT AND POWER FOR A MULTIRACIAL NATION
The Kariba Dam Scheme in the Central African Federation

Erica Wald
VICE IN THE BARRACKS
Medicine, the Military and the Making of Colonial India, 1780–1868

Anna Winterbottom
HYBRID KNOWLEDGE IN THE EARLY EAST INDIA COMPANY WORLD

Cambridge Imperial and Post-Colonial Studies Series
Series Standing Order ISBN 978–0–333–91908–8 (Hardback)
978–0–333–91909–5 (Paperback)
(*outside North America only*)

You can receive future titles in this series as they are published by placing a standing order. Please contact your bookseller or, in case of difficulty, write to us at the address below with your name and address, the title of the series and the ISBN quoted above.

Customer Services Department, Macmillan Distribution Ltd, Houndmills, Basingstoke, Hampshire RG21 6XS, England

Hybrid Knowledge in the Early East India Company World

Anna Winterbottom

British Academy Postdoctoral Research Fellow, University of Sussex, UK

palgrave
macmillan

First published 2016 by
PALGRAVE MACMILLAN

Palgrave Macmillan in the UK is an imprint of Macmillan Publishers Limited, registered in England, company number 785998, of Houndmills, Basingstoke, Hampshire RG21 6XS.

Palgrave Macmillan in the US is a division of St Martin's Press LLC, 175 Fifth Avenue, New York, NY 10010.

Palgrave Macmillan is the global academic imprint of the above companies and has companies and representatives throughout the world.

Palgrave® and Macmillan® are registered trademarks in the United States, the United Kingdom, Europe and other countries.

ISBN 978–1–137–38019–7

This book is printed on paper suitable for recycling and made from fully managed and sustained forest sources. Logging, pulping and manufacturing processes are expected to conform to the environmental regulations of the country of origin.

A catalogue record for this book is available from the British Library.

Library of Congress Cataloging-in-Publication Data
Winterbottom, Anna, 1979–
Hybrid knowledge in the early East India Company world / Anna Winterbottom (research fellow, University of Sussex, UK).
pages cm. — (Cambridge imperial and post-colonial studies series)
Summary: "Hybrid Knowledge in the Early East India Company World presents a new interpretation of the development of the English East India Company between 1660 and 1720. The book explores the connections between scholarship, patronage, diplomacy, trade, and colonial settlement in the early modern world. Links of patronage between cosmopolitan writers and collectors and scholars associated with the Royal Society of London and the universities are investigated. Winterbottom shows how innovative works of scholarship—covering natural history, ethnography, theology, linguistics, medicine, and agriculture—were created amid multi-directional struggles for supremacy in Asia, the Indian Ocean and the Atlantic. The role of non-elite actors including slaves in transferring knowledge and skills between settlements is explored in detail"—Provided by publisher.
ISBN 978–1–137–38019–7 (hardback)
1. East India Company—History. 2. East Asia—Intellectual life. 3. Great Britain—Intellectual life. 4. Knowledge, Sociology of—History. 5. Learning and scholarship—History. 6. Collectors and collecting—History. 7. Authors and patrons—History. 8. Royal Society (Great Britain)—History. 9. Great Britain—Relations—East Asia. 10. East Asia—Relations—Great Britain. I. Title.
DS465.W746 2015
954.02′94—dc23 2015021820

To the memory of my mother Sue and to my father David and sister Emily

Contents

Figures and Tables

Figures

Tables

Acknowledgements

I am very grateful to my PhD supervisors, Lisa Jardine and Felipe Fernández-Armesto, for their support in the earlier stages of this project. I owe particular thanks to Lisa for the inspiration to investigate the lives of Robert Knox and Samuel Baron and to Felipe for his help in achieving a global perspective.

I owe many thanks to Deepak Kumar for acting as my Research Guide during my affiliation with Jawaharlal Nehru University and for his advice about my work and help identifying materials in Indian archives. I also am grateful to Domanik Wujastyk for allowing me to sit in on his MA class at UCL's Wellcome Trust Centre for the History of Medicine. Simon Schaffer and Colin Jones provided invaluable feedback and encouragement during the viva and have been unfailing sources of support ever since.

I acknowledge with thanks funding support for my early research from the Wellcome Trust and Royal Society and, for my research in St Helena, from the Central University of London Research Fund. I have received postdoctoral funding from the Indian Ocean World Centre at McGill under the Major Collaborative Research grant provided by SSHRC and from the British Academy. Both grants enabled me to build on the earlier work I did as well as exploring new directions. While at Sussex, I have been extremely grateful for the kind and inspiring academic guidance and friendship of Vinita Damodaran. At McGill, Gwyn Campbell and several other colleagues offered invaluable help and advice.

I am grateful to the archivists, librarians, curators, and research associates of all the institutions that I visited, in particular Antonia Moon of the British Library, Keith Moore, Rupert Baker, Christine Wollett, and Joanna Corden of the Royal Society, Lacosta McDaniel and Tracy Buckley of the St Helena Archives, and Mark Spence, Charlie Jarvis, and Vilma Bharatan of the National History Museum.

Parts of Chapter 1 have been published as an article in *Journeys*. I am very grateful to the editors of this special edition, Eva Johanna Holmburg and Chloë Houston, for their input. Chapter 2 contains elements of a forthcoming co-authored chapter with Samuli Kaislaniemi and benefited enormously from his insights and findings. Parts of Chapter 4 appeared as a chapter in the work *The EIC and the Natural World*, edited by Vinita Damodaran, Alan Lester, and me. I am grateful to my co-editors for their comments on this piece. Chapter 5 is a revised and expanded version of an article published in the *British Journal for the History of Science*; I owe thanks to the journal editors and reviewers for their help in improving it. Parts of Chapter 6 appeared in *Maritime History as Global History*, and I would like to thank Maria Fusaro and Amelia Polonia for their feedback on this piece.

During the writing of this book, very many friends and colleagues provided invaluable support and encouragement, kindly shared their own work and commented on mine on the way. I am very grateful to Pratik Chakrabarti for sharing his own work ahead of publication and offering very supportive comments on mine; James Delbourgo for very close readings and detailed comments on several chapters; Samuli Kaislaniemi for providing a constant stream of useful and often amusing information as well as reading much of this work and proving indispensable to writing parts of it; Minakshi Menon for close readings of several chapters and friendly advice dispensed over many cups of coffee and pastries; William Poole for many fruitful discussions of Thomas Hyde, Robert Hooke, and much else; and Philip J. Stern for constructive comments and forgiving me for the paella. I'm very grateful to Emily Gray for her help in proofreading early versions.

For being kind enough to read various parts of this work at different stages and offer invaluable comments, sharing their own work and directing me to other sources, or inviting me to present work and providing convivial company at conferences, many thanks are due to Weichung Chen, Albane Forestier, Joshua Herr, Ayesha Irani, M. C. M. Iqbal, Jagjeet Lally, Daniel Margóscy, Vikas Rathee, Samantha Shave, Sujit Sivasundaram, Olivia Smith, and Jenni Thomas.

Without Rohan and Ari, this book might never have been finished.

Note on Transcription and Transliteration

In citations from manuscript material throughout, I have followed the principle of retaining the original spelling and punctuation while adapting the manuscript for printing.[1] Line endings have not been retained. I have silently expanded such common abbreviations as 'wch' for 'which' and placed the missing letters of less common abbreviations in square brackets or footnotes. Superscript letters have been lowered. 'Þ' has been silently expanded to 'th', and the long 's' has been modernised as has the usage of 'u' and 'v' and 'i' and 'j'. Marks over vowels to indicate double letters have been removed and the letters indicated inserted and 'ff' has been changed to 'F'.

For the convenience of general readers, diacritics have been omitted in the transliteration of non-European languages except in direct quotations from other sources and in references.

Introduction: Patronage and the Politics of Knowledge

Hybrid knowledge

This book investigates the creation and movement of knowledge – of people and their customs, of objects, of languages, of plants for medicine and food, and of the topography of land and sea – in and between the settlements of the English East India Company (EIC) during the period 1660–1720. Knowledge is a term used to describe scholarship, to trade, and to personal relations. Knowledge can be transmitted in an oral exchange, a written work, a map or drawing, or an object. In the language of the day, the forms of knowledge I will discuss were classed as 'natural' and 'useful': in modern language, they are encompassed by a broad definition of science, technology, and medicine.[1] In each chapter, relationships between scholars who advised the EIC's Directors in London and those who made collections and descriptions in the settlements abroad are discussed. Many of these scholars were associated with the Royal Society of London and the universities of Oxford and Cambridge and several held governmental or advisory posts. While these men, and occasionally women, played vital roles as patrons, funders, and systemisers of information, the focus in this book is instead on the contexts in which knowledge was generated and on the people who collected information. Through this reorientation, I present a new interpretation of the commercial and scientific 'revolutions' of the period.

Each chapter presents a dedicated intellectual history of a particular work, collection of works, individual, or group of people, while also drawing more general conclusions about a particular genre of early modern scholarship. Chapter 1 examines the genre of geographical and ethnographic accounts, known as 'natural histories', focusing on Samuel Baron's *Description of the Kingdom of Tonqueen* (1686). The text is read in the context of the life of the author; mestizo agent, spy, broker for the EIC, and client of the rulers of Siam and Tonkin as well as collector for the Royal Society. Chapter 2 focuses on linguistic scholarship using the example of English scholarship on the Malay language. English studies of Malay moved from wordlists and manuscript grammars composed during the 1680s to the first bilingual dictionary in English and an Asian language, the *Dictionary English and Malayo, Malayo and English* (1701), published under the name of the trader Thomas Bowrey. Chapter 3 focuses the encounter of Company servants with Hinduism and the practical and intellectual impact of religious toleration in the Company settlements. I examine the translations into English of the *Bhagavata Purana* and the *Samaveda*, produced in Kasimbazar between 1675 and 1677 by the EIC factor John Marshall, based on the oral account of the Brahmin Madhusudana Rarhi. A dialogue in which the two participants discuss their theological beliefs in philosophical terms is also presented. Chapter 4 places the rapidly growing city of Madras at the centre of international networks of exchange of *materia medica* based on the botanical collections (*hortus sicci*) and correspondence of two EIC surgeons and their Tamil and Telugu collaborators. Chapter 5 follows the movement of a travel narrative and 'natural history', Robert Knox' *Historical Relation of Ceylon* (1688), with its author on a series of later voyages, where it was used as the basis of bio-prospecting and agricultural experimentation. Finally, Chapter 6 focuses on the contribution of slaves to the early EIC settlements as well as their resistance. This chapter returns to the sense of knowledge as embodied in people rather than in texts or objects, examining slaves' roles in the transmission of botanical, medical, agricultural, linguistic, and cultural knowledge.

I use the term 'hybrid' to encapsulate what the diverse ways of knowing that I will explore have in common with one another: an admixture of information drawn from diverse sources drawn together to make something new. Hybrid is a word applied to animals, plants, and people. In its original sense, it also has an element of the wild. As Steven Shapin argues, both scientific and social knowledge are in some sense inevitably hybrid, since what we know about natural phenomena is always mediated through our knowledge of the people who describe them, while our understanding of people is conditioned by the explanations they provide for natural things.[2] Hybrid has been used elsewhere to describe institutions such as the botanical garden, which combine aims of science, pleasure, and commerce.[3] Here, the idea of hybridity is also intended to signify human *métissage*, the transformation of natural environments through the movement of plants and the interaction of individual and collective forms of knowledge and power. It also applies to the approaches I will take, which combine intellectual and political histories with cultural readings of texts, objects, pictures, and the lives of individuals.

The term 'hybrid' has also been used previously to describe global encounters, especially in the Atlantic context;[4] indeed, it has attracted some criticism for its ubiquity in this context.[5] Nonetheless, the concept of hybridity has been useful in moving the global history of ideas forward from an earlier language that relied on the concepts of centre and periphery and the assumption that both science and commerce were essentially European creations, exported and modified to a greater or lesser extent.[6] Notably, the term hybridity has seldom been applied to the world of the EIC.[7] Instead, the literature has remained focused almost exclusively on the EIC Court of Directors in London and on English elites abroad. The use of hybridity here is a deliberate challenge to this conventional approach. It is intended to suggest the possibility of translating knowledge across culture but also the clashes and disconnections involved in knowledge networks composed of communities whose interactions can be better classed as 'contained conflict' than as 'partnership'.[8]

The 'world' in which the EIC operated in the late seventeenth and early eighteenth centuries was one of shifting horizons. While the early Directors certainly possessed imperial ambition, their vision was of a series of interlinked island and coastal 'plantations' and trading stations rather than of large swathes of territory inland. This conceptualisation of empire followed previous European imperial experiments in both the Americas and the Indian Ocean. The geographical boundaries of the EIC's world also advanced and retracted based on the political entanglements of each region within which they operated, spilling out of the Indian Ocean and into the South China Seas to the East and the Atlantic to the West, influenced by and interacting with the regional dynamics of areas such as the Bay of Bengal. Each chapter of this book opens with a local history of the settlement or settlements in question, with the aim of understanding how these particular histories informed the knowledge that was produced. Each place is also presented as a node in the overlapping political, trading, and cultural networks formed not only by the EIC but also by its regional and European competitors. The aim is not to present a comprehensive history of the Company settlements – for example, those in western India and Persia are not discussed in detail here – but to achieve an understanding of the early EIC as a fluid, networked, entity.[9]

The chapters in this book travel from East to West, calling at settlements that would have been familiar to most EIC ship crews. They aim to represent the changing geographical orientation of the EIC network over time in the context of regional political dynamics. Chapter 1 deals with East Asia and mainland Southeast Asia, focusing on the efforts of members of the EIC to follow the Portuguese and rival the Dutch in establishing themselves in the lucrative trade between China and Japan during the 1670s and 1680s by establishing bases in Vietnam, Taiwan, Siam and Southern China. Chapter 2 focuses on English efforts to rival the Dutch in Maritime Southeast Asia by establishing settlements in Java and Sumatra from the early seventeenth century and the shifting of English attention eastwards to the South Sea Islands by the early eighteenth. Chapters 3 and 4 both concentrate on eastern India. Chapter 3 deals with the Company's first attempts to establish their 'government' in Bengal during the 1670s and 1680s in the wake of Mughal conquest and Portuguese retreat. Chapter 4 is centred on the diplomatic contacts between Madras and the Mughal Empire, from the abortive conflict of the 1680s up until the more stable period of the 1710s and 1720s, amidst the larger context of the Mughal-Maratha struggle for supremacy in the Deccan. Chapters 5 and 6 have oceanic settings: both examine EIC efforts to rival the Dutch and later French in establishing productive island plantations. Chapter 5 moves from the large Indian Ocean island of Ceylon to the smaller Atlantic islands of St Helena and Cape Verde, and Chapter 6 also examines St Helena, in the light of its connections to West Africa, Madagascar, and the Company settlements in Southeast and South Asia as well as the Atlantic world.

Through an analysis of the circulation of knowledge between these early EIC settlements, this book argues that the seventeenth and early eighteenth centuries were the formative and, in some senses, the most 'revolutionary' periods of Enlightenment scholarship in the early colonial settlements as well as in Europe. Locating the discussion of both scientific and commercial revolutions within a networked understanding of empire allows for a nuanced discussion of the extra-European sites in which knowledge was created, a wider recognition of the range of actors involved in the production of knowledge, and an examination of the functioning of patronage on local and global scales. It also reveals that the processes of codification and redeployment of knowledge were central to the global political interactions of the period, including the process of colonisation, but the ways in which information circulated and was used were determined and limited by external factors, including violence and forced migration.

Revolutions and reorientations

The period 1660–1720 was a time of global flux, intellectual and social. The early seventeenth century has often been described in terms of 'crisis' in parts of Europe and Southeast and East Asia, with falling temperatures ('the Little Ice Age') prompting crop failure; declining populations; stagnating trade; conflict within and between states; and religious and political change, in some cases revolutionary change, as in the conflicts known as the British Civil Wars.[10] Elsewhere, it was a time of prosperity, notably in much of South Asia, where the Mughal Empire was exerting its authority over an ever more vast area[11] as part of a wider move towards centralisation and consolidation of large states across much of the early modern world.[12] By the second decade of the eighteenth century, the tide was turning. Mughal authority had been decimated by a series of succession wars from the death of the Emperor Aurangzeb in 1707. Meanwhile, the great empires of China and Japan, while remaining engaged with the world stage, were closing their harbours to the majority of foreigners and often forbidding their subjects from most foreign trade and maritime voyaging.[13] The period in between, roughly 1660–1720, was one of conflict and creativity, in which centralising states across the world competed for valuable resources and strategic alliances.

The European scholarship of this period – that of the so-called 'scientific revolution' – has been described as 'intoxicating, intellectual and spiritual upheaval': arguably the most radical period of investigation and experimentation in the intellectual movements that have been grouped together under the rubric of Enlightenment.[14] If scholarly trends elsewhere in the world have less often been described in such terms, this might be ascribed to a number of factors, including the earlier initiation of certain scholarly trends in Asia than in Europe, a lesser emphasis on the value of novelty in some cultures than others and an awareness on the part of commentators looking back of the active suppression of many means of producing and deploying knowledge during the colonial period that followed.[15] Nonetheless, and despite the long lack of recognition, a host of recent studies have demonstrated that the scholarly trends that have been considered definitive of 'early modernity' – from voyaging,[16] to collecting (of plants, animals, people, and 'curiosities'),[17] the development of systems of classification, the methodisation of agriculture, calendric reform, mapping, and medical and ethnographic investigations[18] – took place worldwide. While debate remains over questions of periodisation[19] and method,[20] the claim for a 'global early modernity'[21] has gained widespread currency, alongside similar claims for a global understanding of the Renaissance and Enlightenment.[22] Alongside a comparative approach demonstrating parallel trends in the development of knowledge systems, an argument has been made by Sanjay Subrahmanyam and others for writing 'connected histories'.[23] Recent approaches to tracing global connections have included a focus on individual lives[24] and micro-histories of particular works[25] or objects.[26]

An important reason for the designation of the 'early modern period' at the origin of the world of today is because the roots of modern forms of empire, capitalism, and science were put down during this time. In keeping with the re-imagining of early modernity as an era of global exchange, recent studies of capitalism and of natural philosophy have moved away from earlier models that imagined these ideologies as having been created in northwest Europe and exported fully formed to the rest of the world. Instead, such ideas are understood as being formed in their modern incarnations through the process of global encounters, while their claims to be new and unique products of Enlightenment thought have been challenged by demonstrations of the non-European roots of many of the ideas and technologies of the period.[27] The focus of 'new' imperial histories on both local histories and networked global exchanges has parallels with methods in both the history of science and environmental history.[28]

European trading companies are often hailed as the forerunners of modern corporations as well as the ancestors of colonial states, while the 'natural philosophy' developed in the seventeenth century is seen as providing many of the founding principles of modern science.[29] Recent accounts of the Vereenigde Oostindische Compagnie (VOC) and Dutch colonial world adopt a perspective on empire that undermines centre–periphery models and emphasises the creation of imperial identities in the course of contact. Leonard Blussé's study of the VOC's Asian capital of Batavia (Jakarta) places the acculturation of imperial agents and their uneasy coexistence with Chinese merchants at the centre of his characterisation of the early VOC. A number of in-depth studies of the VOC settlements in their immediate cultural contexts have been followed.[30] Connections between the various VOC settlements have also been studied from below. For example, Marcus Vink and Kerry Ward use forced migration to demonstrate the dynamic and contested evolution of the VOC's sovereignty forged between the various Dutch settlements in the Indian Ocean.[31]

The relocation of the making of Dutch identity within an imperial context has allowed for new approaches to scholarship in the 'Dutch Golden Age'. Notably, Harold Cook's work explores the connection between Dutch Empire and the practice of natural philosophy, demonstrating the close connections between the VOC and Dutch scholars in terms of the movement of botanists, physicians, and naturalists as well as plants, medicines, and specimens between the overseas settlements and the Dutch Republic. Cook's work also highlights the close connections, intellectual and financial, between the objects of scientific and mercantile interest, which he describes as the 'passions of body and mind united'.[32]

The conceptualisation of information as produced in multiple sites by a number of actors and circulating through networks of informers around sites of colonial settlement as well as to and from Europe has affinity with the recasting of the British Atlantic world as a 'kaleidoscope movement of people, goods, and ideas'.[33] Scholarship on the interactions that resulted has moved from the concept of a bounded 'contact zone' to understand new hybrid cultural practices emerging from multiethnic, cross-class encounters in multiple locations and situations, in which power relations differed.[34] Examining Atlantic history from below has generated the understanding that capitalism was founded not solely in the transformation of European financial institutions, but in the mobile, multiethnic body of labour that challenged imperial visions of order.

This re-conceptualisation of the Atlantic world paved the way for studies of non-elite contributions to the environmental and medical landscape of the Americas.[35] The neglect of the Iberian experience in the Americas has been corrected by new studies of the networks formed between indigenous medical practitioners, Creoles, slaves, and others, as well as the role of institutions such as the Casa de la Contratación in Spain, formed to collect and test new materials and techniques and despatch lists of queries around the world.[36] This allowed for a recontextualisation of scientific and technical practices. For example, recent studies of mapping in the context of the early modern new world have located it in the context of native American as well as European techniques and spatial concepts.[37] Meanwhile, studies of missionary contributions to botanical enterprise and geographical and ethnographic descriptions have been used to illustrate the formation of knowledge networks based on factors other than nationality and to direct the discussion away from national histories of empire and natural philosophy towards global histories of collecting and exchange.[38] Recent studies have demonstrated the integration of many Jesuits into the pan-European exchange of ideas known as the 'Republic of Letters'.[39] Early Portuguese observations of social, medical, and religious life in India have also received detailed attention.[40]

The EIC is the best known of the early modern trading companies, having eclipsed its Dutch rival in terms of wealth and influence after the mid-eighteenth century. Meanwhile, the early Royal Society and well-known members like Robert Boyle (1627–1691) and Isaac Newton (1642–1727) are often taken to epitomise the 'scientific revolution', notwithstanding recent debates over the validity of this term.[41] Nonetheless, in contrast to the wider trends within the literature, studies of these two emblematic organisations have remained almost exclusively focused on London and on the 'gentlemanly' pursuit of natural philosophy and commerce. The next sections will discuss the reason for this conceptual divergence and suggest some ways in which the origins of science and colonialism in the English context could be re-conceptualised, including by borrowing from recent scholarship relating to imperialism and natural philosophy in other contexts.

Power and politics in the early EIC settlements

Histories of the EIC normally begin with the granting of the charter of 1600 by Elizabeth I and recommence with the Company's reformulation afterwards in April 1661, when the restored Charles II confirmed Cromwell's 1657 charter. The EIC is traditionally regarded as undergoing a transition from an organisation solely concerned with trade to an entity concerned with the rule of large tracts of territory after the Battle of Plassey in 1757. As Philip Stern points out, these conceptual stages have been mapped onto the larger theoretical division of imperialism, consisting of a 'first', primarily maritime, empire based on settlement in the American plantations in the West and 'adventuring' and trade in the East and a 'second' empire of colonial power in Asia.[42]

With few exceptions, histories of the EIC take as their starting point the formal structure of the organisation headquartered in London.[43] There are good reasons for doing so. The Company's organisation was hierarchical: so that the format of the Director assisted by the Court of Committees (or after 1709, the Court of Directors) and a series of subcommittees was replicated in each of the settlements, each of which were governed by 'factories', whose affairs were deliberated by a Council, presided over by a Governor or President. Smaller, less well established 'factories' reported to larger and more secure establishments, which reported in turn to the Directors. The documentation this process produced was immense, famously occupying several miles of shelf space in the India Office, and now in the British Library. The Company produced its own historians with the appointment of Richard Hakluyt (c. 1552–1616) as official historiographer from the first years of its existence onwards.[44] Members of the Company were also avid pamphleteers and polemicists, including Josiah Child (1630–1699): the Company's Director during the 1680s and author of a number of tracts on political economy.[45]

Challenges to the London-centric approach to the EIC came first from economic history. The work of K. N. Chaudhuri, S. Arasaratnam, Michael Pearson, Kenneth McPherson, and others placed the European involvement in Asia within the wider perspective of Indian Ocean history.[46] Focusing on the interconnections between the European and indigenous trade during the 'age of partnership' or 'contained conflict' exploded myths about the 'peddling' nature of inter-Asian and Indian Ocean trade before the period of European expansion and the theory that the decline in this trade was owing to European dominance of the market. These studies also demonstrated the importance of the involvement of the Company in inter-Asian or 'country' trade, both officially and as a result of the private trade of its members. The period between c. 1600 and 1757 was thereby recast as one of 'considerable indigenous political, cultural, and economic life, and a continual jostling between locals and intruders for a share of traditional commerce, not an age marked by the death throes of indigenous enterprise'.[47]

Miles Ogborn and Philip Stern have recently led a reinterpretation of the early EIC from the viewpoints of cultural geography and political history. Both focus closely on the practices of writing and representation of the early Company.[48] Ogborn examines the function of the Company records in creating a conceptual division between public and private and establishing a sense of not only social but also moral order in the minds of the early Company employees.[49] Stern sees the 'Company State' as a veritable leviathan of letters, essentially creating its legitimacy by its appropriation and deployment of stately forms of writing.[50] Stern provides an important reinterpretation of the early Company as a political entity, challenging many of the traditional narratives about the development of the British Empire and questioning distinctions between the public and private and the commercial and imperial. Traditional distinctions between centre and periphery in Company history have also been challenged by Ogborn's emphasis on the global movement of texts[51] and by Stern's point that the Company settlements in India styled themselves much like the Siddis of Janjira: as small tributary polities in relation to the Mughal Empire.[52] Bhavani Raman's recent study takes a different view of the function of writing within the early-nineteenth-century Company settlement of Madras, showing that the impression of legitimacy generated by the mountains of paperwork could be used to blame institutionalised forms of exploitation and malpractice on those lower down in the hierarchy of writers.[53]

These valuable economic and political re-evaluations have been based on close analysis of the records of the early modern EIC. However, there remains a danger of mistaking the EIC's official representation of itself for reality and for overestimating the control of the London Directors over events on the ground. Chaudhuri's systems theory approach led him to conclude that 'Anyone who reads through the general letters written by the Court to the officials in the Indies cannot fail to be struck by the remarkable consistency of the Company's aims and methods'.[54] Stern similarly claims that 'Company leadership were guided by a coherent, if composite, set of political ideas about the duties of rulers and the nature and extent of political authority'.[55] Such an argument leads almost inevitably back to an understanding of the EIC as a political entity centred on London. Even when they focus on the settlements, studies of the early EIC continue to concentrate on elites and their metropolitan links.[56] In this sense, the historiography of the early EIC remains out of step with recent studies of other early modern empires discussed above.

Examining writings generated in, around, and about the Company will be important to this book as well. The aim here is to move beyond the focus of previous studies on official records and widen the scope of analysis to examine the large number of unofficial documents carried on each EIC ship. These writings were produced by people at all social levels of each settlement, from the Governors to the writers, free merchants of all nations and, on occasion, even the slaves. They included petitions and statements directed to the Court of Committees complaining of mismanagement of the settlements or mistreatment of their inhabitants by the Governor or Council and private letters, gifts, and works of scholarship directed to particular correspondents within the hierarchy of the EIC or among its circles of advisors. These materials flowed through networks established on the basis of relations of family, faith, or mutual interest.[57] Their contents often determined the contents of the official records in terms of which individuals were favoured and which policies were pursued. Working with these materials provides an insight into the complex politics of patronage that lay beneath the impression of hierarchical order produced by the EIC's charters and the deliberations of its London Directors.

The traditional framing of the EIC according to its charters and the deliberations of the London Directors means that the impact of the revolutionary period between 1642 and 1660, which immediately precedes the time span discussed in this book, is seldom considered. Following the outbreak of the British Civil Wars in 1642, the Company essentially ceased to exist.[58] The subsequent removal of King Charles I invalidated the Company's charter and licensed competitors during the Commonwealth period included the Courteen Association and Assada merchants, who merged with the original Company in 1650. From 1653, the joint stock lapsed entirely, and by 1656, the remaining Directors were planning a sale of all the Company's rights and properties for £14,000, until the chartering of a new company in 1657.[59] This is normally referred to in the literature of the period as the time of 'interloping' or free trade.

The neglect of this period in the history of the EIC stands in stark contrast to the debate begun by Peter Linebaugh and Marcus Rediker over the effects in the Atlantic world of the explosion of colonists, labourers, sailors, pirates, and slaves, many of them schooled in the revolutionary ideas of the period.[60] In fact, radicals migrated East as well as West: Shafa'at Khan notes that 'many of the Cromwellian soldiers went out to India' and quotes John Evelyn's comment that the refusal of some members to take the oath prescribed by the EIC's royal charter of 1660 was due to 'the "Anabaptists", whose object was to continue their private trade to the East'.[61] When the Company received its new charters, the men chosen to people the factories were those who had already established themselves abroad. This was necessary not only to avoid the very high death rate among new arrivals in the tropics, but also in order to capitalise on the local connections of these men.

Rather than the EIC being ruled by a coherent vision of authority, in the half-century following the civil conflict in Britain, strands of monarchical, oligarchic, and democratic thought were fought out in and between the settlements. An example from the beginning of the period is Sir Edward Winter's seizure of Fort St George, the Company settlement in Madras, in 1665. According to his own account, Winter had deposed Governor George Foxcroft on account of the 'treasonable statement' made by the latter's son. In what sounds like a crude restatement of Hobbesian contract theory, young Foxcroft had apparently claimed that 'he was bound to obey and Serve the King noe longer than the King could protect him'.[62] After Foxcroft had been deposed in the name of the English King, an assembly of EIC servants in Madras took a vote to ask

Winter to resume the government.[63] While this assembly conceded that they were not permitted under the EIC charter to elect their leader, they considered it necessary in view of the perceived threat from the Dutch in collusion with the Kingdom of Golconda, in whose territory the Madras factory was at the time. Winter's ability to influence this situation was swiftly demonstrated by a translated agreement between Mirza Malik Beg, *Diwan* of Orissa, and a Governor of the VOC, who agreed that the Dutch company would not attack or interfere with English ships and would make reparations in case of any transgressions, because 'this is the King [of Golconda]'s Country and nobody can fight here'.[64] Winter's ability to influence local politics was acquired during the time of 'interloping', when he built up a private estate around Viravasaram.[65] The debate over the case in England occupied almost an entire year before ships were sent to remove Winter in January 1668, who surrendered but was permitted to continue living in the settlement without punishment.[66]

As this incident demonstrates, the inability of the London Directors to control events on the ground was due not only to the motley collection of Englishmen they assembled to people the settlements but, more importantly, to the situation these men found themselves in. In this example, Winter's faction represent themselves variously as loyal subjects of the English king and as a quasi-democratic body capable of electing their own leader, in direct contravention of Company policy, but lastly and most decisively as vassals of the ruler of Golconda. Although many previous studies have stressed the difficulty that the Directors had in enforcing their dictates over their settlements, most neglect the basic fact that even the more established settlements such as Madras were constantly at risk of violent eclipse by either the Mughal Empire or regional powers: in the case of Madras in this period, the kingdom of Golconda. The fragile networks of settlements spanning South Asia and Southeast Asia and making occasional inroads into East Asia and Africa that made up the majority of the Company's trading posts in this period were still less able to determine their own fates. Composed of a handful of often sick and disorientated men, many of them enslaved, these settlements were completely dependent on local assistance: linguistic, diplomatic, military, and financial. In many cases, the Company's agents were themselves being used as go-betweens, brokers, or facilitators in local or regional power struggles. This meant that the settlements often acted in ways that were entirely contrary to the intentions of the Company's London Directors, at times even competing rather than cooperating with one another.

As Colin Newbury notes: 'it is abundantly clear that initial contact through maritime trade left the East India Company in a condition of dependency on Indian rulers, brokers, and financiers until the last decades of the eighteenth century'.[67] The impression that the EIC was interested only in trade in their early years has been justly corrected.[68] However, until the factors were able to negotiate or force permission to fortify their settlements and to defend them with armies of slaves and mestizo soldiers – and, as will be argued below, this aim was central from the beginning – it was local rulers and not the Company in London who ensured their safety from local conflicts and judged conflicts between the competing European powers as well as enabling their trade. Thus, the history of the Company at the beginning of this period was directed more by the local political realities in which each settlement was embedded than by commands from and events in distant London.

While having their employees in the service of foreign rulers was dangerous for the Company hierarchy, it was also necessary in order for them to successfully negotiate the settlements' security. Local knowledge was also vital to the ability of their servants to provide them with useful information: commercial, military, and cultural. At the same time, the EIC in London needed to limit and circumscribe these connections in order to retain the allegiance of their servants. At the beginning of the period this was difficult to achieve, and the Company Directors often resorted to breaking their own rules of hierarchy and ordering servants whose actions became problematic to be violently removed by their subordinates. Such men were often denounced as having been involved in private trade, 'interloping', or general immorality. For example, despite being employed by the EIC at the time, Edward Winter had previously been denounced as an interloper after a personal conflict with Mir Jumla[69] became problematic. On this occasion, the Company in London ordered members of the Masulipatnam (Machilipatnam) factory to remove Winter by ambushing him aboard ship.[70] However, all efforts to remove him failed, and he later returned to London on his own terms and was knighted.

While the Company portrayed interlopers as a separate group of independent traders, an impression that endures in later scholarship,[71] the reality was far more complex. Who was inside and outside the Company at any one time was dependent on perspective, locality, and expediency. The success of Company settlements continued throughout this period to rely on recruiting traders who had established useful local connections. A good example is Sir Thomas Pitt, who a biographer compared to Jekyll and Hyde on the basis of his contrasting images as a 'swaggering desperado' and as a respected Governor of Fort St George in Madras and English member of parliament – a biographer compared to Jekyll and Hyde.[72] Denounced repeatedly by Company officials during his successful career in India as a private trader from 1678, Pitt eventually became not only a key member of the Company, but also an important figure in English parliamentary politics. Pitt's effectiveness as an EIC Governor in India was in a large part due to the same local connections that had made him a dangerous rival.

The agents of the 'new' East India Company arrived in India in 1699. Presenting gifts and 'curiosities' to the Mughal Emperor Aurangzeb, Ambassador William Norris denounced the old Company as 'Thieves & Confederates with the Pirates'.[73] Nevertheless, this 'new' organisation – in fact formed mainly from disaffected employees of the 'old' – followed the pattern of its predecessor in recruiting its servants from among those who had previously served in Asia. Despite their commissions from London, 'new' Company agents usually failed to unseat their rivals. As the agent John Beard pointed out to his challenger at Balasore, a commission from the English King meant little without the collection of grants and alliances with Indian rulers held by members of the old Company.[74] Despite imprisoning particular members of the 'old' Company for piracy, Aurangzeb refused to favour the 'new' Company, stating that his ports were open to all.[75] The 'civil war' between the two Companies was conducted in courts and pamphlets in England, but by a truly violent struggle for supremacy on the ground.[76] Once again, those with strong local connections were triumphant. For example, despite the attempts of his cousin, John Pitt, to unseat him, Thomas Pitt retained control of Fort St George under the 'United' Company that eventually resulted from this period of conflict.[77]

In a revealing passage of a letter to London by the 'new' Company's Surat factory, the agents complain with reference to the proposed formation of the United Company that the aim of 'securing the trade from those untoward Hydras of mischief…residing in these parts…may have been glossed and credited by interested men'.[78] In other words, the formation of the new Company in London had failed to slay the alarming 'hydra' formed by those Englishmen already established in Asia. The echo here of the same language that was used to express elite fears about the revolutionary Atlantic world is no accident, but reflects the reality that the EIC Directors in London consistently failed to control events in their settlements before the second decade of the eighteenth century. Instead, they intervened when possible using force or persuasion, or compromised with those who gained power on the ground. Often, like Pitt and Winter, those men who became powerful in the settlements also came to assert influence in London and beyond through an extension of the same networks of patronage they were part of in Asia.

To summarise, the half-century following the re-establishment of the monarchy in England saw a complex struggle to control the sources of knowledge and hence wealth and power in the overseas settlements in Asia as well as the Americas. The alienation of monarchical powers to the trading companies in theory gave them 'stately' power over all Englishmen in the East, a point which they laboured in their petitions to the crown, wining permission by 1686 to use force against both 'native princes and European interlopers'.[79] However, for most of this period, directives from London were insufficient to control the situation on the ground, where power was wielded on the basis of local relations of patronage established by individuals. When they could not be overthrown by force, such powerful individuals had to be comprised with. These men often ultimately became those who wielded power in London. This violent struggle for influence is the context within which the collection of information or 'useful knowledge' described in the chapters that follow should be viewed.

Useful knowledge and the early Royal Society

Like, studies of the early EIC, accounts of the 'scientific revolution' have traditionally focused on European capitals.[80] The Royal Society, or the 'Royal Society of London for Improving Natural Knowledge', which has often been the focus of such histories in the British context, received its first royal charter shortly after the EIC, in 1662.[81] The aims of scholarship and empire are connected within the charter, in which Charles II announces his intention to 'extend not only the boundaries of the Empire, but also the very arts and sciences'. The Society is named as a body 'corporate and politic' and is constituted, like the EIC, of a President and Council as well as its fellows.

In London, the EIC connected with the Royal Society both formally and informally, through overlapping membership and friendships and rivalries that spanned the organisations. From 1682, members of the Royal Society held stocks in the EIC, as well as the Royal African Company, on behalf of the Society.[82] Many prominent members of the Royal Society, including the diarists John Evelyn and Samuel Pepys, also held their own stocks in the Company. Members such as John Fryer, Isaac Pyke, and Joseph Collet were elected to the Royal Society on the basis of their positions in the EIC. The personnel of both the EIC and the

Royal Society also overlapped with members of the English parliament, court, and navy: prominent examples include Samuel Pepys, Sir Joseph Williamson, and William Petty.

The Royal Society shared many other features with the EIC, including its Baconian commitment to keeping detailed vernacular records of all meetings and correspondence. Both institutions kept libraries including numerous works of travel in other European languages and collections of material objects sent from abroad. In London, the EIC had a storehouse or museum of specimens sent to them in this period by their servants abroad. Though references to it are rare, it seems to have been a considerable collection according to a contemporary account of the visit of an Italian duke.[83] The Royal Society also had a 'repository' of curiosities, animal, vegetable, and mineral, sent to them from around the world, a collection that was eventually transferred to the British Museum.[84] It is clear that objects flowed freely between the collections of the two organisations, which in some cases overlapped. Between 1672 and 1675, the Royal Society shared Gresham College with some of the EIC's collections.[85] On occasion, the Royal Society was consulted by agents of the EIC to interpret some of these objects. This was the case in 1680, when the fellows were asked to determine the 'virtues', or useful properties, of a bezoar stone that had been presented to the EIC by the Sultan of Bantam, Abdulfatah Ageng.[86]

Like the EIC, the Royal Society was extremely concerned with shaping its own image: histories of the organisation appeared from Thomas Sprat's work of 1667 onwards, stressing the society's commitment to the experimental method and the generation of 'useful' knowledge in the pursuit of the goal of 'improvement'.[87] Histories of the Royal Society traditionally privileged experimentation as its core mission.[88] However, experiments often failed or proved inconclusive, making it necessary to rely on the reports of others. This was the case with the bezoar stone.[89] Despite efforts to test its alleged effectiveness against poison by administering it to two dogs that had been given *Strychnos nux-vomica*, the experiments proved inconclusive, and, in their report to the EIC,[90] the Society were forced to rely on the reports of accounts of its use in the East Indies and Swahili coast by Jean-Baptiste Tavernier and other travellers, and the account of an Italian virtuoso, Francesco Redi, of his own trials of the stone.[91]

The bias towards experimentation in the scholarship concerning the Royal Society might be redressed by taking note of an early statement about its purpose by its first secretary, Henry Oldenburg. His list of the aims and achievements of the organisation begins with 'Queries and

Directions, they have given abroad', followed by 'Proposals and Recommendations, they have made' and 'Relations, they have received'. 'Experiments they have tried' is given only fourth place in this list of priorities.[92] As Oldenburg implies, the functions of directing travel, collecting and translating works of travel and natural history and making recommendations on the basis of these activities were at least as important as experimentation to the early Royal Society, if not more so. Indeed, the practices were interrelated: although members sometimes travelled themselves, as with Edmund Halley's voyage to St Helena,[93] more often they relied on a network of foreign correspondents to perform experiments as well as record observations.[94] Experiments were often devised on the basis of earlier observations.

Famously, the Royal Society's motto is *nullius in verba*; often translated as 'take no one's word for it'. In fact, the early members often found themselves relying on the words of many others, including a host of unknown strangers. In this sense, the knowledge that the Royal Society assembled was thoroughly hybrid in its origins. At the same time, information was being presented in increasingly standardised ways, including through articles in the journal *Philosophical Transactions*, as well as being incorporated into natural histories, dictionaries, maps, tide tables, and astronomical charts.

Shapin's classic challenge to the Royal Society's claims to reject all knowledge based on authority encouraged a new focus in histories of science on the social basis of scientific practice.[95] His account argues that the exchange of natural philosophical information was based on trust, the social basis of which in the English context was the possession of gentlemanly status. This argument has found an interesting contrast in Cook's location of the scientific revolution in the Dutch context within the mercantile concepts of 'interest' and 'credit' and the emphasis on plain speaking shared by merchants and natural philosophers.[96] These differing approaches are in part due to the difference between an early modern England still dominated by the gentry in social if not economic terms and the merchant oligarchy that held sway in the Dutch Republic. Even in England, however, noble and bourgeois modes of exchange overlapped considerably,[97] and in their relations with the EIC, as well as in their emphasis on 'useful knowledge', repeated throughout the early records, the Royal Society undoubtedly also invested in mercantile concepts of value.

The deep connection of the Royal Society's understanding of what constituted useful knowledge to the projects of colonisation and international trade – including the slave trade – can be demonstrated by both the context and content of many of the early experiments. Numerous examples, including that of the bezoar stone mentioned above, are concerned with testing the value of products from the Indies – camphor, cinnamon and nutmeg to name just a few others – or understanding Indian Ocean technologies, from writing to dyes to weights.[98] Other experiments are illustrative of the Royal Society's connection with the Royal African Company. Early examples include an early trial of pendulum watches, conducted aboard what appears from its route to be a slave ship.[99] Robert Hooke's efforts in 1680 to counterfeit, using glass, a bead of blue stone, 'much esteemed by those of Guiny [Guinea]', also seem to have been directed towards the slave trade.[100]

Some more recent accounts of the Royal Society and similar institutions, such as its French counterpart l'Académie des sciences, have paid attention to collaborations between scholars and the agents of empire, on both theoretical and practical levels.[101] Studies of the scientific revolution have moved away from a sole focus on experimentation and the 'hard' sciences to envisage practices of collection, classification, and description as central to the new natural philosophy.[102] The understanding that such practices did not take place within the enclosed space of the laboratory has led to an interest in practices undertaken by the Royal Society such as the use of travel narratives and the framing of queries for foreign correspondents and instructions for collecting and preserving specimens.[103] However, for the most part, these studies focus on the reception of information in London, rather than its generation on the ground. To fully understand the hybrid knowledge that was assembled by the Royal Society into 'useful' accounts of the natural world, it is necessary to go beyond analyses of the relations between natural philosophers in Europe and the European traders, colonists, and missionaries they relied on for information, to examine in detail the social contexts in which information was gathered.

Detailed studies of the later interaction between science and empire in the context of colonial India have been carried out. Christopher Bayly focused attention on the importance of diplomatic and scientific knowledge to the establishment of imperialism.[104] He demonstrates that the Company partially took over Mughal systems of information and shows that this process was linked with scholarly enterprises, including cartography, astronomy, medicine, and ethnography. A large body of scholarship has now emerged that focuses on specific areas of science and interrogates both the exchanges of knowledge and techniques and the unequal power relations within which they arose.[105] However, with some important exceptions, most of these works of scholarship focus on the period after 1757. Even recent studies of science in the service of empire have imagined a transformation of the nature and purposes of the Royal Society in the mid-seventeenth century from a 'gentleman's club with an interest in nature' to a significant player in the colonial enterprise.[106] This shift is placed at around the same time as the EIC was traditionally assumed to have morphed from an organisation mainly concerned with trade into the ruler of large tracts of territory.

In terms of the failure to engage with the earlier history of empire on the ground, scholarship on the early Royal Society therefore remains out of step with recent writing on the British and Iberian experience in the Atlantic and the world of the VOC, *Compagnie des Indies orientale,* and *Estado da India.* While the idea of science in the service of empire may be apt for the later period, it does not readily apply to the seventeenth and early eighteenth centuries when scientific practices were still being formed and institutions like the Royal Society lacked the extensive funding and personnel required for overseas missions, while empire in the East remained confined to small and relatively powerless settlements. To achieve a global understanding of the scientific and commercial 'revolutions' of the period, it is necessary to understand how new forms of natural philosophy and empire emerged simultaneously and co-dependently during the early modern period.

Recent histories of science have questioned earlier models that assumed that scientific innovations emerged from Europe and were then disseminated to the rest of the world.[107] These studies have not only demonstrated that modern science grew out of complex interactions between the knowledge traditions of different parts of the world, which altered the character of both colonised and colonising societies.[108] The question remains of how disparate forms of local knowledge came of be combined into forms of commerce and scholarship that were able to claim global applicability. Theories include Bruno Latour's theorising of the form of circulation by which such materials were produced as an 'accumulation cycle', in which the concentration of knowledge gathered from various different points at one location makes that point into a centre.[109] Latour's model is helpful in understanding how 'centres of calculation' can arise as well as the importance of transportable forms of knowledge for the exercise of power from a distance.

Latour's account is primarily one seen from a metropolitan European centre, from which perspective the particularities of the exchange are irrelevant. As such, it can tend to obscure the agency of both the travelling collector of information and that of the 'native informants', who are assumed to be essentially passive yielders of information rather than active participants in an exchange.[110] It also leaves untouched the question of the interaction of multiple competing centres of calculation engaged in the collection of people and information and the political reasons why some prevailed over others. If, as I will argue, the accumulation of knowledge was multi-centred in the mid-seventeenth century, Latour's model is insufficient to explain the imbalance in access to information and thus power that began to be felt by the mid-eighteenth century.

Some historians have taken on the idea of circulation as central to the history of ideas while rejecting Latour's Eurocentrism. Kapil Raj argues that the creation of knowledge involved 'reciprocal, albeit asymmetric, processes of circulation and negotiation' throughout the colonial period.[111] His approach has been important in focusing attention on the generation of knowledge in colonial settlements, as well as in highlighting similarities between the French and English encounter with Asia in terms of its relevance for natural philosophy. However, attempts like Raj's to explain how the project of modernity was co-created on the ground often encounter difficulties. One lies in the use of terms such as 'collaboration' and 'circulation' for a range of different encounters and methods of distribution. Such studies also suffer from the related charge that focusing on instances of peaceful cooperation to produce knowledge or the personal entanglements of imperial agents detracts attention from the wider colonial context of exploitation and violence, which meant that knowledge was often lost or suppressed rather than transmitted. As one critic of such approaches in relation to the EIC puts it: 'by looking at the Company through the lens of culture, the underlying purpose of its presence in India is forgotten'.[112]

In this book, I will suggest some concepts that can be used to understand early modern natural philosophy. These can be summarised as *openness* to information and ideas from a wide range of sources and people; *collaborative work*, including composite works such as dictionaries or herbaria produced by organised bodies of scholars and networks of collectors linked by ties of exchange and patronage; *collecting or prospecting* for plants, words, ethnographic objects, and even people; *classification*, the development of systems to mediate between similarity and diversity by grouping languages, plants, and people into broad families while species names proliferated; and *circulation* the processes of copying, translating, stealing information from friends, rivals, and strangers, distributing it for corrections and incorporating it into new works to be circulated once more. These intellectual trends were tied to the large-scale processes of growth in global population and connectivity that took place in the period: being created in the process of encounters that resulted from these developments. In examining the accumulation and transmission of knowledge in global early modernity, however, it is important to underline that it was constantly accompanied by and was often allied with violence, including warfare and forced migration. As Paul Gilroy notes in the Atlantic context, the rationality that lay at the heart of Enlightenment knowledge also underpinned the cruelty and terror at the centre of the slave trade.[113] How circulation was combined with coercion in the world of the early EIC will be discussed in the next section.

Circulation and coercion

As noted above, rather than a territorial power centred on the Indian subcontinent, the early EIC on the ground was made up from a series of small settlements that spanned from the South Atlantic to East Asia.[114] Each settlement was embedded within quite a different environmental, political, and social context, and this affected the types of knowledge that were produced. The ways in which knowledge was circulated between settlements was also dependent on environmental factors including the seasonal monsoon. The pattern of wind and tide meant that the settlements were connected in specific ways, so that for example ships called at St Helena on the way back from India and on the way to Bencoulen (modern Bengkulu) in Sumatra.

The environmentally determined movement of ships between places of trade, refreshment, raiding, and settlement affected the ways in which the settlements were connected to one another in terms of the transmission of people, goods, and information. This point may be illustrated by the story of a woman who came to be known by her Christian name Maria. Before being taken aboard a slave ship on the Guinea coast, Maria wrapped some yams in her clothing. Arriving at St Helena in the South Atlantic, she planted them by a spring. Coming to know of this experiment in around 1717, Governor Isaac Pyke copied her method of wrapping the yams and sent them on the next ship to Bencoulen, Sumatra.[115] Yams became a staple dish in St Helena, even generating the nickname 'yamstocks' for the inhabitants. Another dietary staple determined by the environmental networks in which the island was embedded was Indian rice, and *pilau* (pilaf) remains a favourite local dish.

While, as noted above, previous scholarship on the EIC has concentrated almost entirely on English elites, the populations of each of the early settlements were made up from people of several different nationalities. In India, these included the Portuguese, who with the ebb of opportunities in the Portuguese empire in Asia often turned to the English settlements to find employment in trading or military capacities.[116] The Portuguese population of Bombay was estimated at 60,000 by 1688.[117] The Eurasian population – often also called 'black Portuguese' or simply 'Portuguese', regardless of the source of their European ancestry – also formed a significant proportion of early EIC settlements. The EIC followed the examples of the Portuguese and Dutch in encouraging intermarriage between soldiers and local women.[118] This, the Company declared in 1687, was a matter of 'such consequence to posterity' that a payment should be made to the mother of any child born of such a marriage and brought into the Protestant church.[119] In 1711, Lockyer estimated the Company's servants at Madras to consist of 250 Europeans and 400 Eurasians of mixed European and Asian or African descent.[120]

Armenians were crucial to the EIC's settlements in Persia in particular, where they managed the trades in silk and broadcloth – an arrangement that was formalised in 1688 by a treaty between the Company and the head of the Armenian community in Isfahan. Armenians also migrated to the EIC settlements in India and beyond.[121] Parsis (or Parsees), descendants of Persian Zoroastrian migrants to India, were crucial to the functioning of the Bombay and Surat factories in western India. In the Southeast Asian settlements, independent Chinese communities within the English settlements were crucial to their functioning. In Bencoulen and St Helena, around half of the population of the settlements were enslaved Africans, mainly from Madagascar. Slaves were also crucial to the Indian settlements, especially as soldiers in the Company's armies. All of these people, to whatever extent they were able, drew on the wider knowledge networks in which they were embedded to gain wealth and reputation within the settlements and beyond.[122] This meant that each settlement was embedded in multiple networks of political, economic, and cultural relations outside of the control of the London Directors. These relations differed depending on geographical location.

Within each of the Company's settlements, relationships were forged among the inhabitants that enabled each of the settlements to survive: slaves built fortifications and tended gardens; interpreters negotiated the agreements with local rulers and translated correspondence; spies informed about goings-on in princely courts and impending conflicts; Chinese and Armenian traders established parallel settlements; and missionaries adopted the habits and some of the ideas of other religions. These exchanges were memorialised in the forms of dictionaries, wordlists and grammars; translations of religious texts; collections of dried plants and packets of seeds; natural histories and travel accounts; geographies and maps. As I will demonstrate throughout, these materials were crucial to establishing the local relations of patronage and power on which the survival of the settlements depended. They were also re-circulated within the Company networks and enabled further projects of colonial settlement.

As argued above, the local contexts of knowledge making should not be assumed to have been consensual and must be located within the wider contexts of international competition for territory and profit.[123] At the same time, the agency and agendas of those involved in collecting and transmitted knowledge should not be neglected. While all the inhabitants of the EIC settlement were involved in the creation of knowledge, the extent to which they benefited from it depended on their ability to package and market the resulting information. Slaves like Maria were only able to benefit in an immediate sense from their skills or from the transfer of a particular type of knowledge. In contrast, an individual such as Isaac Pyke, equipped with links to scholarly circles including the Royal Society as well as a circle of elite colonists like himself, was able to forge a transnational reputation and career based on repackaging and circulating the same type of knowledge through writing letters to important correspondents or publishing papers in the *Philosophical Transactions*.

For those inhabitants of EIC settlements who were able to establish relations of patronage, activities such as collecting plants, identifying precious stones, describing territories and peoples, making maps, and compiling wordlists and dictionaries could mean advancement through the official ranks of the Company and membership of institutions such as the Royal Society as well as valuable opportunities for private trade,[124] and lucrative positions in the service of Asian rulers. Those men who became important in colonial settlements were often the same men who later came to influence colonial policy and natural philosophy in London.

This view of the Company settlements brings us closer to recent studies of the other empires discussed above than to the traditionally top-down view of the EIC and English natural philosophy. Learning from studies of the Atlantic world, the *Estado da India*, the *Compagnie des Indes*, and the VOC for the study of knowledge creation and distribution in the EIC's settlements reflects real contemporary overlaps and borrowings. In its early days, the EIC settlements bore close similarities to those of its Dutch counterpart and rival. As shown by repeated instructions issued by the Directors to their servants to follow the examples of the VOC in everything from training slaves to translating works of scripture, the Directors in London were concerned to emulate their more successful rivals. Both the VOC and the EIC inherited from the *Estado da India* everything from settlements such as Bombay to the populations of Eurasian soldiers who were vital to their armies. The French *Compagnie des Indes*, a latecomer to the scene along with the Danish and Swedish trading companies, followed many aspects of the political, economic, and social organisation of its larger rivals, while competing for territory and influence across Asia and for bases in the Indian and Atlantic Oceans.[125]

Overlap with the Atlantic world is evident particularly in the example of St Helena, where ships from Brazil and New York called on slave trading missions to the West African coast and Madagascar, often leaving sick members of the crew behind as well as the slaves that the EIC officially required them to leave at the island. The Directors sent the laws of Barbados to be implemented and despatched West Indian planters to experiment with sugarcane on the island. The close connections between the European trading companies and the savant institutions as well as the traders and settlers of the Atlantic world are also evident in the trajectories of many of the individuals who appear in this account: for example, Thomas Bowrey and Robert Knox both traded to the Atlantic as well as the Indies. Cooperation coexisted with rivalry between nations, companies, and scholars, as demonstrated by the appropriation of cartographic information and guarding of botanical secrets. The forging of an English imperial identity was conducted in competition with other European nations. Indeed, this process of competition, along with the growing importance of the nation state, fostered the later ideas of English exceptionalism that have influenced much previous scholarship on the EIC and Royal Society.

I argued above that, at the beginning of this period, Englishmen did not possess a secure sense of national identity, having emerged from the long civil conflict of the mid-seventeenth century with several competing models for political and religious authority. Neither the sense of colonial separateness that Guha calls being 'not at home' in empire[126] nor the primacy of metropolitan allegiances was inherent in early English imperial expansion. It was the dangerous nature of the men the early EIC relied on to provide them with information that led to attempts from London not only to monopolise and regulate the provision of information, but also to engender in their servants a sense of racial and cultural difference that might bind them closer to a sense of being part of an English 'nation'. This process is observable during the period that I discuss, examples including the Company's demands for ever-increasing jurisdiction over all Englishmen in the East; the erection of boundaries between 'black' and 'white' towns in Company settlements; the growing concern regarding Company servants who spoke local languages too fluently; and the emergence of a racialised discourse of human difference by the end of the period. The circulation of labour, enslaved and indentured, and the construction of fortified and armed settlements supplied by plantations, pioneered in this period in St Helena and Bencoulen, were intended to break the pattern of dependency on the local rulers, brokers, soldiers, and workers described above.

None of these measures to bind the settlements more closely to metropolitan authority and to inculcate a sense of difference went uncontested in the settlements. Go-betweens often succeeded in manipulating the desires of both the Company and Asian rulers to their own ends; the repeated rebellions of slaves and settlers on St Helena as well as the circulation of pamphlets and direct appeals to England often succeeded in resisting and reducing the Company's 'despotic' power on the island. Meanwhile, the boundaries of race that were imposed on the Company's settlements were blurred until rendered senseless in the face of the development of Creole societies, which became crucial to the Company's functioning in both Bencoulen in Sumatra and St Helena in the South Atlantic. While creolisation remained the exception in the British colonial world, hybrid forms of colonial knowledge became ever more important. I will return to the fate of hybridity after 1720 in the concluding discussion.

In what follows, I aim to throw new light on knowledge making in the world of the early EIC settlements. Each chapter addresses a different geographical region and a particular form of scholarship, although there are many connections between both the locations and the practices discussed in each. I show how these works emerged from, reflected, and redirected the multiethnic and multi-directional flows of people, objects, and information between the settlements of the early EIC.

1

Curious Collectors and Infamous Interlopers: Samuel Baron and the EIC Settlements in Southeast and East Asia

Introduction

In 1686, Samuel Baron wrote from Fort St George, the Company headquarters in Madras, to Robert Hooke and Robert Hoskins of the Royal Society, enclosing a draft of his *Description of the Kingdom of Tonqueen*. The manuscript was not published immediately, but appeared in an early-eighteenth-century collection of voyages and is still considered a valuable source of information about early modern Vietnam.[1] Baron informed his correspondents that he would shortly be embarking on a journey to China, from where he would send any 'curiosities' worthy of their notice.[2] Shortly thereafter, Baron left Madras as head of a mission to establish a 'factory', or trading post, in Amoy (modern Xiamen). However, on hearing of his appointment, the London Directors wrote to the Governor and Council of Fort St George, ordering them to dismiss Baron, as 'no servant of ours, but a deserter, the history whereof is too long to tell'.[3]

In this chapter, I tell the story of Samuel Baron, the son of Dutch and Vietnamese parents who became a 'naturalised Englishman'. Baron was central to the EIC's attempts to gain a foothold in the lucrative trade between China and Japan: negotiating their settlement in both Tonkin (Northern Vietnam) and Taiwan and briefly heading the Amoy factory. He also played a key role as a provider of information, not only to the EIC and scholars in England but also in the courts of Southeast Asian rulers. I aim to explore in detail the shaping of the identity of a provider of 'useful' information to the EIC and the Royal Society, to examine the formation of national and ethnic identities in the course of contacts between different cultures and to produce a reading of natural histories or geographic and ethnographic 'descriptions' that goes beyond 'Orientalist' perspectives to consider in detail the circumstances of their production and the ends for which they were intended.

Baron's lifetime, the second half of the seventeenth century, was a time of transition in Southeast and East Asia. The rise of the Qing dynasty and the period of expansionism that followed saw new territories and ethnic groups integrated into China.[4] The trade of East Asia in the 1660s centred on the exchange of high-quality silk from China for Japanese silver. The Ming empire's earlier prohibition of direct trade with Japan had left smaller states and foreign actors – including the European trading companies in Asia – scrabbling for roles as intermediaries.[5] It also encouraged the diversion of the trade through the entrepôts discussed in this chapter: Taiwan (also called Formosa in this period), Siam (modern Thailand), and modern Vietnam (then divided between Tonkin, Annam, and Cochin-China), as well as the port cities of the Malay Archipelago discussed in Chapter 2.[6] During the period of Qing expansion, these areas also provided refuge for Ming loyalists, and thus the maritime world became briefly important to the centre of the Chinese empire.[7] In each place, rulers faced the task of using the interest of foreigners in trade and settlement to their advantage while preventing any one faction from assuming too much power. It was against this background that the EIC, in competition with international rivals, attempted to establish bases in Siam, Tonkin, and Taiwan during the late seventeenth century.

Such experiments in managed multiculturalism meant it was crucial for all these groups of competing traders and rulers to gather information about one another. One way to do so was by amassing collections of foreign 'novelties' or 'curiosities'. These included scientific instruments and weapons: telescopes, watches, burning and perspective glasses, as well as firearms and gunpowder, often sent to Asian rulers by representatives of the European trading companies.[8] Meanwhile, Chinese compasses and Japanese steelyards (weighing scales) were in demand among the intelligentsia of Europe.[9] Maps and pictures were also exchanged,[10] as were animals (alive or dead), plant specimens, and drugs.[11] More bizarre items were also involved: during the 1680s, King Narai of Siam perplexed the English factors by demanding 'A Chrystall branch & amorous representations in wax work',[12] while Hans Sloane later attracted ridicule for his interest in Chinese earpickers.[13] Cook has drawn parallels between the early modern interest in such curiosities and the emerging concept of scientific 'objectivity' as a type of knowledge founded on a detailed acquaintance with material objects.[14]

The display of exotic objects, and even people, was a means of exhibiting one's comprehension and command of the world. For example, the Ming official Zheng Zhilong, at one time commander of Amoy and an important mediator with the Dutch, displayed his power with rarities including a collection of Christian objects and a bodyguard of African slaves.[15] The display of the 'painted prince' Jeoly or Giolo by William Dampier (1651–1715) validated the latter's status as a famous buccaneer.[16] Written accounts were another means of exchanging information: many different cultures produced natural histories, ethnographic writing, and maps as a result of voyages of 'discovery' in all directions.[17] These descriptions functioned as both justifications and manuals for colonialism and, as such, shared many features across cultures, while at the same time being intended to satisfy particular needs and interests.

To manage these exchanges, which were often crucial to the process of negotiating trade and settlements, people who could mediate between two or more languages and cultural systems were essential. Studies of early modern South Asian states and the colonial powers that displaced them have emphasised that in both, the maintenance of power was dependent on accumulating various types of knowledge through 'spies, informants and collators of gossip'.[18] Those who could mobilise and deploy such information have been described as 'brokers', 'go-betweens', or *passeurs culturels*.[19] They bridged gaps between competing imperial networks of diplomacy and trade as well as between subaltern and elite cosmopolitanism in the early modern world.[20] Such people fulfilled a number of roles, being specialists in language, law, navigation, mapping territory, explaining the manners and customs of foreign peoples, or collecting medicinal plants. They often worked in several of these fields and are difficult to define because they cannot be assigned to any fixed social group: indeed, their roles depended on their insight into other worlds, either by changing allegiances or by prolonged contact with another group.[21] In the process of mediating between different systems of knowledge, such go-betweens created new form of knowledge.[22]

Bénat Tachot and Serge Gruzinski's term '*passeurs culturels*' is expressive of the mixing and moulding of personal identities.[23] This brings us close to Stephen Greenblatt's concept of 'self-fashioning', drawn from his reframing of Jacob Burckhardt's 1860 thesis concerning the rise of individualism in the European Renaissance. Greenblatt describes the 'self-fashioning' of writers as an 'artful process', suggestive of 'hypocrisy or deception'.[24] In many ways, this description recalls the common depictions of interpreters or go-betweens as duplicitous individuals, expert in crafting their identities for their political or social ends. Indeed, in his later work, Greenblatt describes the cultural go-between, who managed encounters in the course of travel and colonial conquest, as the ultimate creator of selves: one 'who passes from one representational form to another, who mediates between systems, who inhabits the inbetween'.[25] Anthony Pagden relates Greenblatt's thesis to other developments in the early modern world by arguing that encounters with lands and peoples that could not be explained with reference to traditional authorities forced the creation of new personal identities.[26] The same might be said of the encounter with unfamiliar aspects of the natural world that took place in the same period. Although he does not use the term 'self-fashioning' and while his emphasis is on the creation of truth as opposed to fantasy, Shapin employs a similar approach to Greenblatt in understanding how Robert Boyle's crafting of his personal identity interacted with his claims to present credible information in the realm of natural philosophy.[27]

Despite recent scholarly interest in the go-between, there have been few detailed studies of the self-fashioning of early modern intermediaries of non-European origin. One exception is Natalie Zemon Davis' account of al-Hasan al-Wazzan (Leo Africanus), which explores how the author of *Description of Africa* negotiated his Muslim and Christian identities, using the idea of the 'trickster' to reveal the ambiguities and double meanings of his work.[28] Sanjay Subrahmanyam's case studies of three early modern 'aliens' emphasise the 'friction and discomfort' involved in traversing cultural identities. The role of mixed-race people in the history of early modern Asia has been particularly neglected,[29] especially in comparison to the extensive literature on *métissage* in the American context.[30] However, as an investigation of Baron's life will reveal, these people were in an ideal position to span cultures. *Métissage* was multi-directional. As well as the importance of Eurasians like Baron in both the settlements of European companies and the courts of Asian rulers,[31] contemporaries like Zheng Chenggong (1624–1662, known as Coxinga or Koxinga to the Europeans), Zheng Zhilong's son by his Japanese consort, who bridged Asian cultures were also vital to intercultural contacts.

Focusing on people who spanned cultures can highlight connections as well as disjunctures between the techniques of gathering and using knowledge in different cultures in this period. Using the framework of one man's life to explore the wider history of imperial expansion and conflict in this period can allow us to move beyond narratives focusing on particular nations and their ideologies towards a picture of the connections between the underlying concerns and practices of different competing actors in this period of increasing global contact.[32] Investigating the lines of communication established by the particular connections and spaces of exchange created by go-betweens can also demonstrate the fragility and contingency of both colonial expansion and scientific knowledge.[33] The biographical investigation that follows emphasises both the importance of individual circumstance in determining the connections and misunderstandings that emerged from the multidirectional contacts of this period and to illustrate commonalities in ways of representing oneself and others across national, ethnic, and linguistic boundaries.

Self-fashioning I: Becoming English

Baron's background is obscure, partly, I will suggest, a result of his own efforts to mould his ethnic and personal identity. Previous accounts almost all mistake his origins, and none give an account of his life after despatching his account of Tonkin to Hooke and Hoskins. Salomon Baron, as his name is given in early records, was born in the mid-1640s in the capital of Tonkin: then called Thâng Long and popularly known as Ke Cho (Chachao or Cacho in European documents), now Hanoi.

Tonkin had been a Chinese colony in antiquity, maintaining a tributary relationship, and China remained an important cultural influence and trading partner. Although the *Le vua* (or 'Emperors') were recognised as the rulers of both Tonkin and Annam, since 1590 real power in each region had lain with the *chua* (called the 'General' or 'King' in European descriptions), drawn from the Trinh dynasty in the north and the Nguyen in the south. Informed by neo-Confucianism, the *chua* of Tonkin, Trinh Tac (r. 1645–1682), led a drive towards centralisation and tightened state control over foreign traders.[34]

The primary goods that attracted traders keen to break into the Japan–China trade to Tonkin were raw silk and silver, although other commodities such as tortoise shell, areca nuts, and cinnamon were also profitable. Chinese traders dominated the trade of the capital until the 1670s, although by the time of William Dampier's visit in the later 1680s, they had been banished along with other foreign merchants to Hien.[35] The Dutch had been allowed to establish a trading post in Tonkin in the 1650s in return for services to the Trinh rulers, including attacking Cochin-China.[36] Other groups of foreigners which Baron would have encountered in his youth included small communities of (often-Christian) Japanese merchants and French and Portuguese Jesuit priests.[37]

Salomon Baron was born the 'natural' son of a Dutchman, Hendrik Baron, and a Tonkinese woman in around 1645.[38] Hendrik appears in the VOC records in 1651 aboard a ship that had become stranded on the coast of Annam. As he was familiar with the language and culture of Tonkin, having already lived there some years, Hendrik brokered a treaty, ending a period of conflict since 1639.[39] His linguistic and diplomatic skills later allowed Hendrik to rise to head the Dutch factory in Ke Cho.

The sexual alliances of members of the European trading companies with Asian women are often noted but seldom investigated in terms of their political or economic significance.[40] In Tonkin, such alliances of the VOC agents were extremely advantageous, as women controlled most trade and dominated the profession of money-changing, thus possessing unrivalled knowledge of local markets.[41] While some alliances were temporary, Hendrik Baron maintained a long-term cohabiting relationship with his Vietnamese consort.[42] Baron himself is therefore likely to have been instructed in both indigenous and European trade by both his parents from an early age. He also claimed to have travelled to Japan, Taiwan, and China as a young man.[43]

In 1659 Salomon left for Holland, funded by the VOC.[44] He evidently changed his name at some stage during the European visit, of which few details are known. It is clear from his work, however, that he acquired a thorough knowledge of contemporary and classical European culture: he quotes Luther's speeches, compares the Tonkinese to the ancient Spartans and, to please his English readers, drops in a reference to the defeat of the Spanish Armada. Baron seems also to have read European travel accounts dealing with South and East Asia.[45]

Baron first appears in the Court Minutes of the EIC in London in March 1671, where he is described as a Frenchman: the first of many mistaken statements about his background.[46] Baron had approached an English Ambassador in Paris and suggested to him a scheme for reopening the English trade with Japan. This suggestion was extremely well timed, following up as it did on an initiative led from Bantam (the EIC headquarters in Java) to reopen the English trade with Japan, which had been abandoned almost 50 years previously. Baron was therefore employed on an unusually high salary of £120 per annum as second in command on a voyage to Japan.[47] Although the EIC hoped that Japan would provide a market for the English broadcloths that had been rejected across Asia, they were also aware that any entry into the Japanese trade would involve establishing bases in the entrepôts discussed above, in order to acquire commodities that could be profitably exchanged. It was Baron's unrivalled experience of the local languages and cultures of this region that allowed him to demand a salary and position that would have been unthinkable for most men of his age, particularly an unknown foreigner.

Three ships, *Zant, Experiment* and *Return,* were therefore despatched to Tonkin, Formosa, and Japan.[48] Baron was intended to play a key part in the negotiations in each place. The factory in Vietnam was established in June 1672, with the aid of Tonkinese interpreters, merchants, carpenters, and other workers recruited by Baron at Batavia and sent with the English factor William Gyfford (or Gifford).[49] The agreements brokered by Baron with the Ming loyalist Zheng Chenggong also allowed the establishment of an EIC settlement in Taiwan in July 1672.[50] The *Return* had stopped at Siam on its voyage to Japan in 1674, landing three factors, who settled themselves in the old quarters of the EIC in Ayutthaya, initially meeting with good reception and making agreements for trade.[51]

Despite these relatively auspicious beginnings, the settlements in both Taiwan and Tonkin failed to profit in their early years and were subject to strict control of the authorities and hostility from other groups of traders. On arriving in Tonkin, Gyfford was told by a mandarin that the factors were now 'in the position of a married woman, that can blame no-one but herself for being brought into bondage'.[52] Gyfford apparently agreed: he appealed to the EIC Directors in London not to blame the factors 'who are in reality no better than slaves', for the settlement's lack of profit.[53] War with the Dutch also meant that the EIC was excluded from most regional trade. Faced with these difficulties, Gyfford fell into a depressive illness.[54] The helplessness of the English factors is an illustration of the complete dependence of the early European factories in Asia on men like Baron and their local connections.

The voyage from Taiwan to Japan was disastrous for the EIC. The war which had broken out between the English and Dutch in Europe (1672–1674) meant that two of three English ships that left Taiwan were captured by the Dutch, and Baron was imprisoned in the VOC settlement in Batavia. On reaching Japan, the crew of the third ship were questioned about the English King Charles II and his Catholic Portuguese wife Catherine of Braganza – the Portuguese having recently been expelled from Japan following a Christian uprising. Eventually, they were refused permission to trade and ordered to leave.[55]

Despite the spectacular failure of the mission and the suspicion that must have been associated with his Dutch connections, the EIC intervened with the Dutch commissioners then in London to procure Baron's release.[56] Given his connections with other Europeans in Asia, Baron was also valuable to the EIC as a spy: he informed London about goings-on at the VOC base in Batavia, the designs of the French Compagnie des Indes on Ceylon's cinnamon as well as the allegedly treacherous intentions of the Company's own servants.[57] The use of Baron to provide intelligence to the EIC was not unusual: the archives are filled with factors' reports of the sinister designs of their colleagues and competitors. Which version of events was accepted depended on a web of allegiances running beneath the official Company hierarchy.

Maintaining his position with the EIC depended on Baron moulding his identity so as to attract patronage and assure his patrons of his loyalty. In his initial representation to the Company, Baron stressed his connections to Britain, noting a Scottish grandfather on his father's side and downplaying his Asian identity by describing his mother as 'of the race of the Portugalls'.[58] Baron's early correspondence with the Company stresses his devotion to Protestantism and King Charles II.[59] His facility with English is clear even from the early missives. Furthermore, Baron was evidently a man of personal charm: he was initially described by the EIC as 'an active, intelligent person, with competent abilities for his yeares'.[60] All this helped him fashion an 'English' identity. Although the letter appointing Baron second in command on the mission to Japan notes that he should not succeed as Governor 'being not of our nation', after three years in EIC service, Baron became a 'naturalised Englishman', entitled to hold any Company post.[61] While the Dutch records refer to him as a '*Toncquinese mixtese*',[62] none of the English records makes any comments as to his ethnic identity.

Baron's remarkable success in fashioning an English identity might be considered in conjunction with the creation of an Asian identity by the young Frenchman who called himself George Psalmanazar and succeeded for some time during the early eighteenth century in convincing a number of elite Englishmen that he was a native of Formosa.[63] As Michael Keevak has pointed out in his study of Psalmanazar, his typically European physical appearance was not used to discredit his story, since there was no concept of race as such in eighteenth-century Europe, skin colour being seen as purely a result of environmental factors. Keevak argues that the main factors in determining identity were language, religion, and cultural factors such as eating habits and modes of dress.[64]

Keevak concludes that the failure to recognise Psalmanazar's fraud was a result of European inability to see coherent or internal organisation different to their own in other cultures and thus to define those who were not European.[65] While this may be true, Baron's ability to become English through acquiring mastery of the language and professing allegiance to the King and to Protestant Christianity demonstrates the more general porosity of identities in this period.[66] Other examples of people who were considered to have crossed national boundaries through their linguistic abilities and comportment include the description of Jean-Baptiste Tavernier as a 'Dutchified Frenchman'[67] or the complaints of Robert Knox' rebellious crew members that he was 'an old Portuguese rogue' (Chapter 5).[68]

Self-fashioning II: 'Mr Barron of Funquin'

Baron selected other facets of his ethnic identity to highlight when it suited him. Returning to England in 1677, Baron encountered Robert Hooke in a London coffee house, presenting himself as a native of Tonkin and debating with him on a number of subjects including the process of petrification. Hooke also introduced Baron to Robert Boyle.[69] With Hooke's help, Baron also seems to have defended Gyfford, who by this stage had been recalled to London on charges of mismanagement and private trade.[70] After 'proffering his services' to Hooke as a collector of information, Baron returned to Tonkin. Baron's skill in attracting the attention of natural philosophers in London might also be compared with Psalmanazar's success in attracting the patronage of several scholars. Both did so by combining their claims to special knowledge of Eastern societies with the social aspects that encouraged trust, including Protestant faith, a fluent knowledge of European languages and the right kind of 'curiosity' about the natural world.[71] Baron was not a fake, in that he offered real experience of Tonkin and maritime East Asia. However, as I will show in the next sections, the knowledge that he supplied was carefully packaged to achieve specific ends.

Baron arrived back in Tonkin officially as a private merchant – he was constantly wary of appearing in his homeland in too close an association with the English factors, arguing at one stage that if he went to Tonkin at all, it should be 'in the character of inspectator'.[72] Nevertheless, the records of the EIC show that he was paid to act as a go-between and informer.[73] He seems to have been involved in negotiations with the *chua* Trinh Tac on behalf of the English factory. However, as the English factors complained, Baron had his own agenda. He had entered the service of the *chua's* grandson, who 'adopted' him and gave him permission to trade. In return, Baron brought his patron 'foreign curiosities'.[74] Baron also tried to persuade Trinh Tac to open his own trade with Manila and Siam, telling the *chua* of the vast profit made by foreign traders in his kingdom.[75] Baron thus established relations of patronage in both England and Tonkin, both based on the provision of information and 'curiosities' or physical objects from another culture.

A curious marker of Baron's presence in Tonkin at this time is his autograph, carved on a rock on the entrance to Day River along with the date 1680.[76] Baron prospered during his time in Tonkin in the late 1670s or early 1680s, and it was probably then that he married a local woman. Robert Knox, encountering Baron in Vietnam, reported to Hooke in 1681 that he 'lives splendidly, Is married to a very fair White gentile woman'.[77] A letter addressed to him in 1684 when in Siam presents greetings to 'your Lady',[78] and Baron is referred to as being married to a 'Tonqueener' in the Fort St George census of 1688 (discussed below). Assuming these references are to the same woman, it seems likely that Baron's wife accompanied or followed him during his adventures after leaving Tonkin. This would not have been unusual in Southeast Asia, where women often travelled on trading ships.[79]

Another powerful patron, whose favour Baron briefly acquired, was King Narai of Siam (r. 1659–1688). While the *chua* of Tonkin kept foreign traders strictly under their control during the mid-seventeenth century, Narai and his predecessors had pursued a policy of liberalising trade, allowing foreign merchants to settle in the capital and taking some into his service. In 1605, Japanese merchants were allocated a settlement in the capital Ayutthaya, from where they exported sappanwood and animal skins.[80] An embassy was despatched to the Hague in 1609, and Narai sent Ambassadors to Persia,[81] Tonkin,[82] France,[83] and England[84] and kept up a regular exchange of embassies with the rulers of Bengal and Golconda.[85] By the mid-1680s, the expatriate population in Ayutthaya thrived, with traders and envoys from the Dutch Republic,[86] China, the Malay world, Cochin-China, Lao, Pegu (modern Myanmar/ Burma),[87] and Iran[88] joined by French and Portuguese missionaries and priests. They were encouraged by policies of religious toleration and free trade.[89] The English made two attempts to begin a factory in Siam: in 1613–1623 and 1666–1688.[90] However, they faced a constant problem with deserters. By the 1680s, at least 11 English naval commanders served King Narai, while a contemporary account put the total number of English in Ayutthaya before 1688 at 120.

The friction generated by such intercultural contacts could have creative, but also explosive, consequences.[91] For example, astronomical instruments were taken from France to Ayutthaya.[92] Musical and artistic exchanges also occurred in the course of the embassies between France and Siam.[93] Other contributions of foreign settlers to Thai affairs were less welcome: for example, Japanese settlers were massacred in 1632 after supporting the usurper Prasad T'ong. A similar fate awaited Europeans who interfered in the internal affairs of the kingdom.

Baron traded privately from Tonkin and Bantam to Ayutthaya during the late 1670s. In 1682, he persuaded King Narai to appoint him as the commander of a ship to Tonkin with a cargo of rice.[94] However, arriving in Tonkin, he found Trinh Tac dead and his patron apparently mad and had his own cargo of rice confiscated.[95] Furthermore, in 1683, three Chinese Ambassadors had arrived in Ke Cho at the court of the vua Le Hi Tong (r. 1676–1705).[96] Although his references to this period are obscure, Baron seems to have been punished for his refusal to lead a military force against Cochin-China and perhaps for his association with enemies of the Qing like the Zheng rulers in Taiwan.[97]

Baron therefore fled back to Siam, arriving in February 1684 in a state of destitution.[98] Arriving after a space of two years without goods or profits from his voyage, Baron came into conflict with Chinese traders and the King's officials, and took refuge with the English factory.[99] By this stage, one of the most influential men in the kingdom was Constantin Hiérachy, popularly known as Constantine Phaulkon, Phaulcon, or Falcon, a Greek who had been in the service of erstwhile EIC employee George White during the 1670s,[100] but who had later entered the service of King Narai. Phaulkon soon began to court alternative trading partners, including the French, who had established a factory in the capital in 1680, aided by the Société des Missions Étrangères.[101] The celebrated exchanges of embassies between Ayutthaya and Versailles followed during the 1680s.[102] Similarly to Baron, Phaulkon was fluent in European and Asian languages;[103] he moved between cultures with ease. An English account of 1684 depicts him at home 'entertaining the Chineses with a Comedy'.[104] Phaulkon's popularity aroused jealousy from various quarters: the chronicler of the Persian embassy to Siam notes that he had displaced the Shi'ite Persian merchant Aqa Muhammad Astarbardi (Okpha Sinnaowarat) as Chief Minister, complaining that 'the Frank has endeavoured to weaken the King's character and encourage the royal fickleness'.[105] As was perhaps inevitable, given the similarities between Phaulkon and Baron, they became rivals, and in 1684 Baron wrote a scathing report of the Greek's 'monstrous' character and his plan to use his influence to displace the EIC in favour of 'interlopers'.[106] The reference to interlopers here is pure polemic, as Baron himself was in Siam in the capacity of a private trader rather than an EIC employee. He urged the Company to send a military force to remove Phaulkon and his associates. By 1686, Phaulkon was also calling for European military intervention in Siam, urging Louis XIV to seize Mergui for France.[107] Despite his rivalry with Phaulkon, Baron built up strong enough networks with local officials and Portuguese priests in Ayutthaya to secure the release of two English factors who were imprisoned after coming into conflict with Phaulkon in 1684.[108]

Despite Baron's warnings and an explicit command in 1684 from London to break off all relations with Siam,[109] the private interests of members of the Company meant that EIC ships continued to be sent from both Surat and Madras to trade in Siam.[110] As was often the case, local politics triumphed over orders from London. In 1685, the Siamese Ambassador was received with ceremony along with his Persian counterpart in Fort St George.[111] Eventually, a feud broke out between Elihu Yale, William Gyfford's successor as President of Fort St George, and Phaulkon over the proposed sale of some jewels by the former to the latter.[112] In October 1686, another errant Company servant, Samuel White (brother of George) – whom Narai had appointed *syahbandar* (harbour master) – attacked the King of Golconda using Siamese warships with English captains.[113] Golconda was an important patron of the Madras settlement, where the instructions of the Directors to make war on Siam were finally acted on. King James II supported the declaration of war, with the aim of forestalling Louis XIV's ambitions for a settlement in Siam to open up trade with East Asia.[114] The second French embassy to Siam arrived in 1687 with secret instructions to seize the ports of Bangkok and Mergui.[115] Like the Japanese before them, the European traders were punished for interfering in Siam's affairs, and about 200 were massacred in Mergui in 1687.[116] This was followed in 1688 by the execution of Phaulkon[117] and the expulsion of all foreign traders from Ayutthaya by Okphra Phetracha, an army general to whom the mandarins and *sangha* (Buddhist monastic community) had entrusted the affairs of the kingdom after King Narai fell ill in May 1688. After Narai's death in July 1688, a further round of violent expulsions targeted Catholic missionaries. Meanwhile, in England, James II was overthrown in the Glorious Revolution of 1688–1689.

The Siam incident demonstrates the impact the intrigues of go-betweens could have in this crucial period of change. Diplomatic and scientific exchanges were fragile, built as they often were on chance encounters and the relations of a few individuals. While some foreign missionaries and visitors continued to gain entrance to the kingdom,[118] the aftermath of the so-called 'revolution' of 1688 in Siam[119] ended the period of intense contacts with the outside world for some centuries. By the 1930s, the country, renamed Thailand ('land of the free'), was the only Southeast Asian country not to have been colonised by a European power. The unceremonious ejection of the Europeans from the Siamese kingdom, like the humiliation of English in their 'war' against the Mughal Empire that followed (Chapter 4), demonstrates that whatever their naval prowess, Europeans forces were by no means assured of victory against established Asian polities at this stage. The period of contact with Siam continued to have echoes in Europe, from Louis XIV's construction of enlightened absolutism[120] to John Locke's use of the kingdom of Siam to exemplify a functioning society of atheists.[121]

Fashioning 'others'

Describing Tonkin

Baron missed the culmination of the drama in Siam, having arrived by 1686 at Fort St George in Madras, where he completed his *Description of the Kingdom of Tonqueen*. The *Description* had begun as an exercise in correcting the account of Tonkin by Jean-Baptiste Tavernier, which had been translated into English in 1680[122] and was presumably sent by Hooke to Baron in Tonkin. The practice of asking one correspondent to correct the account given by another was a common strategy among members of the Royal Society for gathering information at a distance. Baron, however, soon departed from the text of Tavernier's account, deeming it 'fabulous and full of gross absurdities'.[123] Nonetheless, he continued to use Tavernier's structure as a basis, his chapters covering the geographical position and size of Tonkin; its nature and produce; its riches, trade, and money; military strength; the manners of the people; marriage customs; socialising and pastimes; scholarship; physicians and diseases; government, law, and policy; the court of the *chua* and his officials; ceremonies; funerals; and 'idols' and 'superstitions'. Baron's account fulfils the requirements for a natural history set out in Boyle's general guidelines, in giving an in-depth description of the natural and human geography of the country.[124]

I used the idea of 'self-fashioning' above to describe the process of creating and maintaining personal identities: a concept that has been applied both to the writing of fictions and diaries and to the management of cultural boundaries in the process of mediating between cultures. While the broker has been cast as the ultimate constructor of selves, 'Orientalist' works, including European travel narratives, geographical and historical descriptions, works of literature and art, and analyses of language, have been viewed as tools for the construction of 'others' as part of the project of imperialism. Like human brokers, descriptions like Baron's could serve to enable colonialism on a practical level: functioning as manuals for colonialism that instructed potential colonists on the geography, languages, and customs of the land to be colonised. Such narratives have also been argued to provide the theoretical underpinnings of colonialism via 'demarcations of identity and difference often based upon ideological and mythical distinctions between civilization and barbarism and tradition and modernity'.[125]

Critiques of post-colonial approaches to works of 'Orientalism' have often pointed out that the assumption that European writings about the rest of the world unconsciously echoed a set of primal prejudices derived ultimately from the encounter with Islam constitutes the same sort of essentialism of which Orientalists themselves are accused of perpetrating.[126] In addition, the uniqueness of European constructions of the 'other' is questionable, given that in the age of global contact, people from all over the world encountered, described, and attempted to subjugate one another.[127] Furthermore, the understanding of the cultural hybridity of the go-betweens who provided the information on which ethnographic accounts were based is difficult to reconcile with the idea of such texts as monolithic purveyors of the prejudices of an insular culture. Nonetheless, it is difficult to deny the essential claim of post-colonial studies that such accounts often function to create and perpetuate stereotypes.

Baron's text presents a particular problem since it contains several tropes taken to be emblematic of the European colonising imagination, including negative assessments of the civilisation of Tonkin. If, as I have argued, Baron retained a Vietnamese identity that was at least as strong as his identification with European modes of thought and representation, how can these negative comments be explained? Mary Louise Pratt advances the concept of 'auto-ethnography', defined as 'instances in which colonised subjects undertake to represent themselves in ways that *engage with* the colonizer's own terms' (emphasis in original).[128] Such works involve partial collaboration and appropriation of the idioms of the coloniser. While the definition does not quite fit Baron's case (for one thing, he urged rather than submitted to colonial intervention in Tonkin), it is helpful in recognising that while certain stereotyped views of societies do emerge from 'Orientalist' texts, their authors often employ them knowingly. In this chapter and those that follow, I will argue that rather than straightforward carriers of European prejudices, the works produced in the process of early modern globalisation were, like the broker him or herself, often hybrid products of encounters that had specific aims, which can only be understood by reading them in the contemporary context in which they were produced. Information is revealed, hidden, or distorted, and comparisons and translations of concepts are made depending on the aims of the author(s). Reading the *Description* with Baron's self-fashioning in mind thus allows us to achieve a fuller understanding of his work.

The *Description of the Kingdom of Tonqueen* was dedicated to Baron's patrons within the EIC, in particular his old friend Gyfford, then President of the Madras factory.[129] As he had earlier done in person with the help of Robert Hooke, Baron used his text to defend Gyfford against charges of mismanaging the Tonkin trade. Instead, he credits him with increasing the English understanding of the country, 'as was very advantageous to commerce'.[130] The *Description* was also intended to generate funds for Baron's patron. In his letter to Hooke and Hoskins, Baron asks that his work should be published and that the 'money the said description will yield' should be delivered to Charles Chamberlain to be passed on to William Gyfford.

Natural histories like Robert Knox' *Historical Relation of Ceylon* (Chapter 5) had the potential to raise significant funds and patronage for their authors. They could also be of practical use to the EIC, and this aim is demonstrated by Baron's statement in his letter to Hooke and Hoskins that the account is intended to equip a new commissioner to conduct business in Tonkin on his first arrival in the country. Baron's account contains features that are clearly designed to entice the EIC to invest further in the Tonkin settlement, a project from which he would have benefited. For example, Baron claims that European broadcloth was in demand in Tonkin.[131] As he must have known, this statement had not been borne out by the experience of the factors, who complained that broadcloth could not be given away, much less sold.[132] Rather, it was designed to appeal to the Company, which was constantly engaged in a search for a market for the unpopular fabric during the first century of its existence.

The production of ethnographic descriptions of foreign peoples for the purpose of colonialism was not confined to European powers, but was also practised by the Chinese as part of the process of assimilating different ethnic groups into the Qing empire.[133] Like their European counterparts, these Chinese accounts had practical purposes, as manuals for the governors of new territories as well as demonstrations of the achievement of knowing and understanding the 'nature' of people. During the period of transition from Ming to Qing, less formal accounts were composed by those who used their knowledge of the areas on the fringes of the empires to attract patronage in the scramble for territory. While there are undeniable differences, reading such accounts alongside their European counterparts serves to undermine the claim to uniqueness of many of the features that have been considered to comprise the 'Orientalist' viewpoint.

Comparing Baron's account with the late-seventeenth-century description of Tonkin by a Chinese trader, Pan Dinggui (潘鼎珪),[134] reveals several similarities both in the tone of the ethnographic description and in the types of natural historical information that is included. Little is known of Pan Dinggui, a merchant-scholar active in Gaozhou, Guangdong Province, as well as Taiwan and Annam. His descriptions of several of these places appear in the Shuoling collection of geographical descriptions published in 1702.[135] It is unclear how Pan's texts came to be included in this collection, and we know nothing of their intended audience. Nevertheless, it seems that Pan lived among the foreign merchants from the 'four cardinal directions' at Pho Hien.

Both Baron and Pan start with histories of Tonkin, beginning with the period of Chinese conquest, and both offer anthropological accounts of the differences between Tonkinese and Chinese, using a story about the former's 'converging toes'.[136] Pan makes similar observations to Baron about the influence of Confucianism in Tonkin and the systems of government, education, and honours. Like Dampier, Pan notes the custom of women conducting the majority of trade and notes with disapproval women's participation in public life and ability to choose husbands freely. His description of the natural history of the country pays particular attention to the unusual shapes of the limestone cliffs that rise from the water along the long river journey to the capital, a feature that had also been remarked on by Robert Hooke in an account he received from one of the Bantam factors.[137] He also describes the isle of Hainan, which produces pearls and was inquired after by Robert Boyle.[138]

Both Baron and Pan denigrate the culture and values of the people of Tonkin. For example, Baron writes, 'the Tonqueenese are not altogether so fradulent, and of that deceitful disposition as the Chinese; it may be, by reason they are inferior to them in craft or cunning'.[139] At times, Baron takes pains to be less complimentary about the Tonkinese customs than Tavernier had been in his narrative. For example, he writes in the chapter on learning that the Tonkinese are 'wholly ignorant of natural philosophy, and not more skill'd in mathematics and astronomy', adding that 'their poesy I do not understand, and their musick I do not find delightful or harmonious; and I cannot help but wonder by what faculty Monsieur Tavernier has discovered them to be the most excellent of all the oriental people in that art'.[140]

Ibn Muhammad Ibrahim, the Persian chronicler of the embassy to King Narai, pokes fun at the Siamese army, depicting it as merely an assembly of the peasantry.[141] Similarly, both Baron and Pan downplay the significance of the Tonkinese army and the manliness of the people in general. Pan states incorrectly that there was no permanent militia and argues that despite its large population and territory, Tonkin was poorly defended. Baron mocks the eunuch mandarins and describes the people as 'cowards'.[142] Such claims acted to reassure potential colonialists that they would not face significant opposition.

Pan concludes by lamenting the loss of Chinese territory under the Ming empire.[143] Although we can be certain of neither the intentions of Pan's text nor its reception, this statement suggests that the merchant-scholar was urging Qing intervention in the former Chinese colony, the rulers of which had been unwilling or unable to curb the 'piratical' activities of Ming refugees.[144] Although further verification is necessary to support the suggestion, Arnold Vissière believes that Pan's account eventually influenced the expedition to Tonkin of the Qianlong Emperor in 1788–1789.[145] As this brief comparison demonstrates, many of the techniques of both describing and 'othering' societies that were regarded as potential targets for colonial settlement were shared between the powers competing for territory in the unstable maritime world of Southeast and East Asia in this period.

Just as Baron fashioned his own 'English' identity, he presented to his English patrons an image of Tonkin that consciously used the tropes of the travel narrative to entice the Company to further invest in a settlement, from which he might have benefited, by offering evidence for the ultimately inferior and docile nature of the people to be ruled. This picture should not, of course, be equated with his real view of the society. Baron may have altruistic reasons for urging the English and other foreigners to invest in Tonkin, where he had experienced a period of increasing poverty since the introduction of tighter state control.[146] The cargoes of rice that he brought from Siam in 1682 were perhaps intended to relieve the famine that was ravaging the country at the time as well as for his personal profit.[147] In a later account of the famines of Bengal, Baron writes, 'there can be no advantage more uncomfortable than that which arrises from the poverty and misery of the poor tho it may be as well Charity as Interest to deal with them sometimes'.[148] Although Baron's background was unusual for the author of an 'English' natural history, I argue in the chapters that follow that many of those who filled similar go-between roles presented only certain aspects of the other worlds they were familiar with to their European correspondents, often exhibiting expected attitudes rather than simple reflections of real prejudices.

'Not according to art': The illustrations to the *Description of the Kingdom of Tonqueen*

Ibn Muhammad Ibrahim notes that, being eager to learn about the rulers and customs of other countries, King Narai 'sent everywhere for pictures depicting the mode of living and the courts of foreign kings'.[149] The exchange of images alongside other 'curiosities' had become a standard feature of early modern diplomacy as well as an occasional focus of trade. The Mughal Emperor Jahangir had received paintings of the English nobility, presented to him by Ambassador Thomas Roe, with enthusiasm, while a depiction of the naked Venus brought to Japan had aroused interest as well as confusion.[150] Jean-Baptiste Tavernier, who reports conversing with the Shah of Persia on the subject of paintings of women,[151] also relates that his brother Daniel had 'stor'd himself with a considerable number of small curiosities, to present the King [the *chua* of Tonkin] and his Nobility', including pictures of courtesans.[152]

A consensus was evidently developing about the types of subjects that should be produced for the benefit of curious foreign observers. George White, brother of Samuel, who accompanied a Siamese Ambassador who crossed from France to England in 1684, showed to scholars in London maps of China, the world, and Siam, along with a depiction of the King's palace, naval parades ('triumphs') on the river in the capital, the funeral procession of the King, and 'curious figures of several machine relating to the fire works'.[153] Very similar scenes were depicted by Engelbert Kaempfer on his later visit to Siam as well as in Baron's images of Tonkin.[154]

The early modern artistic exchange between Asia and Europe[155] was already leading, by the late seventeenth century, to an export trade in paintings and wallpaper from China, and, by at least the 1730s, paintings were produced specifically for European markets.[156] The eighteenth century also saw the emergence of 'Company school' painting in India and the large-scale production of illustrations for works of natural history.[157] As will be discussed in Chapter 4, however, the practice of Asian illustrators collaborating with European collectors on works of natural history had already begun in the seventeenth century. The fact that the illustrations for ethnographic accounts were often based on original illustrations made in the place being described has often been neglected. However, several of the illustrations published in Engelbert Kaempfer's *History of Japan* are clearly based on originals by Japanese artists.[158] As Baron notes, Tavernier also boasted that his map and illustrations of Tonkin had been 'drawn upon the place'.[159]

The illustrations that accompany the text of the *Description* were reproduced from drafts that Baron sent to Hooke and Hoskins, which have recently been recovered.[160] There are a total of twelve illustrations in addition to the map, depicting 'Cha Cho', or Ke Cho, as seen from the river, with the palace, arsenal, and English and Dutch factories marked; a bridal procession; entertainments; sports and feats of agility performed at New Year; the process for selecting scholars or 'literadoes'; the *chua* giving audience; martial exercises; the court and inner courtyard of the *chua's palace*; the *chua* taking part in a religious procession; and a funeral procession. Baron notes that the images he had sent for the printer to work from were copies of originals made by 'a Tonqueener of Eminent Quality' and were 'true and exact, tho' not according to art'.[161] Seventeenth-century Ke Cho was a centre for art, paintings being sold around the modern Hoan Kiem district, and Vietnamese painters incorporated some Western conventions such as the 'snake' layout for processions.[162]

Baron's illustrations, particularly those that depict the court of the *chua*, are not dissimilar in style to a series of Hang Trong paintings that depict the ruler of the heavens on a throne surrounded by military mandarins in the same manner as an earthly ruler.[163] These paintings were normally made using woodblocks and ink washes. It would therefore have been straightforward for Baron to commission a local artist to produce his illustrations. Baron's copies are on European paper, using black pen and ink washes in grey, yellow, and red. Numbered keys describing the activities being depicted are added to the images. The manuscript images and labels contain some corrections, presumably made by the printer. The 12 illustrations include the detailed illustrations of costumes, buildings, and vessels that were common to Western and Chinese ethnographic depictions of the period.[164] The depiction 'The City of Cha-Cho, the Metropolis of Tonqueen' (Figure 1.1)[165] indicates the situation of the city using a compass rose of the type used in European maps,[166] but provides an illustration rather than a map of the city. A similar style is apparent in a late-seventeenth-century map of Jambi in Sumatra, produced in the region.[167]

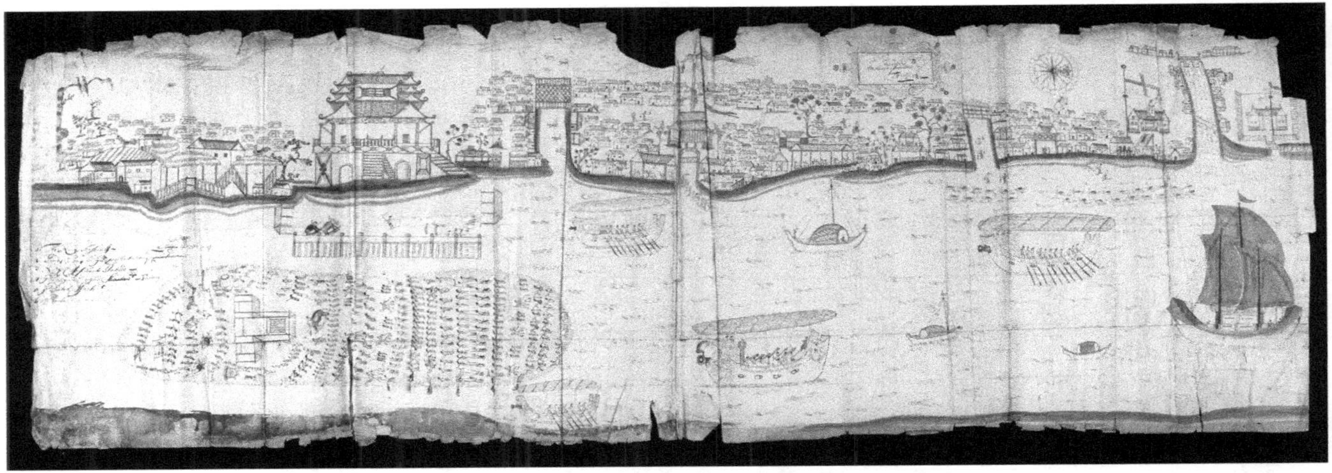

Figure 1.1 Draft illustration for Baron, *Description*, Plate 2: 'The City of Cha Cho, the Metropolis of Tonqueen', manuscript copy, courtesy of the Royal Society

The perspective is an interesting feature of several of the illustrations. In particular, the dais in the illustration of the 'Bova or King of Tonqueen' (Figures 1.2a and 1.2b) and that on which the judge is seated in the illustration 'The Manner of Their Dancing upon Ropes and Other Sorts of Play' as well as the buildings shown in several of the illustrations demonstrate the use of axonometric or parallel perspective as employed in Chinese and Japanese painting, so that there is no vanishing point: the view is always a 'bird's-eye' one, and the size of the figures and architectural elements in the foreground and background remains constant. The lack of a fixed optical plane means that in many cases the side and front of buildings can be seen simultaneously.[168]

The intaglio process of transferring sketches like those Baron made to a copper plate meant that the images were copied again by the engraver.[169] Some differences are therefore observable between Baron's originals and the printed illustrations. Notably, the European copyist has altered the perspective of the drawings, whilst retaining it in others. While the use of parallel perspective is made use of in some European maps of the period, it is rarely employed in depictions of landscapes, except in cases such as Kaempfer's work that draw directly on East Asian models.[170] As well as altering the perspective, the European copyist has given the people depicted in the illustrations more European features, whereas in Baron's originals they are distinctively Asian. That the copyist was probably unconscious of making this alteration supports the idea that racial difference was largely invisible as a marker of difference even in mid-eighteenth-century Europe.

The draft version of the map that accompanies the printed *Description* has not yet been recovered. Baron says that it was 'drawn and computed' from two others in Tonkin. As this comment implies, maps were produced and circulated throughout the region. At Batavia, the Dutch had established a cartographical office by the 1620s, and by 1630 had also established a hydrographical office. Baron's map (Figure 1.3) bears a strong resemblance, though is not identical, to Blaeu's map of the same region produced in around 1663: perhaps one of the two sources he mentions.[171] Although the EIC were less organised in their production of maps, it is clear that charts were produced and circulated within Asia, as Baron's reference to 'our sea draughts' shows.[172] The practice among all the European companies of copying and incorporating information from maps produced by their rivals was common, and will be discussed further in Chapters 2 and 4.

A close reading of the text of the *Description* and attention to the images that it includes demonstrate its hybrid nature. It includes several

(a)

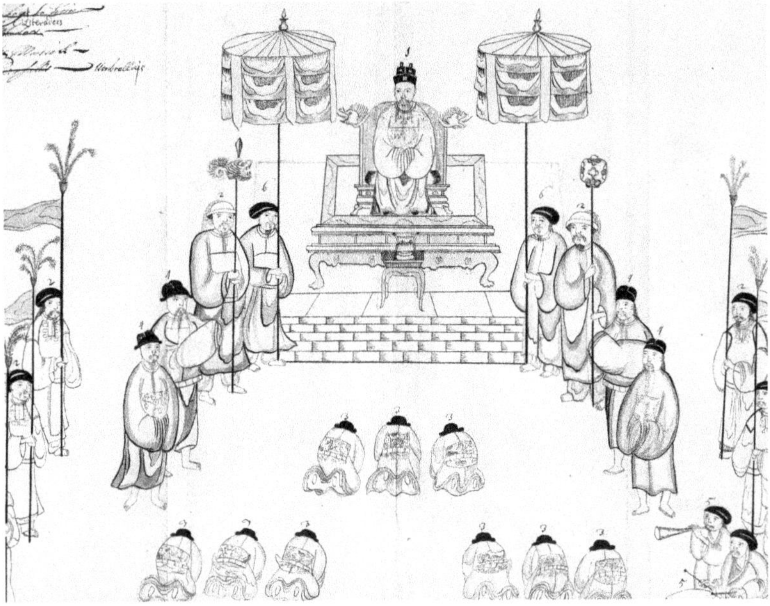

(b)

Figure 1.2a and b Baron, *Description*, Plate 8: 'The Bova or King of Tonqueen When He Gives Audience': in manuscript, courtesy of the Royal Society and published version, courtesy of the British Library

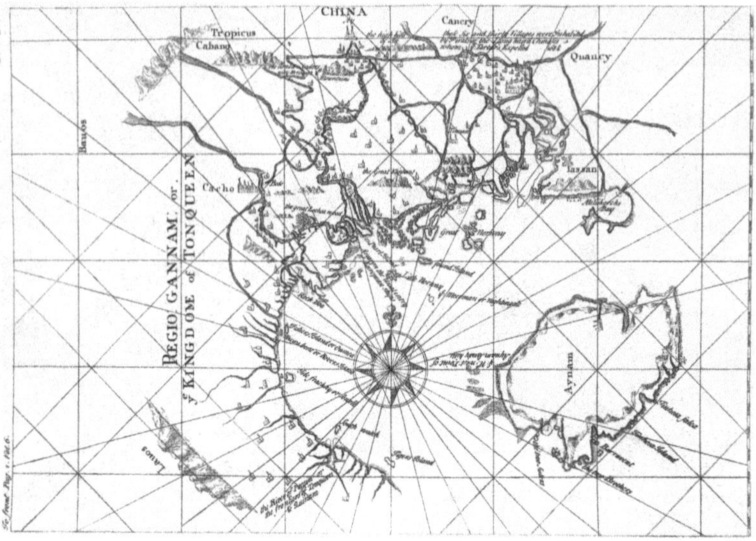

Figure 1.3 Baron, *Description*, Plate 1: 'Regio Gannam or the Kingdom of Tonqueen', courtesy of the British Library

features that are shared with Chinese descriptions of Vietnam and draws on illustrations made in Tonkin by local artists. Closer attention to the process of compiling natural histories like Baron's challenges any straightforward reading of them as repositories of European prejudices, suggesting that derogatory tropes about other societies were both shared by several colonising powers of the period and employed knowingly by their authors in order to achieve specific aims. While Baron's other identities are often hard to discern under the English persona he cultivates in his writing, the mixture of styles Baron uses in his illustrations provides a small demonstration of his different perspectives on the country of his birth.

Other selves: Amoy and after

Having despatched the *Description* in 1686, Baron was preparing for his next diplomatic assignment, in Amoy. Gyfford needed Baron's help to negotiate a new set of power relations that had emerged in the region as a result of the southward expansion of the Qing forces into the coastal areas and outposts until recently held by the Ming loyalists.

In Taiwan, where Baron had negotiated the founding of a factory in 1672, the English ambition had been to inherit the profits gained by the

VOC from exporting deerskins and sugar before their 1661 expulsion.[173] As in Tonkin, however, the English factors had failed to break into the regional trade in any profitable way during the 1670s.[174] Nevertheless, they were granted permission to trade in Amoy, which at this stage was also controlled by allies of Zheng Jing, son of Zheng Chenggong. A factory was opened in Amoy in 1677 with Taiwan placed under its control the following year.[175] In 1680, however, Amoy fell to the Qing, and the English factors fled after the Zheng leaders to Taiwan. In 1683, the Qing forces led by Shi Lang (known to the English as Sego or Seco) conquered Taiwan,[176] exacting tribute from the English factory there.[177] However, under the new regime, they were once again permitted to trade with Amoy, where some factors were despatched, meeting with those men who with Baron's help had fled there from Siam after being imprisoned by Phaulkon.[178] Given his inguistic and diplomatic skills, Baron was the ideal candidate to head a voyage to Amoy and consolidate the trade there.[179]

At Amoy, Dutch, Siamese, and Portuguese traders were taking advantage of the new Qing rulers' decision to liberalise trade at the port. Although two English ships that arrived there in 1685 had been allowed to trade, they had not been able to procure exemption from the customs duties imposed on all merchants.[180] Baron arrived in May 1686 and entered into negotiations with Shi Lang. However, he fell out with his subordinates, who complained of his closeness to the Chinese and neglect of them, while he complained that his 'malicious and refractory' underlings had refused Shi Lang's proffered agreement.[181] Baron eventually boarded a Chinese ship, apparently to act as navigator, translator, and mediator on a voyage to Fort St George. However, Captain Burton, another English trader in Amoy, wrote to the Council in February 1687 to report Baron's 'fatal accident'. The Chinese vessel had capsized and sunk with all the crew, including Baron, aboard. Burton adds that without a 'wonderfull act of providence', Baron must be dead, and then goes on to defend him against allegations of deserting to the Dutch, adding that 'were Mr Baron now alive to hear this story, it would Even Scare him to his grave'.[182]

Baron had been neither drowned nor scared to death, for he appeared in Malacca some time later.[183] His explanation for this 'wonderful act of providence' is not recorded, and the whole episode is subject to multiple interpretations: did Baron defect to the Dutch once more, as his accusers claimed?[184] Was he scared enough by the accusations to lie low until he could assure himself of at least some welcome in Madras? Did the 'expensive and extravagant wanderer', as the Directors called him,[185] simply decide to engage in some private trade with the profits of the China voyage? Or did Baron really have yet another miraculous escape via one of the fishing boats Burton mentioned nearby the wreck?

By February 1688, Baron had returned to Fort St George accompanied by three Chinese representatives from Shi Lang, to negotiate with the Council for a 'mutual trade'. Baron had apparently encountered these men in Malacca, where they had ended up after escaping from some English pirates who had taken their ship. On their arrival, 'they were handsomely received in the fort'.[186] Baron himself received a less warm welcome given the directive from London to sack him, the testimonies of his fellow merchants in Amoy, and the disappearance of the ship, profits, and documents from the voyage.[187] The replacement of his friend William Gyfford with Phaulkon's then ally, Elihu Yale, as President at Fort St George is also likely to have affected his prospects.[188] However, Baron remained too useful to the Council to be dispensed with altogether and was called on to advise concerning the treatment of the Chinese representatives.[189]

After this incident, Baron apparently settled for some time in Madras: he and his wife appear in a February 1688 census[190] and in the survey of houses in 'white town' in August 1688 as the tenants of one Bernado Medom.[191] Baron then disappears once more from the Fort St George records until 1693, when he was arrested for 'tampering with some of the soldiers in the garrison to draw them with him to Siam'. For this purpose, he had left white town and taken up residence with a mestizo army drummer in a nearby fishing village.[192] There, as he admitted, he had met secretly with the Siamese Ambassador (perhaps he who had arrived in 1685) and spoken to some of the inhabitants and soldiers about leaving to form a new colony under his leadership in Ayutthaya. This incident was Baron's last (recorded) attempt to establish his own trading settlement: something that appears to have been his ultimate goal throughout his adventures. It is notable that in this instance he drew on his mestizo identity to attract followers. Although there is little mention of them in the literature, aside from infamous examples such as the '*firingi*' pirates of Chittagong,[193] semi-autonomous communities of Eurasians inhabited many of the cosmopolitan cities of early modern Asia.

In 1695, Baron gave the EIC one final piece of advice, which has survived in the copybook of an eighteenth-century governor of Bombay.[194] This interesting document provides a bird's-eye view of the Company's trade in that year. By the time Baron dispensed this last piece of advice, however, he had disappeared from the lists of European residents and seafaring men of Madras, and his burial is not recorded among the lists that remain at St Mary's church. Indeed, in his last years, he seems to have abandoned the 'English' identity he had earlier taken such care to create. Did Baron remain in the fishing village or San Thome after the incident in 1693? Did he ever undertake his intended voyage to Siam? Or did he return to Tonkin? In his *Description*, in a passage concerning the presents of food made to the dead, Baron writes, 'I have, in jesting, told some of them I would not like to die a Tonqueeneese, were it only because the custom of the country, whilst living, allowed me three meals a day, but when dead they would feed me but once a year'.[195] Nevertheless, it is possible that Samuel Baron returned to the place of his birth, 'to die a Tonqueeneese'.

In the final analysis, Baron failed to construct robust enough networks to consolidate his wealth and power through the EIC. The uncertain fortunes of his patron Gyfford were one reason for this. Hooke's apparent neglect of Baron's manuscript *Description* was another. In addition, Baron's construction of a European identity was perhaps not quite complete enough to survive into a new century. Despite the official support for racially mixed marriages in the early governments of the trading companies in Asia, attitudes that stressed 'purity' of blood were already becoming common among both Europeans and Asians of the period. In language that echoes that used for the go-between and the 'interloper', people of mixed backgrounds were often described as threatening, 'disorderly', or 'turbulent' because of their dual allegiances.[196]

By 1700, while the idea of 'race' was still not fully formed, the boundaries between 'white' and 'black' were hardening. In Madras, the walls dividing two sectors of society were complete by 1710,[197] and by mid-century, Armenian and Portuguese inhabitants were ejected from 'white town' despite their Christian faith.[198] While Baron might have been able to become English, in this new climate any children would have struggled to do so, and perhaps this is one reason for his eventual disappearance from 'white town'. In contrast, European contemporaries of Baron's, such as Elihu Yale and Thomas Pitt, who were involved in the same process of brokerage through local and international networks of patronage, became rich and influential, ensuring a place for their posterity among British and Anglo-American elites.

Conclusions

Baron's extraordinary life provides a demonstration of the vital importance of *passeurs culturels* in this period. Baron had personal contacts with all the crucial figures in the complex configurations of power that were being played out during this period: he negotiated with both Zheng

Chenggong and Shi Lang; founded the EIC settlements in Taiwan and Tonkin; and acted as a merchant, adviser and diplomat to the rulers of both Tonkin and Siam. As well as being promoted to one of the most important positions available to a factor in the Company's service abroad, as putative head of the factory in Amoy, he maintained contacts with the Dutch companies and Portuguese and French missionaries. In addition to the skill in languages and negotiations required to straddle these different allegiances, this required constant manipulation of his own identity.

As well as shaping his identity as a diplomatic go-between, Baron created an reputation as a collector of useful natural historical information in his exchanges with Hooke, Boyle, and Hoskins. In doing so, certain ways of engaging with the prevalent culture of curiosity were useful: frequenting the right coffee shops, using the right sort of language to discuss processes such as petrification, and understanding the components of a written 'natural history', in his case the *Description of the Kingdom of Tonqueen*. As the example of Psalmanazar shows, the ability to participate in natural philosophical discourse was often more important than the veracity of the information that was provided. Like Psalmanazar, Baron offered the unrivalled knowledge of a native couched in the language of a philosophical traveller and combined with a profession of Protestantism. While 'gentlemanly' status has been regarded as important in the creation of an identity as a natural philosopher,[199] it was less important for collectors of information: others in the Royal Society's circle included William Dampier, who is often described as a pirate, but whose ability to package the results of his globetrotting as 'useful' information won him considerable patronage. The information that was supplied to natural philosophers was directly relevant to trade; even pieces of information that may seem frivolous, such as the fondness of the Tonkinese for birds' nests, were employed by the EIC in its search for profitable goods to be exchanged.[200]

All the figures I will introduce later in this book as providers of information – botanical, medical, linguistic, cartographic, and ethnographic, as well as political and economic – demonstrate similar trajectories to Baron's in and out of the service of the EIC and Asian rulers. Their lives demonstrate that collecting was not a leisured or peripheral activity, but a dangerous competition for supremacy in the competing networks of patronage that linked Europe and Asia in the struggle to control the trade of South and East Asia. Whether they died wealthy men like Robert Knox or slipped into obscurity like Samuel Baron depended on the strength of the ties that they managed to build up through the distribution of information, be it in the shape of travel accounts, physical objects or 'curiosities', or verbal reports. The authors of such works shaped their representations to conform to the expectations of their audiences, but also to their own advantage: revealing some facts and concealing others.[201] However, the works were also themselves hybrid products, part of both the process of mediation and the definition of difference.

During Baron's lifetime, the exchange of Ambassadors and curiosities flowed in all directions, and the creation of new multicultural trading emporia was determined by the shifting political currents of the maritime world, rather than directed from the distant centres of the Chinese court or directors of the European companies in Amsterdam, London, or Paris. This period was one in which the occupation of the go-between assumed a particular importance. Those who could build up international networks of patronage through their skills in brokerage could rise to the highest positions in the Company. As a young man, Baron was able to make demands on the Company for a high salary and position as second in command in the Japan factory on the basis of his status as one of the few people capable of negotiating settlements in Tonkin and Taiwan. However, such men could also become the most dangerous and therefore had to be either integrated into the Company or denounced as 'traitors', 'deserters', and 'interlopers' to justify their removal. This was the purpose of the type of writing about other EIC servants as well as their European rivals that Baron produced and that was directed against him.

The period of openness did not last. Japan had already banned all foreign travel by Japanese by 1635,[202] and by 1717, the Qing emperor had also prohibited all Chinese from trading with the Southern oceans, while the Chu I-kuei rebellion on Taiwan in 1720 reinforced the Qing tendency to see the maritime world as a centre of subversion.[203] After the territorial expansion into India following the victories of the mid-eighteenth century, the EIC was better able to regulate the work of its European agents and was more assured of their continued loyalty. As Raj notes, the eighteenth century saw the professions of go-betweens assume some regularity, with training in languages as well as classics, sciences, law, and other subjects provided for both European administrators and the indigenous literati.[204] While still crucial to the expansion and consolidation of colonial rule, the standardisation of their professions meant that these men were ultimately replaceable. Asian powers' need to use foreigners in their conflicts during the late seventeenth century had placed the EIC in an advantageous position in terms of the regional trade.[205] Go-betweens such as Baron and the descriptions they provided were essential to this process.

2

Linguistic Landscapes: Early English Studies of Malay and the EIC in Maritime Southeast Asia

Introduction

On his way to the Malay Archipelago, the future founder of Singapore Stamford Raffles (1781–1826) carried an 'antediluvian book', from which he learnt his first words of Malay.[1] The book in question was probably the *Dictionary English and Malayo, Malayo and English* (1701), attributed to the trader Thomas Bowrey (c. 1655–1713).[2] The publication of this work – the first complete dictionary in English and an Asian language[3] – was the culmination of a period of interest in Malay among English scholars that also saw the publication of the four gospels in Romanised Malay and the compilation of several manuscript grammars. These Malay materials, like other works of scholarship discussed in this book, were produced by traders with practical experience abroad, in this case in maritime Southeast Asia, aided by English scholars, and encouraged by patronage from members of the EIC.

Why should Raffles be carrying a work that was already over 100 years old at the time of his first voyage east? Bernard Cohn states that no significant number of the EIC's servants learnt Asian languages until the 1740s or 1750s and that the first English linguistic forays were conducted largely in ignorance of the earlier investigations by the European powers who had settled and traded in Asia before them.[4] Here, I will contest both claims, arguing first that the EIC Directors were extremely eager for their servants to acquire Asian languages from the outset and that during the turbulent first century of the Company's existence, some achieved a degree of fluency that would not be surpassed for some time, necessitating the continued reliance on older works by Raffles and his contemporaries. Second, I will demonstrate that, far from being independent of earlier endeavours, early English scholarship on Asian languages developed as part of a competitive engagement with other European powers. Notably, English scholarship on Malay during the seventeenth century can be envisaged as part of the struggle with their Dutch rival, the VOC, for territory, trading privileges, and supporters in Southeast Asia, with its coveted spices and entrepôts. By the time the *Dictionary* was published, the Spanish were the new rivals and English ambitions had begun to focus on the South Seas: an interest that would eventually lead to the infamous 'bubble' of 1720.

The study of Oriental languages is the original sense of the term 'Orientalism'. As one author puts it: 'the Other is located most fundamentally in language, the medium for representing selves and others.'[5] This chapter examines how these English linguistic works on Malay represent both contemporary European understandings of language and the actual interactions of European travellers with contemporary speakers of Malay. Studies of dictionary-making and language learning manuals often argue that these scholarly endeavours act to define the boundaries of languages through factors such as the selection of script in which a language should be represented, the inclusion or exclusion of loanwords, the definition of 'high' and 'low' forms, and the formulation of theories about the origins and spread of languages.[6] As I will demonstrate through a study of three manuscript Malay grammars, English scholars often applied to Malay ideas drawn from their understandings of other languages – notably Latin and the European vernaculars – as well as contemporary ideas about the potential for a 'universal' or 'philosophical' language. While the materials are rooted in seventeenth-century debates, they also point the way to later works on comparative linguistics.

Materials produced to teach languages invariably reveal something about the mental and social worlds from which they have emerged and into which they expect the learner to enter.[7] The *Dictionary English and Malayo, Malayo and English* includes an extensive list of words with their meanings in English and Malay, running to 468 pages in all, as well as several 'miscellanies' or short sentences, and a series of dialogues. Additional materials include a table of conversions between the Gregorian and Hijri calendars and a Chinese compass. Here I use these materials to build up a picture of the varied interactions that English traders like Bowrey had in the seventeenth-century Malay Archipelago.

Linguistics and politics

Trade and diplomacy

Maritime Southeast Asia was of key strategic interest in the early modern world because of the commercial and diplomatic potential it offered.

The commodities that attracted traders included black pepper from Java, cloves from Maluku, nutmeg from Banda, and camphor from Borneo. Merchants from the Middle East and China arrived in the region from the fifteenth century[8] and Italian, Spanish, and Portuguese missionaries and traders from the sixteenth.[9] The Dutch and English trading companies joined them from the seventeenth century. These newcomers in the region quickly realised the necessity of acquiring 'Malay', an umbrella term describing the Austronesian languages spoken across the Philippines, Malaysia, Indonesia, and parts of Vietnam and widely used elsewhere in the region as a language of trade.

Relations between Indonesia and England began with the EIC's first voyage, when James Lancaster established trading agreements in Aceh in Northern Sumatra and Bantam (or Banten), a thriving port on the north-west of the island of Java close to the Dutch settlement of Batavia (modern Jakarta, Indonesia). The earliest Malay-language materials in England are royal correspondence, dating from 1602 onwards.[10] The Bodleian library received two old Sundanese and old Javanese palm leaf manuscripts in 1627.[11] Archbishop Laud presented two Malay manuscripts in 1633 and 1635. The first, MS Laud OR 291, is a copy of the *Hikayat Sri Rama* – a Malay version of the *Ramayana*.[12] How the infamous archbishop acquired this manuscript is unknown, but a Chinese stamp within its pages suggests it once belonged to a Chinese bibliophile. MS Pococke 433, bought for the Bodleian in 1693, is the *Hikayat Bayan budiman* or *Hikayat Khoja Mimun*, a collection of tales told by a parrot to a merchant's wife. This was probably based on the sixteenth-century Persian *Tuti Nama*, itself a version of an older Sanskrit work.[13]

Competition between the English and Dutch for control of the Southeast Asian spice trade was fierce throughout the seventeenth century. Although the 'accord of 1619' divided control of the spice trade of the Moluccas, Banda, Amboina, and Java between the two nations, the agreement soon broke down.[14] The infamous Amboina Massacre of 1623 became the subject of a war of pamphlets, plays, and paintings in Europe.[15] While Dutch supremacy was widely acknowledged in the region by the 1630s, English merchants continued to trade during the British Civil Wars, particularly for pepper on the Sumatran coast. The need to fend for themselves prompted some to acquire more than a smattering of Malay. The Courteen Association established a factory at Aceh in Sumatra but it was short-lived and impoverished.[16] The three Anglo-Dutch Wars (1652–1654, 1665–1667, and 1672–1674) were fought abroad as well as at home. The VOC's drive to bring the Malay world under the control of their base in Batavia involved enforcing trade monopolies, collaborations with enemies of Aceh like Johor and Ternate, and the use of military force to exclude Asian and European competitors. The Dutch imposed an embargo on Bantam between 1651 and 1659, proving detrimental to the English traders there but leading them to found some smaller trading settlements in Sumatra and Java.

At the time of the reformation of the Company in 1658–1662, English trade in Sumatra centred on Aceh,[17] where Thomas Bowrey wrote 'all masters of English ships and vessels are very Nobly Entertained'.[18] Despite this warm welcome, the port was firmly under the control of the Sultana Taj al-'Alam Safiyyat al-Din Syah (r. 1641–1675), and the EIC was never permitted to build its coveted fortified settlement. Perhaps for this reason, diplomatic overtures made by the Sultana to King Charles II were largely ignored.[19] By the 1660s, Bantam had developed into a thriving port under Sultan Abdulfatah Ageng (1651–1682), with a population of over 100,000,[20] exporting opium and textiles throughout the region to the detriment of the Dutch factory at Batavia.[21] However, by the mid-seventeenth century, the English presence in both Java and Sumatra was again under threat from the Dutch. The English therefore stepped up efforts to forge diplomatic ties in the region and, from the 1660s, supplied arms to the rulers of Bantam.[22]

In August 1675, two visitors from Bantam arrived in London, accompanied by an elephant. Less information is available about the men than about their exotic charge, but they stayed in London until February 1676.[23] More is known about the next visitors from the Malay world: an embassy from the ruler of Bantam, Sultan Abd al-Kahar, in London from April to July 1682.[24] The circle around the two Ambassadors included members of court and parliament and interested scholars. EIC agents with experience in the Malay world were present to translate and advise.[25] The embassy delivered a letter in Arabic from Abd al-Kahar, the *raja muda* or 'young King', as the EIC referred to him, to distinguish him from his father, Sultan Ageng. Abd al-Kahar's letter called for urgent aid against the Dutch, who he claimed have 'invaded the territories of my neighbours and are now entertaining the same designs on my own'. The Sultan requested large-scale military assistance, including the supply of 4,000 muskets and an expert in manufacturing cannons.[26]

To translate the letter, the court employed the Orientalist Thomas Hyde (1636–1703), at this time Bodleian librarian as well as Secretary and Interpreter of Oriental Languages to the court and who, from 1701, would become Professor of Arabic and Hebrew at the University of Oxford. Hyde's interest in comparative linguistics and unrivalled access to the Malay manuscripts and letters in the Bodleian and at court rendered him central to all the seventeenth-century English projects on Malay.

Hyde's notes on the embassy highlight the reluctance of the EIC to cooperate with the Sultan's requests.[27] Hyde's comments were probably intended partly to enhance his own employment prospects: he stresses that the Crown should accept his services, rather than relying on the 'coarse' translations of merchants. However, it is clear from EIC correspondence that efforts were made to limit the size of the Sultan's embassy[28] and to discourage a return embassy from Charles II.[29] Despite the Dutch threat, the EIC were perhaps unwilling for the English King to augment the already formidable military might of the Sultan.[30] Whatever its reason, the EIC's failure to take the message borne by the embassy seriously had consequences. The returning Ambassadors were met off the coast of Bantam in early 1683 with the news that the English had been expelled by a Dutch-backed rebellion of the 'young' Sultan after his father, Sultan Ageng, had tried to resume power. The ship was diverted to Batavia, where the factors had set up a temporary base.[31] The overthrow of Sultan Ageng saw the end to any credible alternative to Dutch power in Sumatra.[32]

The next English attempt at diplomatic relations with the Malay world came with the mission of Ralph Ord and William Cawley to Aceh in 1685.[33] The factors were again denied permission to build a brick fort, on the grounds that they might enslave the inhabitants were it granted. However, while there, the rulers of Pariaman – on the west coast of Sumatra – approached them and offered a settlement. At the last minute, however, the mission sent from Fort St George to settle in Pariaman was diverted to Bencoulen, 300 miles south.[34] This met with fury from the London Directors, who, as will be discussed in Chapter 6, considered the settlement an unhealthy place. Over the next few years, subordinate English settlements were begun in Sumatra: at Indrapura, Manjuta, and Sillebar. By 1700, factories had also appeared in Triamang, Ketahun, and Seblat.[35]

By the early eighteenth century, the intense and violent rivalry between the English and Dutch had largely abated, and Spain was the new enemy from the time of the War of the Spanish Succession (1701–1714). By the time Bowrey's *Dictionary* was published, English ambition in the Malay Archipelago had begun to shift eastwards, towards the South Seas and the lucrative galleon trade from Manila. English interest in the Pacific was also piqued by the adventures of William Dampier, who sailed the Southern Seas, mapping the South Seas islands, and had landed in Australia twice, in 1688 and 1699. These adventures were described in his *New Voyage around the World*.[36] A powerful illustration of the allure of the Pacific was provided by the 'painted Prince' Giolo or Jeoly, whom Dampier brought back with him from the island of Gilolo in 1691.[37] Before his death, probably in 1682, Giolo visited Oxford and conversed with Thomas Hyde, perhaps in Malay. This encounter resulted in Hyde's anonymous account of Giolo's life and religion.[38]

The EIC and language

The dedication page of the *Dictionary English and Malayo, Malayo and English* is divided in two: containing the names of the 'Governor, Deputy Governor and Committees of the Honourable East India Company' (the 'old' Company) on the right-hand side, and on the left the 'Directors of the English East India Company' (the 'new' Company, containing several disaffected members of the 'old', including Streynsham Master, former methodiser of EIC affairs,[39] and George White, brother of Samuel, who had orchestrated the Siam embassy to London mentioned in Chapter 1). By 1701, these rivals were about to begin their consolidation into the United Company, a process in which Thomas Bowrey took an interest.[40] Bowrey says that the work was undertaken for the promotion of trade in the Malay-speaking world and that the Directors and Governors of the two companies had encouraged the publication of the *Dictionary*. In practice, this probably meant that individual members had subscribed in advance, providing funds for printing the work.

Why dedicate a linguistic work to a trading company? In fact, Bowrey was not unusual in doing so, but followed both Dutch and English precedents. Dutch scholarship had moved swiftly from a Malay wordlist of 1599[41] to the 1603 publication in Amsterdam of a book of Dutch and Malay dialogues compiled by Frederick de Houtman[42] and dedicated to the 'seventeen gentlemen' of the Dutch East India Company (VOC).[43] The four gospels were translated into Malay between 1612 and 1646,[44] and a work entitled *Spiegel van de Maleysche Tale* or 'A Looking-Glass of the Malayan Tongue' was produced in 1612 to instruct young converts to Christianity.[45] Malay dictionaries were published in the Hague in 1623[46] and Batavia in 1677,[47] along with several texts printed in the Dutch headquarters. These dictionaries, training manuals, and scriptures were produced by alliances of VOC merchants and Dutch scholars and deployed to bolster VOC power by attracting local Dutch-speaking allies and converts to Protestant Christianity. The spread of Malay along with Christianity in Dutch Southeast Asia often had a deleterious effect on the indigenous languages of the region.[48]

From the beginning, the EIC – like its Dutch rival – was acutely aware of the need for its servants to acquire competency in languages spoken in Asia, particularly regional lingua francas like Arabic, Persian, Malay, and Portuguese. While the resources and organisational structure to adequately support language learning were lacking in the first century of its existence, the EIC tried various methods to promote language acquisition in all their Asian factories, including the appointment of teachers[49] and offering rewards,[50] and occasionally some more draconian means, including isolating young writers from English-speaking communities to force them to learn other languages[51] and even threatening them with punishment for speaking English.[52] Realising that children were more likely than adults to acquire proficiency in other languages, three 14-year-old 'hospitall boys' were sent as writers to the Bencoulen factory in 1688.[53] An oft-expressed aim was to minimise the use of interpreters and go-betweens, who, as discussed in Chapter 1, were widely regarded as devious and untrustworthy. Bowrey writes in his preface that he had learnt Malay in order to dispense with the service of 'a prevaricating interpreter, as they generally are'. These efforts bore some fruit: by 1690, William Dampier reported that the chief at Indrapura spoke very good Malay.[54]

Unlike the VOC, the EIC were slow to set up printing presses in their settlements. To encourage the production of language-teaching materials in England, the EIC partnered with scholars. In such cases, Company members stressed their pious intentions to promote conversion to Protestant Christianity in Asia above the more mercenary aims of dispensing with brokers. In the case of Malay, members of the company had partnered with scholars and churchmen on several previous occasions to produce linguistic materials. Each time, their interest coincided with the diplomatic relations described above. In the later part of the seventeenth century, English scholars tended to rely on redeploying Dutch materials. This was the case with Augustus Spalding's version of Frederick de Houtman's Dutch-Malay dialogues,[55] produced via a Latin translation. Spalding's work was dedicated to the EIC and financed by Richard Hakluyt, author of the *Principall Navigations* (first published in 1589). Copies of the dialogues were sent out with the EIC fleet. While the work was of questionable use, it established a precedent for collaborations between merchants and scholars to produce linguistic materials, encouraged by incentives from the EIC's Directors.[56]

In the later seventeenth century, Robert Boyle, gentleman scientist and religious adviser to the EIC, was an important source of patronage for translations of Christian texts. In the 1640s, he financed the Arabic translation and distribution to EIC factories of Hugo Grotius' *De Veritate Religionis Christianae*.[57] Several other translated works on Christianity were sent to the Bombay, Madras, and later Calcutta factories,[58] and the Surat factory library contained bibles in several Oriental languages by 1664.[59] Robert Boyle also made ambitious plans for English factors to receive training in Arabic and mathematics and suggested that Asian youth should be instructed in English and 'European Learning'.[60]

In collaboration with the bishops of Oxford and Canterbury[61] and with the support of Josiah Child, then EIC Director, a committee was formed in 1682 to raise money for these ends.[62] Although the committee failed to provide ongoing training for young men in Oriental languages, it began an association between the EIC and missionary groups.[63] That the EIC factors had some success in sowing the seeds of both their language and religion on the ground is indicated by the mention of 'Malay Padres' in both Bencoulen and Tryamong factories by 1695. These men doubled up as interpreters and 'writers of Malay' to the English factories.[64] Bantam served as a central point from where converts to Christianity could be sent as intermediaries and translators could be sent out to serve more remote factories.[65]

Boyle also financed the *Jang Ampat Evangelia* (1677),[66] a Malay translation of the four gospels published by Thomas Hyde and the philologist and churchman Thomas Marshall (1621–1685). Marshall learnt Malay in Rotterdam, where he spent 24 years as chaplain to the Company of Merchant Adventurers before returning to England in 1672.[67] Marshall was an important source of Oxford University Press' collections of materials (matrices, types, and punches) for printing in exotic languages, which he procured from Dutch type founders.[68] Hyde's knowledge of Arabic and Persian meant that decoding the Jawi script was relatively straightforward. Despite Marshall's and Hyde's ambitions, the text of the *Jang Ampat Evangelia*, in Romanised Malay, was simply lifted straight from the Dutch scholars Ruyl, van Hasel and Heurnius' 1651 text. How useful the text was for the English readers is questionable, as is its relevance for Malay Christians. However, copies of the *Jang Ampat Evangelia* were sent to the Bantam factory.[69]

Original English works on Malay: Grammars and poems, 1680–1695

Thomas Marshall and Thomas Hyde continued to collaborate after the publication of the *Jang Ampat Evangelia*. In 1680, an advertisement of 'Books making ready for the Press' at Oxford listed a 'Grammar of the Malaian Tongue'. Among Marshall's papers, there are two manuscript drafts, probably by him, of a Malay grammar: the first original English work on Malay.[70] Despite their originality, Marshall's continuing reliance on Dutch scholarship is evident from a comparison between the versions of the Lord's Prayer in Malay given in different Dutch texts, from which he derives his own new version.[71] Both drafts of the grammar give the Jawi alphabet, with a discussion of the 'extra' letters that are included in Jawi, as compared with Arabic. An interesting feature of the grammars is that they give an archaic form of one of these letters, making it probable that the author had consulted some Malay-language literature.[72] In one of the grammars, Marshall also copied out a section of text from the Bodleian copy of the *Hikayat Sri Rama*.[73] The second draft of Marshall's grammar was sent to Gilbert Burnett, another clergyman who was interested in the EIC's sponsorship of Malay-language materials, and was annotated by a contact of his with experience of the Malay world.[74] In the event, despite the advertisement, Marshall's Malay grammar was never published.

The renewed interest from the EIC shortly before the 1682 Bantam embassy perhaps prompted a new attempt to publish a Malay grammar. One of the translators associated with the embassy, William Mainstone, became involved with Boyle, Marshall, and Hyde. A manuscript grammar under Mainstone's name remains among the Ashmole manuscripts at the Bodleian library.[75] Although it draws heavily on Marshall's drafts, it is more complete and includes a preface based on Marshall's 'Address to the English Reader' in the *Jang Ampat Evangelia*. Mainstone's grammar provides a comprehensive discussion of the grammar and parts of speech of the language. It gives the Malay words in Jawi script with transliterations and includes samples of the scripts of Java and Makassar. Towards the end are some materials in Chinese, apparently appended later.[76]

In May 1683, Mainstone wrote to Boyle, enclosing for his 'judicious and impartial censure' the final version of his Malay grammar.[77] Mainstone also mentions his intention to produce a Malay dictionary. No response is recorded, and Mainstone died shortly after his second letter to Boyle. Although Boyle recalled the grammar in 1684 and wrote to Hyde about it, the latter dismissed it as a 'pitiful trifle, not of any worth': an opinion that seems surprising considering that the grammar was in fact by far the most comprehensive work on Malay yet available in English.[78] A later request from Hyde to Boyle to recover the dictionary that Mainstone had compiled also proved inconclusive.[79]

The dismissal end to the embassy put an end to the EIC's interest in financing Malay-language projects for some time. English interest in Malay had not entirely dissipated, however. At some stage, probably before his work on the observatory of Ulugh Beg (1394–1449),[80] Hyde had a set of types made as supplements to the Arabic alphabet to supply the 'extra' letters in Malay, Turkish, and Persian. These were rather poorly made 'portmanteau types' from which parts must be trimmed off to make characters.[81] Their advantage for Hyde was that they served for all three languages. He used them to print a poem in Malay in the *Pietas Universitatis Oxoniensis*.[82] The poem, describing the laments of the King at his wife's death, was probably composed by Hyde himself, demonstrating that he had a basic mastery of Malay. However, it obeys English rather than Malay conventions, following a simple rhyme scheme, AA, BB, CC, DD.[83] As Annabel Teh Gallop notes, the poem contains some strange features, including a variable formation of the letter 'nga',[84] perhaps a result of the unreliable portmanteau types. After OUP had purchased Hyde's types in 1693, another Malay passage was printed in their *Specimen*.[85] This is a biblical passage, describing the visit of the Magi to Jesus in Bethlehem.[86] During 1688–1689, Hyde and Boyle corresponded about a project of printing the gospel of Luke and the *Acts* in Malay using Jawi script:[87] the passage printed in *Specimen* is perhaps an isolated fruit of these efforts. Hyde was soon to be involved in a new project, however, the *Dictionary English and Malayo, Malayo and English.*

Compiling the *Dictionary English and Malayo, Malayo and English*

Thomas Bowrey's career exemplifies the routes by which English traders made their fortunes in the half-century following the re-establishment of the EIC in 1657. Company bases in the Indies provided safe havens, but the real money was to be made outside the official sphere. This meant acquiring local contacts and know-how, which could facilitate participation in the 'country' or inter-Asian trade and be traded for patronage within and beyond the EIC circle. Bowrey arrived in India in 1669, assisting with the Company fortifications at Madras that year.[88] Despite maintaining close relations with many EIC employees, owning several East Indiamen, and holding stocks in both 'old' and 'new' companies at various points, Bowrey made his fortune as a private trader.

Bowrey travelled widely in the East Indies, Africa, and the Americas and left an extensive set of charts of the Malay Archipelago, the River Hugli, and the Persian Gulf and Sri Lanka.[89] He was involved with several schemes to set up banks and 'plantations', offering his expertise to potential patrons including the King of Prussia[90] as well as to the EIC and South Seas Company, to which he bequeathed his journals of Africa and the South Seas, respectively.[91] His account of his travels and observations around the Bay of Bengal from 1669 to 1679 was published in the twentieth century.[92] His only publication during his lifetime was the *Dictionary English and Malayo, Malayo and English*. While he was also credited with the authorship of the *Dictionary of the Hudson's Bay Indian Language*, this attribution has been questioned.[93]

Bowrey claimed to have begun the *Dictionary English and Malayo, Malayo and English* during idle hours on his voyage home when he sailed as a passenger back to England in 1688 and enlarged it with 'some helps I have attained since'.[94] One of the manuscript notebooks among his papers does appear to represent an early draft of a Malay-English vocabulary,[95] but the words it contains are a tiny fraction of those included in the final work as well as diverging from it in the spelling of particular terms and lacking the diacritical marks that are used in the final version. Thus, while the *Dictionary* is presented as the sole work of Thomas Bowrey, like the other works of scholarship discussed in this book, the story of its creation is more complex.

The main source of the help Bowrey mentions was Thomas Hyde. Despite the disparagement Hyde had earlier expressed about the linguistic skills of merchants, he was deeply involved with the EIC, having relied on their networks around Surat to acquire the materials for his magnum opus *Historia Religionis Veterum Persarum*.[96] Bowrey and Hyde corresponded frequently between 1700 and 1702, collaborating on the *Dictionary* and planning further linguistic projects.[97] As well as Malay, they had copper plates engraved of alphabets and some basic words of vocabulary in 'Tartar', Sinhala, Sanskrit, Telugu, and Tamil.[98] These plates were distributed among EIC factors for correction and additions and eventually printed in Hyde's posthumous collection.[99] Hyde supplied Bowrey with further vocabulary, including a list of Malay words in Jawi script that were finally printed at the end of the *Dictionary*.[100]

A more serious challenge to Bowrey's claim to authorship of the *Dictionary English and Malayo, Malayo and English* was put forward by one Henry Smith, the author of several corrections to a proof copy of the work that ended up among the papers of William Marsden (1754–1836) in the archive of the School of Oriental and African Studies (SOAS).[101] Temple identifies Smith as the brother of Bowrey's uncle, who was ordered to be sent back from India in 1669 for being a 'ne'er-do-well',[102] but seems to have disobeyed this command, leaving EIC employment to trade privately and acquiring competence in Malay.[103] In one section of the SOAS *Dictionary*, Smith adds a complaint:

> My Dictionary, which the foregoing should have bin onley the Coppy off, is so strangely perverted thro' [deleted: Ambition] Ignorance of the genuine Elegence and Meaning of the language, that it would have puzled a learned Malay to have pickt out the meaning of the short sentences, for they are very concisive in there discourse, using noe circumlocations or tautologie.[104]

Assuming that Smith's claim is true, how did Bowrey come to acquire and enlarge Smith's original dictionary, and what of the 'ambition' that had prompted it? The reason Smith was apparently willing to consign his own work to Bowrey is revealed in the correspondence between the two men. During summer 1700, while the dictionary was being prepared, Smith was awaiting trial at Newgate prison, probably on charges of piracy.[105]

Bowrey helped Smith out of his difficulties and later employed him as an agent, or 'mole' as Smith referred to himself, in Scotland, where Bowrey was attempting to recover compensation for his ship, the *Worcester*.[106] In return, Smith presumably consigned his linguistic work to Bowrey, who enlarged it with the aid of his own memories of the Malay language and the other materials supplied to him by Hyde. While Marsden claimed that Bowrey had 'derived... no advantage whatsoever from the preceding publications',[107] in fact, the *Dictionary* drew heavily on earlier material, whether acquired directly as in the case of Smith's dictionary, or mediated through Thomas Hyde.

The 'ambition' that prompted Bowrey to enlarge on Smith's original work was associated with Bowrey's own plans for a new English intervention in the Malay world. He writes that Malay will be of use in the 'South-Sea Islands, in which Countries so great a part of the Trade of India is negotiated'.[108] Bowrey, along with a Captain Rossey, had been called before the EIC Court of Directors in 1699 in order to discuss the prospects for settlements in the Southern Seas.[109]

Bowrey's draft proposal for a 'considerable settlement' on an (unnamed) island near the Straits of Malacca ruled by the King of Johor remains among his manuscript collections.[110] As a free market, he wrote, the island would also attract the trade of other colonial powers, including the Spanish, who would bring bullion from the Pacific galleon trade to the island. In essence, Bowrey's South Sea plan sketches the outlines of the settlement that Raffles founded at Singapore just over a century later. In fact, while the island in question was never named, it is not unlikely that it was Singapore Bowrey had in mind, given his description of its situation and his careful depiction of Singapore in one of his charts.[111] Bowrey perhaps hoped that the *Dictionary* would advance the cause of his project for a settlement, having requested a fee of £500 for each scheme adopted by the EIC.[112] Although it seems that his proposal was not given further consideration, Bowrey's plan was certainly not ill timed, given the contemporary English interest in breaking into the Spanish control of the Pacific trade.

Mapping Malay

Bowrey's map 'Of the Countries Wherein the Malayo Language Is Spoken' gives a reasonably accurate account of the distribution of Malay (Figure 2.1). Bowrey's use of rhumb lines and the compass rose, survivals from the old portolan charts, was unusual by the early eighteenth century, and Dampier omitted both elements in his similar, although far less detailed, map of the region.[113] The map gives place names clustered around the coasts, reflecting the European ignorance of the hinterlands, and Borneo's blank centre is filled by the compass.

As noted in Chapter 1, the English lagged behind the Dutch in the art of map making. However, Bowrey produced several charts, some of which he bequeathed to the EIC. They include maps of Gombroone, the

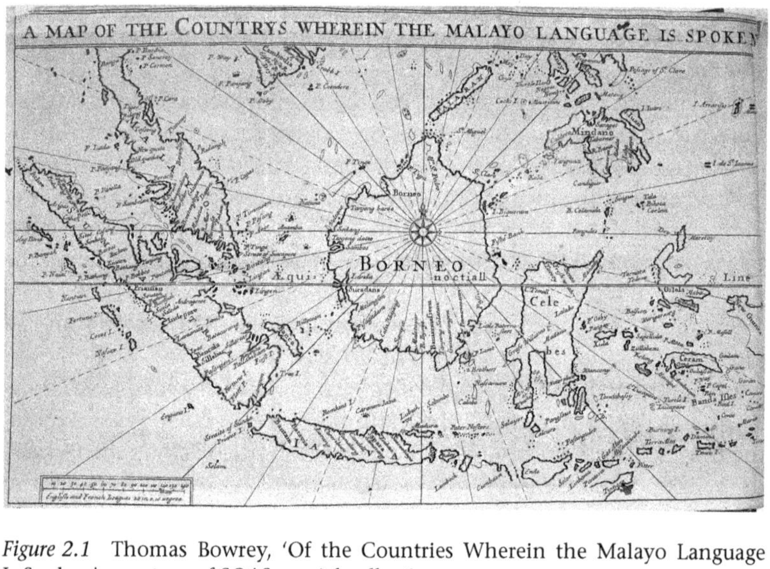

Figure 2.1 Thomas Bowrey, 'Of the Countries Wherein the Malayo Language Is Spoken', courtesy of SOAS special collections

English factory in Persia;[114] of Java;[115] of the Mindanao group of islands in the Philippines;[116] a coloured map of 'the straits of Sincapura', showing the point of Johor and the island that would later become known as Singapore;[117] and a chart of Ceylon.[118] While at Fort St George, Bowrey made maps of Amoy,[119] and of a stretch of the Hugli river.[120] A map of Australia showing Dutch voyages is said to have been extracted from several ships' logs.[121]

Shortly after publishing the *Dictionary*, Bowrey wrote to the EIC's Court of Directors, noting the 'many faults of our sea charts of India' and proposing that each ship, along with its logbook, should record details of the longitude and latitude, weather, tides, depths of water, and coastal features. Bowrey also provides an insight into his own method of map making, recounting his practice of buying and copying sea charts and amending them during his own voyages.[122] While there is little surviving evidence for the indigenous tradition of Southeast Asian map making in the seventeenth century, charts used by Javanese pilots are referred to in some European sources, and may have provided information for maps like Bowrey's.[123] The oral exchange of geographical information suggested in Bowrey's dialogues would certainly have been incorporated into maps and charts.

Bowrey's method of map making by adding to existing sources thus overlapped with the way in which he had compiled his *Dictionary*, using Smith's original and enlarging it with materials supplied to him by Thomas Hyde. In the same letter in which he requests the collection of maps and geographical features, Bowrey also urges the EIC to order their servants to acquire 'Alphabets and Vocabularys in all the Languages and Characters by them procurable and returne them to me'.[124]

As I will argue throughout this book, a cumulative and collaborative approach was a common feature of early modern scholarship. In fact, different types of information were often collected simultaneously. For example, as well as linguistic materials, Hyde requested Bowrey to procure specimens of coffee trees and sago, samples of medicinal seeds, the picture of a jackal, and even the hand of a giant mermaid.[125] The method of collecting words and scraps of linguistic information from a number of written works and oral sources explains many of the characteristics of the Malay materials produced in England during the seventeenth century: rather than being representative of a single place or time, they are a record of the experience of English contacts with the Malay world over the course of a century. The next sections of this chapter will ask what they reveal about these contacts.

Language as mental mirror

Grammar and etymology

The *Jang Ampat Evangelia* begins its discussion of Malay by noting the dissemination of the Arabic-derived Jawi script along with Islam.[126] This was based on Hyde's observation that Malay contained a number of Arabic loanwords, often being employed to describe monotheistic religion.[127] Hyde's interest in comparative linguistics and the collections of languages and scripts that he made in association with Bowrey led him towards making the sorts of historical connections between languages that were also pioneered by Gottfried Wilhelm Leibnitz (1646–1716) and Josephus Justus Scaliger (1540–1609).[128] This approach was an important, if often overlooked, precedent for the later work of linguists including William Jones.

The *Jang Ampat Evangelia*'s 'Address to the English Reader' was a common source for Marshall, Mainstone, and Bowrey. However, it seems that Bowrey was unaware of the manuscript grammars, stating that he was the first English author to have attempted a Malay grammar.[129] While Bowrey's brief grammatical sketch is intended merely to enable conversation, the three manuscript grammars make more serious attempts to elucidate the structure of the Malay language. While Bowrey's outline deals largely with the spoken or 'low' form of Malay, the Mainstone and Marshall grammars follow the majority of the Dutch materials in attempting to engage with 'high' or literary Malay.[130] In the terminology of the day, Bowrey's is a practical grammar, while Mainstone and Marshall offer a philosophical approach.[131]

As early commands from the EIC to confine young writers within another linguistic community demonstrate, it was understood that immersion in language was an equally if not more effective way of learning language than through the analysis of grammatical forms. This argument had also been made in seventeenth-century European debates over the proper methods of teaching Latin and the vernaculars.[132] However, studies of grammar had philosophical as well as practical purposes. Linguists including Sanctius (1523–1600), the Port-Royal grammarians and, later, Francis Bacon (1561–1626) and René Descartes (1596–1650) argued that languages displayed the thought or natural reason of humankind.[133] Hence, they could theoretically be reduced to universal rules or patterns. Attempts at universal grammar like that of Charles Sorel (1602–1674) or a universal language like that of John Wilkins (1614–1672) were based on the idea that existing languages were a better or worse approximation to a universal and underlying thought, and that studying and comparing them would therefore yield an ideal form.[134] On the other hand, 'natural languages' were considered by Leibnitz and others to be a reflection of the external environment and therefore to condition rather than represent ways of thinking.[135]

The influence of these debates is evident in some of the introductory material to the manuscript Malay grammars. Mainstone begins his dedication with the claim that all languages have implanted in them by divine providence a 'native elegance and harmonious congruity' and that 'the worst of Tongues hath some secret rule or Clavis, to unlock its most recluse Mysteries'. Mainstone regards the *ars grammatica* as the most important of the liberal arts because 'without it, men may gabble (like parrots) but can prescribe no true rule for the speech'. He adds that while providence fits the 'souls and tongues' of a people to their own language, others can only imitate unless they learn the rules of the language.[136] Mainstone also implies here that once a rule has been established to explain a language, it may be applied consistently, to smooth out the confusing diversity found in spoken language through deliberate selection of words or grammatical usages. Marshall expresses a similar idea, noting the real variation in Malay grammar, but stating in reference to his exposition that '[t]hese rules are taken from the dialect which seems best and most used & will give occasion to others to cultivate & adorn these Rudiments more accurately'.[137]

Some modern commentators have noted in passages such as these the willingness of European linguists to impose rules on the exotic languages they encountered, whether as part of a wider appropriation of culture or as a result of anxiety over difference.[138] This tendency is certainly evident in the case of Malay, which was altered greatly by the use of Roman script and the Dutch efforts at enforcing the learning of a standard form of Malay.[139] However, the tendency to search for and apply linguistic rules was also part of a more generalised early modern European attention to the structure of vernacular languages. Whereas classical languages such as Greek and Latin were clearly structured according to grammatical rules, dictionaries and grammars of the European vernaculars had only begun to be produced from the late fifteenth century.[140] Orientalist scholars were aware that languages such as Arabic and Persian had their own systems for explaining grammar: this was part of the reason they were also ranked as classical languages. While 'high' Malay might have some claim to join the ranks of classical languages on these grounds, 'low' Malay fell definitively into the ranks of the vernaculars. This may explain why, like the European vernaculars, it was considered a candidate for standardisation.

Bowrey notes that since his readers will all be familiar with Latin grammar, he has used this as a basis. Bowrey's grammar section in fact only discusses nouns, adjectives, and verbs and does not deal directly with the problems of explaining Malay according to grammatical rules derived from Latin, except to note that nouns are not declined.[141] In comparison, the Mainstone and Marshall manuscripts exhibit a more sophisticated way of explaining Malay grammar that borrows from but does not rely exclusively on the principles of Latin grammar. While these grammars also refer to the noun classes of Latin, they note the problems of matching them consistently to the grammatical rules of Malay and make appropriate adaptations. In his later work, Marsden took the point further, dismissing as 'absurdity' any attempt to explain the rules of Malay according to those of a European language.[142] Nevertheless, his division of the parts of speech in Malay does not significantly differ from that given in the Mainstone grammar.[143]

'Malay' encompasses a large degree of regional variation as well as the 'high' and 'low' forms.[144] Based on their observations of this variation, the seventeenth-century English materials give various opinions concerning the origins of the language. Based on its name and its continental location, Bowrey identifies the Malay Peninsula as the original home of Malay.[145] He lists the other islands where Malay is spoken, including Java, Sumatra, Borneo, and Bali, but notes that these islands have one or more of their own languages and therefore use Malay as a trading language, in a variety – called 'Basadagang' (in modern Malay, *Bahasa dagang*, 'the language of foreign trade'[146]) – that differs from 'true Malay', but is nonetheless well understood. Mainstone shared Bowrey's conjecture that Malay originated in the Malay Peninsula.[147] However, Marsden later identified Sumatra as the home of Malay, noting that the Malay-speaking people of the peninsula were likely to be immigrants.[148] While modern studies of seventh-century inscriptions locate the first Malay empire of Srivijaya in Sumatra, the question of the origin of Malay remains a matter of debate.[149]

The Mainstone and Marshall grammars and Bowrey's *Dictionary* all link their assessments of the origins of the Malay language to the perceived level of civilisation of its speakers. Drawing on a passage originally from the Dutch travel writer Jan Huyghen van Linschoten (1563–1611) and cited by others including John Ogilby (1600–1676),[150] Mainstone links the inclinations of the people of Malacca towards international trade with 'better laws, milder government' and a higher general level of civilisation than their neighbours. This, he claimed, was reflected in the 'elegance of style' of their language.[151] William Dampier makes a similar connection between trade, the Malay language, and civilisation, describing the Muslim 'Indians' who spoke Malay as 'the trading and politer sort'.[152]

Such assessments of the 'civility' of certain peoples or linguistic groups were fluid, often shifting along with the allegiances of the EIC. For example, in Bowrey's earlier description of 'Janselone',[153] he shows less appreciation for the Malay speakers, 'a very roguish Sullen ill natured people', preferring the 'natural Siamers' he met further inland and described as 'a very Civil good humoured people'.[154] Nevertheless, the early arguments made for both the 'civility' of Malay speakers and the argument for the origin of the language in the Malay Peninsula persisted into the colonial period, when Malay speakers were granted recognition as the 'indigenous' people of the peninsula at the expense of other groups such as the Orang Asli, regarded as less 'civilised'.[155]

The grammatical materials produced to explain Malay in seventeenth-century England display a 'philosophical' interest in language that was emblematic of seventeenth-century ideas about universal language and pointed the way towards comparative and historical studies of linguistics. The early discussions are significant in demonstrating an understanding that an analysis of the development of language and the inclusion of particular loanwords might be used to map the historical movements and interactions of people. Their legacies for colonial approaches to language are evident both in the urge to standardise Malay and in the efforts to draw connections between language, participation in commerce, and civilisation.

Vocabulary, miscellanies, and dialogues

If linguistic materials reflect the mental worlds of their compilers, they also mirror some aspects of the situations in which their authors learnt languages. The body of the *Dictionary English and Malayo, Malayo and English* contains 468 pages of vocabulary, with somewhat more attention devoted to the English-Malay section than the Malay-English.[156] While wordlists jotted down by travellers tend to focus on the basics needed for trade, Bowrey's *Dictionary* reflects a wide range of situations, formal and informal. Some vocabulary is undoubtedly derived from earlier materials, including the *Jang Ampat Evangelia*, Spalding's versions of Houtman's dialogues and the Dutch materials that Bowrey refers to in his preface, as well as Henry Smith's original dictionary, and earlier merchant wordlists. Three of the dialogues are taken from the work of Spalding, in turn copied from Houtman, but with some alterations in the detail and updated orthography.[157] As such, the *Dictionary* should be regarded as a composite work rather than as that of a sole author. Nonetheless, the vocabulary and contents of the dialogues can be used to paint a general picture of the world of early modern English traders in Southeast Asia.

A large amount of vocabulary relates to the sailing of a ship (Bowrey: *cāpal*), demonstrating the common practices of using Malay crews on European ships and European traders' use of local shipping. Words relating to pirates (*ōran roompak*) indicate the threat they posed at sea, while vocabulary relating to rivers (*soongey*) and canals (*simpanggam ayer*) indicates participation in inland trade. Miscellanies related to sailing include technical details: 'I am amazed to see a little Ruther steer a great ship too and fro' and 'Tis the loadstone that causes the compass to traverse'. The eighth dialogue is a detailed account of a voyage from Bantam to Persia including the kind of geographical detail that Bowrey would have incorporated into his charts.

A number of trade goods also feature in the *Dictionary*. Spices mentioned are both local and international: they include camphor of Borneo (*bārroos, cāphoor bāroos*) and Japan (*tōhooree, cāphoor tōhōree*); cardamom (*kāpālaga*), cumin (*jintam*), and turmeric (*coonhet/coonhir*), all of which were produced in the Malay world; long pepper (*tābee, chīabee*), of which some species are endemic to Java; cinnamon (*camcāmoon*), brought from Sri Lanka; olibanum or frankincense (*lūbaun*), originating from Somalia or Southern Arabia. Opium (*āpheon*), the name derived from Arabic, was becoming an important commodity in the region when Bowrey was compiling the *Dictionary*. From the late seventeenth century, the VOC claimed a monopoly in the Malay Archipelago on the opium they brought from Bihar ('Bengal opium') and distributed via Batavia.[158] Opium would later be grown in large quantities in Java and elsewhere in Southeast Asia, to meet demand from expatriate Chinese communities.[159] Coffee (*cāwa*) and tea (*cha*) were also imports at this stage: the Dutch began Java's first coffee plantation in 1707.[160] Other trade goods that are mentioned include various jewels, textiles, and ivory (*gadang*). Dialogues 2 and 3 refer to buying nutmeg and other spices. The fourth dialogue, adapted from Houtman, is between a merchant, a king, and the *shahbander* or port official. The fifth, also from Houtman, is about demanding a debt.

The first and most intriguing dialogue is a conversation between two friends, Malay and English. It begins with a discussion of the suicide of a mutual acquaintance, 'Joseph' (whose nationality is not specified). The two friends walk together and drink a cup of wine with water although the Malay friend refuses to accept the suggestion of smoking some tobacco as 'I am not yet used to it'. The reference to suicide here might hint at the concern for the mental as well as physical malaise that the climate of settlements like Bencoulen was feared to visit upon both Europeans and locals, an issue which will be discussed further in Chapter 6.[161] Tobacco and alcohol, as well as other intoxicants including betel and cannabis, provided revenue for the Company, which issued licenses to those who vended them in their settlements.[162] While chewing the indigenous areca (*pēnang*) and betel (*seeree*) were still more popular intoxicants, as well as important social rituals, tobacco smoking was introduced to Java sometime in the sixteenth century and gradually caught on across the Malay Archipelago, only later being added to the betel chew.[163]

Perhaps unsurprisingly, given the emphasis on proselytisation that motivated many previous European studies of Malay, a large number of religious terms are included in the *Dictionary*. As Bowrey notes, most of the vocabulary for monotheistic religion is derived from Arabic. As was usual in missionary practice of the day, Allah is used for the Christian God. Similarly, *mēsājit* and *moskit* are applied to churches as well as mosques. What is more surprising is Bowrey's inclusion of a translation of Christ as *nabbe ēsa* or 'the prophet Jesus', following the Muslim designation. Hints of popular beliefs from outside the monotheistic religions appear in words for a pagan god (*berallah*), witch (*ōran hōbatan, swangee*), and shadows or spirits (*bāyang bāyang*).[164] 'Devout' is translated as *bactee*, signifying the adoption of the Sanskrit-derived word *bhakti* and the devotional practices it connotes into Islam as well as Hinduism in South and Southeast Asia.[165]

In the ninth dialogue, two travellers, a Christian Englishman and a Malay Muslim, ask one another about their religions. The Christian refers to the Trinity and recites the Lord's Prayer and the Ten Commandments – elements of the *Dictionary* that are probably taken from Dutch materials. Despite the initial promise of reciprocity, Islam is not in fact described. While the dialogue is brief and formulaic, it gives a sense of the sort of interfaith debates that will be discussed in Chapter 3. The English factor Joseph Collett reported holding debates with Malay elites on the subject of religion.[166] The references to the spread of Islam along with trade in this dialogue demonstrate the EIC's awareness that they were in competition with Islam as well as Catholic missionaries to attract converts and trading partners.

Conversion to Islam in Java was often associated with the cure of disease. Indeed, the word Bowrey gives for doctor (*tābeib*) is derived from Arabic. Christian missionaries competed by offering medical services to potential converts.[167] The vocabulary for sickness (*sākit*) and medicine (*ooabat*) suggests the local adoption of some elements of European practice including bloodletting (*booang dara*), cupping (*bēskam*), and dissection of corpses or anatomisation (*bēla*). Even the concept of the circulation of the blood (*mengāleh dara*) is translated, suggesting that this fairly recent European discovery was under discussion in the Malay world by the late seventeenth century.

Like any successful private trader, Bowrey would have needed to spend time in palaces and courtrooms negotiating trading permissions. For Europeans in the Malay world, the shorthand for such documents was *chap* or *chop*, which Bowrey defines as 'a signet, seal, an impression or mark with a seal'. Bowrey's vocabulary includes a court or palace (*astāna*), a council of state (*mantree*), and a council house (*rooma bēchara*). The Malay word for 'council' was borrowed by the English factors for the 'bitchars' they held with local rulers.[168] Characters associated with official circles include an emperor (*sooltawn, sooltawn chejālam*), an interpreter (*juree sālin, jurre bāsa*), a judge (*juree hookoom, alsaki*), a eunuch (*sēda sēda, cambiree*), and an executioner (*pemboonoos*). The production and distribution of Malay literature is notable by the appearance of a bookseller (*ōran penkillip kītab*), a bookbinder (*ōran berjewal kītab*), and a dictionary (*kītab bāsa*).

While Bowrey does not seem to have been able to write Jawi himself, depending on Hyde for all the instances given in the *Dictionary*, we might imagine that Bowrey had his letters of requests (*soorat pēmintaawn*) written out by a 'compositor of letters' (*pengārang ooroof*) and then presented them with his own oral explanation. At the end of the dialogue section, Bowrey gives two model letters. The first is from a King of the North (David) to a King of the East (Abdullah), beginning with an appropriate exchange of compliments and recommending a merchant (Agent White) to him. The example was probably modelled on royal correspondence. The second letter, from a merchant named Soliman Golasmith, requests specific quantities of trade goods.

Words for personal relations account for a large proportion of Bowrey's vocabulary, particularly in relation to women. 'Pillow dictionaries' were an important means of language learning for Chinese and European merchants.[169] Relations ranged from brief encounters with prostitutes to formal long-term cohabiting partnerships like that of Baron's parents discussed in the previous chapter. Bowrey gives vocabulary to cover all these situations: from a whore (*soondal*) to a learned wife (*bērālim*). The reader can learn how to discuss love (*cassee*) and betrothal (*toonang*). We get a snatch of dialogue here and there that seems to have come straight from one of these bed companions, in one instance: 'Because of you, Men do scorn me' ('*Carna angkaw ōran mēlēat mooda Champa*').

Several phrases in the miscellanies indicate an interest in settled agriculture, for example: 'Reap the Rice in the east field, sow the west field with beans, plow the field by the river side, and mend the banks'; and 'Rice is now dear, therefore tread out the paddy in the barn, winnow it well and a munday I will sell it'. Agriculture had become increasingly important in Southeast Asia during the seventeenth century, as the drawbacks of relying on a few high-value cash crops such as pepper and nutmeg became evident following the deprecations of the Dutch. In fact, the urban world of commerce that emerges from the numerous terms that appear in the vocabulary Bowrey gives concerning the court and its trade had almost disappeared by 1700.[170]

Vocabulary for food, plants, and animals also appears in the *Dictionary*. Some American transplants like the potato (*ooby kēcheel*) and pineapple (*ānānas*) occur alongside the mango (*ampullum mango*) and jackfruit (*nanca*). Animals that are noted include a camel (*oonta*), a novelty in maritime Southeast Asia and perhaps an indication of the early modern craze for the exchange of animals that was also evident in the King of Bantam's urgent request for some 'great dogs [i.e. mastiffs] of the biggests [*sic*] that can be found in England' that he included in his 1682 letter to Charles II.[171] Descriptions of two animals occur in the final dialogue. One is the 'Ōrang ootan' or 'man of the woods', whom Bowrey describes as 'very ready to learn any actions, but never heard to speak'. This description is part of a contemporary European myth developing around the beast. The Dutch physician and natural historian Bontius had described 'a curious monster with a human face... [that] has the human habit of sighing as well as shedding tears'.[172] This description filtered into other contemporary travel works: for example, an orang-utan couple is depicted in Ogilby's translation of Montanus with human faces, the male offering the female a flower.[173] In fact, as Dellios points out, the Malay word for the creature was *mawas*, but the word 'orang-utan' persisted and is now commonly used,[174] one example of how the linguistic writings of this period with their often comical misunderstandings could affect language. The other creature is described as resembling a hare but with a transparent horn in the centre of its head, which it is said to leave in a tree. While this animal, whose name Bowrey gives as *tadoc palandock*, sounds mythical, it is probably a reference to the horned Bornean yellow muntjac (*Muntiacus atherodes*; modern Malay: *pelandok*).

What should we make of Smith's claim that his dictionary had been 'perverted' by ambition or by ignorance? Some *Dictionary* entries are certainly bizarre – in one example, corrected by Smith, Bowrey gives the word for 'slavery' and the general word for 'slave' as '*lascar*', a word normally used to refer to a free sailor (although, as will be discussed in Chapter 6, this was not always the case). Smith substitutes '*peramba*' for slavery and '*amba*' for slave. Bowrey also gives '*amba*' for a slave born in the house, '*tāwan*' for a prisoner enslaved in wartime, and '*tooboosan*' for a slave who had been bought. Marsden confirms Bowrey's use of '*amba*' and '*tāwan*' and confirms Smith's corrections by giving slavery as '*per-ambā-an*'.[175] In another example, Smith corrects the second-person singular pronoun given as '*Joo*' to '*Kamoo*'. Kader suggests that the original '*joo*' perhaps resulted from the borrowing of the English word 'you' in the Malay ports visited by Bowrey.[176] However, the use of '*jū*' is also noted in the later work of Marsden, although he describes it as 'provincial, vulgar, and not to be found in good Malayan writings'.[177] Thus, while some of Bowrey's word choices are clearly mistaken or seem surprising, the majority are confirmed by other sources.

Appendices: Place and time

Bowrey provides a table of conversion between the Gregorian and Hijri calendars for 30 years from 1701. He notes that the Islamic system of reckoning dates begins with the Prophet Mohamed's exile from Mecca to Medina in 622 AD and that the lunar calendar makes the year around 11 days shorter than the Gregorian calendar, decreed by the Pope in 1582. The transition from the Julian calendar in Europe was slow, and alternatives to the Gregorian calendar were long debated. This stimulated inquiries into other calendars: including the enquiries of the Royal Society into the observatory of Ulugh Beg in Samarkand, which resulted in Hyde's translation of 1665.[178]

Hyde also researched Chinese systems of reckoning time,[179] which were reformed with the aid of Jesuit priests under the Kangxi Emperor (r. 1661–1722).[180] Hyde realised the connection between the ancient Persian and Chinese solar calendars, and this was perhaps another reason for his interest in Giolo and his religion, in which he saw a link between Persia and China.[181] Powerful Southeast Asian rulers also instituted calendric reforms: in the 1630s, Sultan Agung promulgated a new calendar harmonising the Islamic calendar with the Javanese Indic calendar starting from 78 AD.[182] Amid this confusion, Bowrey opts for the two calendars with the best claim for international comprehensibility.

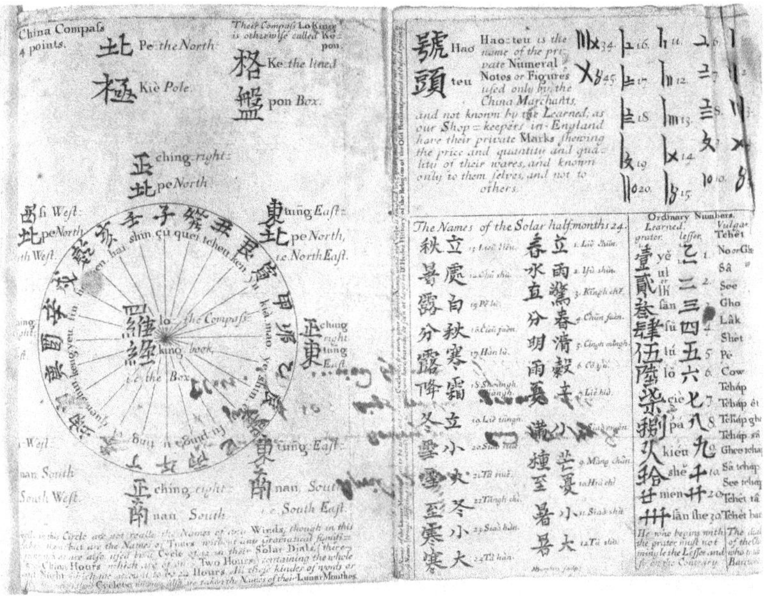

Figure 2.2 Thomas Bowrey, 'the dial of the Chinese who trade at Bantam', courtesy of the British Library

The Chinese compass that appears in the final pages of the *Dictionary English and Malayo, Malayo and English*, labelled 'the dial of the Chinese who trade at Bantam' (Figure 2.2), was the product of another collaboration of Thomas Hyde: with the Chinese Christian Michael Shen Fuzong (c. 1658–1691).[183] Shen visited Oxford during the summer of 1687 and helped catalogue the Chinese books in the library. The two men also pored over the huge map of China that had come into the possession of the lawyer John Seldon (1584–1654).[184] The Seldon collection also contained a Chinese compass. While Seldon believed that he had acquired a sea compass, in fact his was a compass for use on dry land.[185] It is not clear whether Hyde and Bowrey were aware of this difference. Shen and Hyde examined the Seldon compass as well as the map closely: Shen transliterating all the characters shown on the compass with explanations.[186]

The compass that was engraved by Michael Burgess (d. 1693) in April 1701 for Bowrey under Hyde's direction has the 24 compass points marked with their characters,[187] although not the further subdivisions that appear in the notes made by Shen and Hyde.[188] Bowrey had originally wished to match the 24 winds with their names in other languages, but, as the Chinese division would not match the 32 winds or directions used by European navigators, Hyde advised that the Chinese must be given alone.[189] On the facing page, the names of the solar half-months are given in Chinese characters along with transliterations. Next to the solar half-months are given the numerals used by 'those Chinese who trade at Bantam'. Shen wrote out these numbers and their sounds for Hyde.[190] Above this are given 'Hao-teu' or the private numerals used by merchants. Again, these appear in draft form among the notes made by Shen and Hyde.[191]

The EIC, like other Europeans, was heavily reliant on Chinese pilots in Southeast Asia as well as the South China Seas.[192] Whether Malay sailors themselves made use of the Chinese compass is debatable. While Ludovico di Varthema claimed to have observed the use of a compass aboard a Malay ship travelling from Borneo to Java, early Dutch reports from the Malay world insisted that the compass was unknown, based on their failure to market European compasses in Indonesia.[193] Reid's conclusion – that the need was met instead by Chinese compasses – seems to be supported by the inclusion of the Chinese compass in Bowrey's *Dictionary*. Although there were also Malay terms for the eight points of the compass, they are not given in the *Dictionary*.

The inclusion of the Chinese elements in the *Dictionary* is revealing in terms of the type of society the EIC envisaged creating in their settlements. The Chinese presence in the Malay Archipelago had been important since the eighth or ninth century, with fluctuating levels of contact.[194] From the 1680s, a new wave of Chinese settlement in the region began, especially following the lifting of the imperial ban on trade of 1684.[195] Most Chinese traders and settlers in the Malay world came from Southern China, and those who settled in Sumatra and Java spoke the Hokkien, Hakka, Teochew, or Cantonese dialects rather than Mandarin Chinese.[196] This is reflected in the dialect used in the materials on Chinese that were appended to Mainstone's grammar.

By the 1600s, the Chinese were already the largest group of foreign traders in Southeast Asia, with seventeenth-century populations of up to 3,000 at Bantam, 23,000 in Manila before the massacre of 1603 and 5,000 in the Cochin-Chinese port of Hoi An in the 1640s.[197] These expatriates established themselves as craftspeople and tax farmers. They were also important middlemen, sourcing pepper in the hinterlands and in the pepper trade carried on in local craft around the centres of pepper production on the east coast of Sumatra and elsewhere, and by the early eighteenth century they were moving into the mining industry.[198]

As Leonard Blussé has shown, the VOC factory at Batavia was reliant on Chinese collaboration.[199]

Following the Dutch example, the Bencoulen factory was eager to recruit Chinese merchants to live in their settlement, writing in 1686 that the settlement must be fortified in order to protect the Chinese 'who are the only trading people'.[200] As Farrington notes, Bencoulen was a less attractive area for settlement in terms of this trade, and the Chinese population only ever reached an estimated 1,000 by the mid-eighteenth century.[201] Nevertheless, the Chinese were an important part of the settlement, opening bazaars and market gardens and selling arrack. Like their compatriots in the Dutch settlements, they tended to maintain an identity separate from the local population and lived in a Chinese quarter separate from, although adjacent to, the English settlement.

In Bowrey's proposal for a South Seas Company, the increased participation of the Chinese in Southeast Asian trading settlements 'since the English left Bantam' (in 1682) is noted, the number settled in a part of Borneo having increased 100-fold.[202] Bowrey advances the prospect of Chinese settlement and collaboration as a powerful incentive for the proposed settlement. Southeast Asian traders from across the region would also, he writes, flock to the island to exchange their goods for those from Surat, the Coromandel Coast, Bengal, and China. Here, as in many of the other materials included in the *Dictionary*, Bowrey both reflects the contemporary situation and points the way to future imperial projects like the Singapore settlement.

Despite Bowrey's hopes, the EIC did not pursue the idea of establishing settlements in the South Seas. Nonetheless, Bowrey's advocacy of the cause of settlements was taken up by other voices, and the South Sea Company was chartered in 1711,[203] resulting in the infamous 'bubble'. The EIC's settlements in Sumatra were continued, despite the temporary expulsion of the factors from Bencoulen between 1719 and 1724, but the Company's interests, both commercial and linguistic, were turning away from the Malay world and towards the Indian subcontinent. Bowrey's *Dictionary* remained the standard linguistic work on Malay in English until the early nineteenth century.[204]

Conclusions

In a letter dated only 'Thursday evening', the man who called himself George Psalmanazar[205] wrote earnestly to Thomas Bowrey about the system of transliteration he had devised for 'Formosan' – a language that he had in fact devised in its entirety. To 'pronounce the language well and prevent confounding the significance of words', he wrote, one should observe the guttural pronunciation of 'Rh' and 'Kh', 'ph' should sound like 'pf', long and short vowels are marked with microns, while 'f' is 'pronounced unlike any other letter I know of in any language'.[206]

It seems incredible that Bowrey, who produced a detailed sea chart of Taiwan, was taken in by the pretender. The key to Psalmanazar's deception here was perhaps his presentation of his fabricated language using the terminology that had begun to be developed to discuss how to represent unfamiliar scripts and sounds in English. Psalmanazar continued to be taken seriously by colonial linguists in the eighteenth and nineteenth centuries despite his own admission of fraud.[207] While this example is extraordinary, it is emblematic of the long legacy of early modern linguistic works, complete with errors and misunderstandings.

While Bowrey claimed to have compiled the *Dictionary English and Malayo, Malayo and English* from personal memory, it was in fact a composite work based on information gleaned from a number of disparate sources. As such, it can be seen as a testament to the English experience in the Malay Archipelago in the late seventeenth century. The experience included formal encounters in palaces, courtrooms, and customs houses; informal encounters in punch houses and private houses; and living, trading, sailing and fighting alongside Malay, Chinese, Dutch, and Portuguese collaborators and rivals in the hybrid trading world of the late-seventeenth-century archipelago.

The *Dictionary* contains words and phrases drawn from all of these situations – haphazardly mingling 'high' and 'low' Malay and the dialects and expressions common to different regions. Its hybridity, complete with misunderstandings, is an apt reflection of the intimate, varied and changeable interactions that an early modern foreigner would have had in the rapidly changing world of Southeast Asia. As for how far the *Dictionary* represents 'Malay', Smith was probably correct in assuming that a Malay intellectual would have been puzzled by the work. The use of Roman script distanced it considerably from anything that would have been locally accepted as written Malay. Nonetheless, as a guide to spoken language, the terms it contains are still recognisable by modern speakers and scholars of Malay.[208]

The collaboration between Bowrey and Hyde to produce the *Dictionary English and Malayo, Malayo and English* was emblematic of the scholarship of the period, in which merchants and scholars formed alliances around the acquisition and interpretation of information from the extra-European world in the hope of profit and prestige. Such knowledge was valuable to the EIC's Directors, who surrounded themselves with advisors and often acted as individual patrons by trading gifts and preferment for information or publically by subscribing to books, thus assisting in the cost of publication. Nonetheless, throughout the seventeenth century, the EIC as a corporate body lacked the resources to institutionalise language learning among their servants or to follow Hyde's suggestion that Asian scholars should be brought to England in order to teach languages.

The Company's interests in Asia were fluid, following the fluctuations of local politics, and support for language projects was contingent on their perception that a particular language would be of use in establishing trade and settlements. Nonetheless, the advantages for EIC servants of learning languages and recording their experiences in the form of grammars and dictionaries were clear. Almost every such work produced in the century and a half following the *Dictionary English and Malayo, Malayo and English* was dedicated to the EIC. These works provide a testament to the experiences, relationships, and aims of the Company's servants that is often more revealing than the more consciously expressed impressions that were recorded in official documents, travelogues, and letters.[209]

3
Toleration and Translation: English Versions of Two Hindu Texts from Bengal

Introduction

> [I]n reason or objects of sense all religions agree, therefore it is not a small wonder to me why the religions in the world should be so straingly noxious & opposite to each other, whereas did men really consider & throw away prejudice and all tradicion not grounded on the two aforesaid grounds (vizt) Reason & Senses, there would not be such contencions in the world about Religion as there is.[1]

On 25 June 1674, Madhusudana Rarhi Brahmin and a young EIC factor, John Marshall, secluded themselves within the English factory at Kasimbazar (or Cossimbazar) in modern West Bengal to begin a project that must have absorbed much of their spare time for the next two years, continuing after Marshall's promotion to governor of the Balasore factory in 1676 and until his untimely death on 30 August 1677. This plea for inter-religious understanding based on reason and the evidence of the senses opens the manuscript account of a 'familiar & free' dialogue between the two men. The dialogue serves as framing story for the first English version of the *Bhagavata Purana* (BhP), a popular Hindu text, itself composed of multiple religious dialogues leading into illustrative stories.[2] This remarkable document, composed over a century before the founding of the Royal Asiatic[k] Society, runs to 230 large folios, closely written on both sides.[3] It appears to be the first surviving attempt to render a Hindu text in English.

The text, probably drawn from an oral performance, is free-flowing and full of contemporary references, but manages to convey the central religious, philosophical, and cosmographical concepts of the BhP in the lucid but down-to-earth English of an educated merchant. Marshall and Madhusudana also produced an interpretation of the *Samaveda,* which floats still freer of the versions of this ancient collection of verses generally accepted as authoritative. Both texts and the framing dialogue provide a fascinating exposition of the faith of Madhusudana, a Vaishnava – or devotee of Vishnu – after the Gaudiya tradition of Bengal, and the interpretation of his words by Marshall, an English Protestant who during his university studies had become drawn to the group of broad-minded Christians known as the Cambridge Platonists.

How should we approach these texts, which are presented as translations of ancient Sanskrit texts but which constantly draw us down into the world of the two participants in the dialogue, complete with warehouses, factories, and brokers? In one sense, the interaction between Madhusudana and Marshall is the product of a personal meeting of two inquiring minds from two religious traditions that, while very different, converged around some specific questions about the nature of the divine and the physical world. We might therefore choose to look first at the authors' immediate situations and backgrounds, and then at the texts in the light of their form, language, and content, asking what they tell us about the two men and their particular concerns.

If we zoom out from this particular encounter and look out across the network of geographically dispersed settlements of the EIC and their European rivals, we can see other conversations taking place about the similarities and differences between religions and the potential or otherwise for toleration between their followers. In this sense, the conversations between Marshall and Madhusudana were a product of the common situation that these two men found themselves in: of having to negotiate trust in the face of religious difference and political upheaval.

The early settlements of the EIC and the other European trading companies in Asia were composed of Protestant, Catholic, and Armenian Christians; Parsi brokers and businessmen; Hindus and Muslims of many different persuasions and origins; and some Buddhist traders from China and mainland Southeast Asia. They all had to make business deals, swear oaths and design their towns to accommodate religious buildings and sites, and, on occasion, they fought side by side to defend these settlements. This situation prompted interest in other religions. Such interactions were not always peaceable, and some inquiries into religious belief became tools of coercion in the hands of Company officials – a process that was to continue in the colonial period.

How to interpret these interactions? One tale we could tell would involve pushing back the beginnings of British 'Orientalism', in the sense of the study of Oriental religions – normally commenced in the

1760s, around the same period that the EIC was traditionally considered to have transformed itself from a collection of 'mere merchants' into a governing body. We could argue, then, for an approach that considers both the EIC's ambitions of stateliness[4] and their interest in Asian religions and cultures to date from a much earlier period. Marshall's text could be placed in sequence after the earlier work of Henry Lord, on Hinduism and Zoroastrianism, alongside his Dutch contemporary Philip Angel, who similarly composed a text on Vishnu's avatars,[5] and before William Jones and the Royal Asiatic Society.

Another story we could tell about these documents would place them in a tradition of texts actually composed by Christians but drawing on Hindu literary styles that begins with the *Kristapurana* of Thomas Stephens, a Jesuit and perhaps the first Englishman in India,[6] and takes us up to the *Ezourvedam*, the 'French Veda' that fooled Voltaire with its claims to represent a translation from a Sanskrit original.[7] Such texts employed Hindu idioms for aesthetic, playful, or polemic ends, sometimes with the aim of enhancing interfaith understanding and at other times intending to highlight difference. They coexisted with the works of Indian Christians and with translations of Christian works made by Indians of other faiths for their own purposes.

This story, however, forms only part of a longer process of accommodation between different religious communities in India. We could also build a narrative about how this translation of the BhP fits with the history of the Puranic literature and their performance to audiences, which could include untouchables, women, and *mlecchas* (foreigners). In this story, we might locate Madhusudana within a long tradition of Brahmins who constantly retold and adapted Puranic histories to serve contemporary purposes, while serving under various rulers of different faiths. By the eighteenth century, some historians argue that these clerical elites had begun to establish a sense of identity that spanned the subcontinent through narratives of identity based on and woven into textual traditions.[8]

The fate of particular texts at different places and points in time is related to the perceived relevance of the ideas that are contained within them. The metaphysical ideas that the interpretations produced by Madhusudana and Marshall contain were relevant to seventeenth-century European debates about the concept of atomism versus that of mechanism; the existence and significance of a vacuum; the relationship between spirit, matter, and motion; the age of the universe; and the nature of time. As such, they attracted the interest of English philosophers including John Locke and Damaris Masham. By the early eighteenth century, however, with the rise of Newtonian physics, such debates seemed less relevant, and Marshall's manuscripts were forgotten in England. Nonetheless, similar efforts to find support for contemporary European philosophical concerns in the teachings of other faiths, often considered to be expressions of natural theology or natural law, continued to be important, notably to a group of French Jesuits and Orientalists.

The authors and their worlds

Politics and religion in seventeenth-century Bengal

The Bay of Bengal was a coveted site of trade, long famed for its exports of rice, sugar and fine cloth.[9] Duarte Barbosa, who visited in the 1520s, described Arabs, Persians, Abyssinians, Portuguese, and Gujaratis trading with southern India and Southeast Asia.[10] This cosmopolitan trading population was matched by religious diversity and dialogue. Bengal had a long tradition of mother goddess worship, and outsiders in the medieval period had remarked on the region's reputation for heterodoxy, including the blurring of caste barriers.[11]

Islam arrived in Bengal from the thirteenth century, and Sufism became an important force from the fourteenth, borrowing from Hindu religious practices such as yoga and merging the identities and worship of local gods and Muslim saints.[12] Marshall refers to one such practice with his story of a tiger who emerges at night to 'salaam to a fauckeer's [fakir's] tomb', perhaps in reference to the conjoined cults of the tiger god Daksin Ray ('King of the South') and a Muslim pioneer, Badi Ghazi Khan.[13] Elements of Christianity were also absorbed within a religious landscape in which ideas and rituals were readily translated. An example is the 'the lay of Saint Anthony' (*Sadhu Antanir gan*), a ceremony performed at Nagori, Bhawal, near Dhaka, in which Muslims and Hindus as well as Christians have taken part since the seventeenth century, when Antonio de Rozario converted to Christianity following his capture by *firingi* (from 'Frank', in this case meaning Eurasian) pirates.[14] During the seventeenth century, the Baul mystic movement attracted a significant following.[15]

Bengal had been integrated into the Mughal Empire, beginning in 1576, but only reaching as far east as Chittagong in 1666. The Emperor Akbar (r. 1556–1605) was famed for his policy of *sulh-i-kul*, which enshrined equal respect for all religions based on the idea that all were different routes to the same end. Akbar had personally granted a *farman* (a Persian term for 'an order, patent, or passport') to the Portuguese captain, Pedro Tavares, permitting the preaching of the Christian gospel in Bengal and the conversion of anyone who was willing. Jahangir (r. 1605–1627) assembled Jesuit as well as Hindu and Muslim theologians and philosophers learned in metaphysics to debate questions of religion and ethics in a format that drew on Sufi *malfuzat* or 'table talk'.[16]

Revisionist accounts of the reigns of Shah Jahan (r. 1628–1658) and Aurangzeb (r. 1659–1707) have modified their reputations as champions of conservatism and questioned the degree of impact their personal piety had on religious and social policy within the Mughal Empire.[17] Nonetheless, religious toleration was officially more restricted than in previous years, and Aurangzeb reinstated the unpopular *jizya* tax on non-Muslims, a revival that endured until 1720 and was also applied to Christian members of the European companies.[18] Although Hindus continued to be prominent in the Mughal nobility, the reinstatement of the controversial tax raised concerns in some quarters.[19] Meanwhile, with his coronation and symbolic rebirth as a universal dharmic monarch, the Maratha leader Shivaji (c.1627–1680) used Hinduism as a rallying call.[20] Puranic histories began to be employed in a rethinking of Brahminical identity in the period.[21]

Jesuit syncretism

Marshall was not the Christian to be involved in the production of a Puranic text. This may have been the English Jesuit Thomas Stephens. Born in 1549 in Wiltshire, he arrived in India in 1579, remaining until his death in Goa in 1619. He learnt Konkani and Marathi. With the aid of an unnamed Brahmin informant, Stephens composed the *Kristapurana*, 'Christ's' or 'the Christian' Purana, apparently at the request of the local population. The work was first printed in 1616 using Roman script. A handwritten version in the Marsden collection at SOAS employs Devanagari script and is written in more elegant literary language. The text follows the epic style of the Puranas and tells the story of mankind from the Creation to the time of Christ.[22] Many Hindu terms are borrowed to describe Christian concepts, including *karmadosa* for sin and *moksa* for heaven. The language of the text is said to be heavily influenced by the poetry of the Bhagavad sect of saints, whose writings were important to Marathi literature from the twelfth century onwards. The *Kristapurana* refers to God as 'the pervader of the universe' and 'the ocean of all happiness... without beginning or end'.[23]

Several other Christian *puranas* followed Stephen's, including the *St Peter purana*, composed by Étienne de la Croix in Goa and published in 1634, the *St Anthony Purana* by Antonio Saldanha, and the *Thembavani*, composed in Tamil, by Constanzo Beschi.[24] Such texts formed part of the Jesuit technique of 'accommodation' – an offshoot of a long theological tradition going back to early Bible exegesis in which 'adiaphora' (indifferent things) were introduced to mark off the 'human' sphere from the 'divine'.[25] The missionary Alessandro Valignano, who arrived in 1575, was the first to develop the concept in the Indian context. Roberto de Nobili (1577–1656), who famously adopted the dress and eating habits of a Brahmin, further developed the strategy that became known as 'accommodation', in which local forms were adopted for a Christian message. He also took on concepts including caste-based ideas about purity, attracting criticism from fellow Jesuits.[26]

The 'dressing up' of Christian ideas extended to religious buildings: architectural syncretism could be observed between Hinduism and Christianity with churches in St Thomas near Madras resembling temples and Hindu deities engraved in the wooden churches of Goa.[27] Indian Christians also produced texts that employed the idioms of other Indian religions. One such text was Antonio de Rozario's *Brahman-Roman Catholic-Sambad*, a dialogue between a Brahmin and a Christian, published in Bengali and Portuguese.[28] The practice of accommodation prompted fierce debate, generally known as the 'rites controversy', both within and outside the Jesuit order, and eventually became a target of the Portuguese Inquisition set up in Goa from 1560.[29]

Politics and religion in Europe and the early colonial settlements

The question of religious toleration had also arisen in England during the revolutionary years of Marshall's youth. John Milton's *Areopagitica* of 1644 and the Levellers' 'Agreement of the People' of 1647 both called for freedom of religious expression. The English Republic came closest to realising such ideals in 1650, with the repeal of an Elizabethan act requiring attendance at church and the readmission of Jews in 1655. The activities of religious sects regarded as extreme, including the Ranters and Quakers during the 1650s and later the Pantheists led by John Toland (1670–1722), caused further concern. Jews continued to be admitted to England after the Restoration, and the Declaration of Indulgence of Charles II in 1672 allowed separation from the Church of England. However, the radical suggestions of the earlier period were quietened in the decade after the Restoration.[30]

John Locke, who became influential in English politics during the 1680s, put forward a political argument for religious toleration. His *Epistola* makes the case that true belief could only be instilled by 'inward persuasion', and never by coercion.[31] However, Locke's version of toleration was restricted in comparison to that propounded by earlier thinkers, including those known as the Cambridge Platonists. While some radical Whigs hoped the Glorious Revolution of 1688 would provide a more satisfying solution, the Toleration Act enacted under William and Mary excluded Catholics and those who denied the existence of the Trinity from the right to worship.[32]

Where toleration was concerned, the Company settlements were a different matter. Stern notes Josiah Child's comparison between the attempts to enforce religious unity in England and the 'Amsterdam liberty' normal in the English 'plantations'.[33] While Holland's toleration and the English experience in America were likely to have been templates for EIC policy, the impact of the direct exposure to religious toleration in Asia on both European nations should not be overlooked. Despite the efforts discussed in Chapter 2, the EIC made few converts to Christianity in Asia in its early years, and was in no position to acquire them through coercive means. Therefore, the early EIC settlements were heavily dependent on Hindus, Muslims, and Zoroastrians to manage their trade, diplomatic, and agrarian relations.[34] The Company derived revenue from the religious buildings of other faiths that were erected in their settlements.[35] Thus, by default, a degree of religious toleration that would have been unthinkable in contemporary Europe existed in these small enclaves of Company government.

Despite occasional complaints from Christians who considered themselves wronged, the oaths of people of all religions were accepted as of equal weight in courts set up in the early Company settlements. The records of the Madras court described the oaths of the Portuguese defendants administered by the Catholic Portuguese padre and Hindus taking their oaths at the 'Gentu pagoda' (Hindu temple).[36] Accepting such oaths in court implies the recognition of some concepts of honour in other religious systems by the Protestant judges. In Europe, similar systems for adjudicating legal disputes between people of different faiths existed only in parts of Italy and the Netherlands.

In some cases, Hindus and Parsis sought written assurances of freedom of worship when they took up residence in Company settlements. One example lays out the conditions of that a prominent Hindu trader from Diu in Gujarat stipulated when some of his family and representatives settled in Bombay in 1677. These specify the wearing of crimson at weddings and the cremation of the dead, as well as the understanding that no animals would be harmed within the family's compound and that no Christians or Muslims should enter without invitation.[37]

The necessity of cohabiting with people of different religions prompted some inquiries into Indian religions to take place in Company settlements. One of the earlier accounts of Hinduism and Zoroastrianism in English is Henry Lord's *A Display of Two Forraigne Sects*.[38] As Lord notes, he had been encouraged to write his account by the President of the EIC factory in Surat, where he served as a clergyman. The two 'sects' are Hindu merchants, referred to here as 'Banians', and Parsis. Lord claimed to have consulted the 'shaster' (shastras or works of holy law; which particular works he had access to is not clear) with the aid of translators. He also drew on contemporary informants.[39]

Lord's prejudices regarding Indian society[40] and his stated aim of providing an 'indictment' of the religions he describes can overshadow the more radical implications of the work. One notable aspect of the text is that it highlights the Mughal policy of religious toleration. Lord describes the migration of the Parsis from Persia, attracted by the 'liberty of conscience' offered in India.[41] Despite Lord's initial claims that the Banian's religion is heretical, he describes the 'shaster' as containing moral laws, framed according to the familiar model of the Ten Commandments. The fourth principle is to be truthful and not to break one's word in dealings, bargains, or contracts, and the eighth forbids theft and dishonesty. He stresses that the 'merchant men' were those most strictly enjoined to live by these precepts and concludes that 'they seeme to be well grounded, and to carry truth and good reason with them'.[42] Thus, Lord's work acted in its immediate setting to reaffirm the necessity for religious toleration in the Company settlements and to reassure the English traders that they could trust the word of their co-habitants and business partners, as their moral principles were founded in reason, if not in true religion.[43]

Surat was also the probable site of composition for the account provided by the maverick painter and sometime employee of the VOC, Philip Angel (1618–c. 1664), in his *Deex Autaers* (*Dasavatara*), compiled with the assistance of one or more Indian interpreters and illustrators. He presented the richly illustrated manuscript, containing a detailed account of the ten avatars of Vishnu in Dutch, to the VOC Director General Carel Hartsinck in 1658.[44] Although it seems evident that Angel relied heavily on an Indian informant, neither an individual nor a textual source for the work is identified. Despite Angel's claim to have translated the account directly from an Indian-language source, several passages suggest a less direct process of gathering ethnographic information.[45]

Relations of patronage established between merchants or travellers and scholars – such as those of the French traveller François Bernier's with Gassendi[46]– meant that examples from other cultures were increasingly used in European philosophical works, often to demonstrate that the argument being made had a basis in reason or natural law. Individual accounts were drawn, often unattributed, into compendia. Samuel Purchas used de Veiga, and Philippus Baldaeus and Olfert Dapper drew on Fenico and on Angel's manuscript for their descriptions of Hinduism.[47] Rubiés argues that by the 1660s and 1670s, a struggle was already under way in Europe over the authoritative interpretation of the religion of the Hindus.[48]

Athanasius Kircher's widely read work *China Illustrata* (1677) contains information from his Jesuit colleagues about Hinduism. The work gives drawings of nine of Vishnu's avatars, supplied by the German Jesuit Heinrich Roth, and five pages of Devanagari script.[49] There were also attempts in these early European accounts to map Indian religious practices onto those known elsewhere in the world. Lord's work portrayed Indian religions as degraded forms of an original Christianity, and Kircher, da Valla, and others speculated about connections between Chinese religions, Hinduism, and other world religions, including the ancient religion of Egypt.[50] As Findlen notes, Kircher drew criticism for his conviction that Christian truths were echoed in the practices of pagans, but his emphasis on commonalities rather than differences between faiths and cultures would provide a basis for later comparative investigations.[51]

To summarise, within the settlements of European companies, enforced liberty of conscience led their inhabitants to engage in investigations into other religions with a degree of openness that would have been difficult to parallel in Europe given contemporary restrictions on religious toleration. Some accounts found their way back to Europe as individual works or compilations.

It should be emphasised that religious toleration was practised in the early Company settlements out of necessity, not because the early EIC factors were particularly enlightened in their thinking. While cooperation, cultural as well as commercial, did exist in the pre-colonial period of European trade and settlement in Asia, Sanjay Subrahmanyam's substitution of 'the age of contained conflict' for 'the age of partnership' is as pertinent to the exchanges of knowledge as commodities.[52] Early modern Europeans believed that supernatural powers could be harnessed by the practices of other religions as well as by witchcraft, and they were as eager to discover and exploit these secrets as they were hungry for the commodities of Asia. An example of the forcible appropriation of religious knowledge is given in the journal of Isaac Pyke for 1706–1709, in which he describes how a Captain Baker held a Hindu holy man hostage, demanding to be instructed in divination.[53]

EIC presence in Bengal

At the time of Madhusudana and Marshall's conversation, the EIC's Bengal factories, so important in its later history, were in their infancy. The English trade in Bengal had been originally established in 1633 – the same year the Portuguese were finally expelled from much of the region. The Company received a *farman* from Shah Jahan, and a settlement at Hugli, by this stage the most important port in the region, was established after a further grant from Shah Shuja in 1651.[54] However, the Mughal succession struggle had disrupted the EIC's trade during the 1650s, as did the conflict with Mir Jumla over the seizure of his junk by Edward Winter.[55] Therefore, it was only in 1660 that the EIC officially re-established themselves in the region. Relations with the Mughal Empire and its regional representatives continued to be tense throughout the 1660s and 1670s. Despite the EIC's attempts, the factors were not able to procure an imperial *farman* allowing free trade in the region.[56] In this unstable situation, the English factors in Bengal were keen to recruit local supporters. As Job Charnock the founder of Calcutta wrote in 1688, the 'serenity' of the English government was hoped to be a key attraction.[57]

In the case of Kasimbazar, an important centre for silk weaving[58] where the English had been granted a factory in 1658, the most important relationship was with Shaista Khan, who had succeeded Mir Jumla in 1664 as Nawab of Bengal and ruled the area until 1677.[59] Despite conflicts over the supply of saltpetre, he renewed an earlier permit granted to the English in 1672,[60] although he would later close down the factory between 1685 and 1701.[61] Marshall mentions visiting Shaista Khan's residence and the Mughal mint at Rajmahal on EIC business.[62]

The Brahmin and the factor

Madhusudana's presence within the factory suggests that he played some part in its functioning. This was common in the early EIC settlements, where Brahmins served as political advisers, bankers, informers, representatives, and go-betweens in dealings with regional powers.[63] Streynsham Master lists three wealthy Brahmin bankers to the Kasimbazar factory in addition to several officials concerned with legal affairs, revenue collection, accounting, and cash-keeping.[64] Relations between the English factors and these officials were not always examples of peaceful cooperation, as demonstrated by the 1675 trial of Matthias Vincent for murdering Raghu, the factory's *poddar*, or cash-keeper, and extortion of the factory's Indian writers.[65]

Madhusudana himself is absent from the Company records. All that we can say about him must therefore be deduced from his name, his choice of texts, and his expositions of them. Marshall names his partner in conversation 'Muddoosoodun Rauree, Brahmin'.[66] Khan interpreted this as 'Madhusudana Radha', but it seems more likely that the correct rendering would be 'Madhusudana Rarhi', 'Rarh' roughly corresponding to the geographical area that is now West Bengal.[67] The *Bhagavata Purana* (BhP) – the first text that Madhusudana presented to John Marshall – who rendered it 'Seebaugabatpoorum' (Sri Bhagavata Purana)[68] – is particularly important to the followers of Krishna Chaitanya (1486–1533). Chaitanya was considered by his followers to be the combined avatar of Radha and Krishna. The Chaitanya school placed great emphasis on *bhakti*, or loving devotion to Krishna, which is laid out as a coherent philosophy for the first time in the BhP.

By the late seventeenth century, Chaitanya's disciples were divided between the intellectual traditions of the Gosvamis of Bengal and of Vrindavan. Meanwhile, various popular Vaishnava movements focused on goddess worship and erotic practices.[69] The use of Puranas to bridge the gap between Brahminical and popular religion was notable in Bengal,[70] and goddess worship was brought into the textual tradition earlier on, as is evident from works such as the *Devi Bhagavata Purana*, a work which is closely related to the BhP, but which stresses the role of Lakshmi, Vishnu's consort, as Mother-Creator.[71]

About Madhusudana's partner in conversation, we can say a little more. John Marshall was born in Yorkshire in 1642, amid the outbreak of the British Civil Wars. He studied at Christ's College, Cambridge, during the 1660s, where he was exposed to some of the most radical religious thought of the period: including the work of the famous heretic John Milton.

During his studies, Marshall became close to Henry More (1614–1687). More, one of the so-called Cambridge Platonists, was known for his rejection of Calvinist concepts of predestination and emphasis on man's free will and innate goodness. More corresponded with Descartes from 1648 and with Ralph Cudworth, Cromwell's one-time advisor on religious affairs. He was influential in introducing mechanist philosophy to England, although he later became critical of many of its elements.[72] Combining an atomistic natural philosophy with Platonic metaphysics, he rejected the extremes of mysticism and scepticism and strongly advocated religious toleration.[73] Another close friend of Marshall was John Covel (1638–1722). Covel was a fellow of Christ's when Marshall was a student, but left in 1670 to become chaplain to the Ambassador to Constantinople. Covel left an account of the voyage, paying particular attention to observations concerning Judaism.[74] On his return, Covel met John Locke in Paris, and the two corresponded from 1679 onwards.[75]

Marshall would have liked to stay on at Cambridge, but as he wrote, 'his countryman' Henry More was already a fellow, and his college had a rule against employing two men of the same county.[76] Marshall's was becoming a 'Company' family, and Lord Craven and his own elder brother Ralph Marshall recommended him for EIC employment. John Marshall was appointed a factor with a salary of 30 pounds per annum.[77] He sailed for Masulipatnam (modern Machilipatnam in Andhra Pradesh) in September 1668 and stayed for eight months before continuing to Balasore in July 1669.[78] Thereafter, his diaries give detailed accounts of his movements in Bengal, where he spent some time in Hugli and Dhaka, before being employed at Patna.[79] Marshall became second in command at the Kasimbazar factory in 1675, having arrived in around April 1672.[80] Not unusually, Marshall acted as both a Company factor and a private trader, selling sword blades, fabric, and rose water in Balasore, Hugli, and Patna.[81]

Marshall was apparently planning to return to England at some stage. In his will, drawn up during a bout of sickness in 1672 while in Patna, he leaves money for a mourning ring to Elizabeth, daughter of John Atwood, perhaps the woman he intended to marry.[82] In a 1675 letter to Covel, enclosed with a jar of 'Mango Acchar' (pickle), he hopes to see his friend in England shortly, but writes: 'I am now so habituated to the customs of these parts that nothing now seems strange to me.'[83] Marshall recorded his impression of India assiduously and perhaps intended to publish his accounts on returning to England. In the event, his only published work was a letter to Nathaniel Coga, another Cambridge associate: 'Giving an Account of the Religion, Rites, Notions, Customs, Manners of the Heathen Priests Commonly Called Bramines', which appeared posthumously in the Royal Society's journal, *Philosophical Transactions*, in 1700. This paper gives a summary account of some of Marshall's observations and conversations in India.[84]

Dharma for the factory

We might guess from their frequent meetings within the factory that Madhusudana and Marshall worked together on some aspects of the governance of the settlement and its delicate relations with regional powers. Why, then, did they diverge from the daily affairs of the factory to discuss questions of religion and philosophy?

The goal of preserving dharma, in the sense of good, balanced government, is central to the BhP. The establishment of enlightened royal order is the perpetual problem that Vishnu must solve. At different times, he appears in the world in order to destroy corrupt kings, to protect righteous kings, or to himself become a king and demonstrate ideal leadership.[85] Recitals of the BhP and the other Puranic texts had long taken place in sites of administration, including law courts and army camps, to Muslim as well as Hindu rulers.[86] Wealthy merchants of all faiths provided another source of patronage for mobile Brahmin intellectuals in early modern India.[87]

At the time of Marshall's untimely death in 1677, he was third in line to head the Company factories in the Bay of Bengal.[88] Contemporaries praised him for his modesty and honesty, and he seems to have been unusually conscientious, refusing bribes and recording small favours in his diaries and will. He managed to prosper in the early Company settlements without becoming the target of the usual accusations of corruption and mismanagement.[89] Giving lessons in government to the young factor, who could potentially have acted as a patron later in his career, might therefore have seemed apt to Madhusudana.

As for Marshall, he may have seen an opportunity to test the doctrines of natural religion he had imbibed at Cambridge, as well as to exercise his skills in governance. The factors of the early Company did not view themselves as 'mere merchants'.[90] Like Marshall, many were university educated and retained an interest in philosophy and theology and links of patronage with scholars in Europe.

The texts and their composition

Text and performance

In his will, John Marshall 'of Pattana in the Province of Hindostand in East India merchant being weak of body but of perfect mind' refers to 'all my manuscripts of what I have taken concerning this part of India'.[91] He requests that his brother Ralph show them to his two former mentors, Henry More and John Covel.[92] In the will, Marshall specifies that there are double copies of some of his manuscripts. In addition to some copies made later,[93] some of these doubled originals survive among the Harley manuscripts now in the British Library. Some of these surviving manuscripts are in Marshall's hand[94] while others are not.[95]

Harley 4256 contains the dates on which the translations of the BhP were made, on vellum insets by which the text is divided up into five books (A–E). These dates indicate that translation of the large work was begun on 25 June 1674 and that the final book was begun on 29 May 1677, just a few months before Marshall's death. The framing 'familiar & free dialogue' (BL MS Harley 4254) is dated 18 March 1675, shortly before the completion of Liber B. The *Samaveda* is dated in the manuscript version to 24 July 1676, thus having been composed during the interval in the recording of the longer work. The manuscript copy of the *Samaveda* begins with the statement that it represents Madhusudana's translation into 'the Hindostand language out of the Sinscreet & translation from his into English by John Marshall'.[96]

The use of 'translation' in some of the manuscripts initially suggests a process involving three texts: a Sanskrit original, a written translation into a vernacular language, and finally the English version. That at least one original text was consulted in the version of the BhP given by Marshall is indicated by a passage at the end of Liber D in MS Harley 4256 (the most complete version of the BhP), in which Marshall notes, 'Hither writ 160 pages left 63 to write of that book called Serebaugabut poran'.[97] Marshall seems to have been relatively gifted in languages and had a collection of Arabic and Persian manuscripts that he bequeathed to a fellow factor, Matthias Vincent, in his will.[98] However, there is no evidence that he was able to read Bengali script, which appears nowhere in his notes, and the example of Devanagari script he gives appears amateurish.[99] While John Haddon Hindley supposed that Marshall had translated the text from Persian,[100] this seems unlikely, given the lack of Persian terms (aside from a few common loanwords like *wakil*) in the English text, as compared with a number of both Bengali and Sanskrit words.

A slightly different process is suggested at the beginning of the truncated copy of the BhP in Harley 4253 (not in Marshall's hand), which reads: 'related by John Marshall'. 'Related' has been crossed through and 'translated' substituted. 'Translated' also appears in Harley 4256 (in Marshall's hand). What the original use of 'related' and the two hands present in the manuscript copies might suggest is a somewhat different process involving only two texts but three people. In other words, a possible mode of composition is that Madhusudana recited the texts in Bengali, perhaps loosely based on a Sanskrit text that formed the basis for a free translation.[101] Marshall translated the spoken Bengali into English and he, or an amanuensis, wrote it down. Finally, either Marshall or the amanuensis copied out the text from the rough draft.[102]

Bhagavata Purana

The manner and order of the production of the texts explain some of their characteristics. In particular, the freedom of interpretation that they demonstrate might be attributed to the element of improvisation present in an oral performance. Although Marshall apparently wished to claim a textual origin for his translations, oral transmission would not have been a deviation from custom. The word 'purana' means 'old', and the works are largely commentaries and amplifications on the older Brahminical literature. Nevertheless, they are fluid works that began as oral recitations and they are still often recited in the vernacular.[103] The BhP is normally considered one of the eighteen major Puranas (*mahapuranas*). The bulk of this material was stabilised by the Gupta period of the fourth or fifth centuries CE, but individual works are difficult to date. Some authors locate the BhP as early as the third century CE,[104] while more recent authors favour a later date, up to around 950 CE.[105] While the BhP is specifically named as the text being translated, Madhusudana's oral rendering might have incorporated elements of the vernacular Bengali Krishnavijaya or Krishnamangala literature beginning with Maladhara Basu's *Srikrishnavijaya* of c. 1498, which focuses on the tenth and eleventh chapters of the BhP.[106]

The Marshall-Madhusudana version of the BhP is recognisable in that it contains the stories of many of Vishnu's avatars recounted by the sage Shuka (which Marshall gives as 'Sookdeeb') to King Parikshit ('Porakeet'). Varaha (or in Bengali, *baraha*, from which Marshall derives 'Burrawhoos'), the boar, lifts up the earth on his tusks;[107] Vishnu fights Hiranyaksa ('Hernaik') and Hiranyakasipu ('Hernecusseepoo'), and Narasimha ('Nursing'), the man-lion, appears from a pillar to defend the devotee Prahlada ('Purrand').[108] We learn of Sita's abduction by Ravana and her rescue by Hanuman and his monkey army,[109] and we see Krishna open his mouth, revealing the universe to his astonished foster mother Yashoda ('Jessoodah').[110] More than half of the book is taken up with the life of Krishna, including his childhood in Vrindavan ('Sere Brindabund'), the 'city of gardens', and his triumphs against a number of demonic enemies and his dalliances with the young *gopis*, Radha among them,[111] and with the Brahmins' wives.[112] The inclusion of the character of Krishna's consort Radha in the story, from which she is absent in Sanskrit textual versions accepted as authoritative, indicates her importance to Madhusudana.[113] It also suggests the possible influence of the Krishnamangala genre on the presentation of the text.[114]

Despite its relative faithfulness to Sanskrit versions of the text in terms of the arch of the story, its protagonists, and meaning, the text is also replete with references that ground the stories in nearby places and contemporary concepts for Madhusudana and Marshall. For example, one passage describes some of God's power being transferred to a river that runs down a hillside 'near Dacca in Bengal... And to this day the waters which come from those hills is esteemed much of & held dear by the Drugsters & apothecaries as a thing with great virtues.'[115] Such mapping of Puranic stories onto familiar places was a common strategy during early modern recitals and could be linked to specific claims over the places mentioned.[116]

Madhusudana also tailored his story to his merchant audience. In recounting Hanuman's mission to Sri Lanka ('Lunka') to recover Sita, Hanuman is accompanied by 'his lascars' and Ravana is told that 'Hunneman was come from Ram as a bakeele or Soliciter', employing the anachronistic but familiar Persian/Urdu term *wakil* (Anglo-Indian: '*vakeel*').[117] The characters in the BhP behave like early modern merchants. Having determined to set Hanuman's tail on fire, Ravana takes out 'all the cloth in his warehouse' for kindling[118] and Krishna employs an accountant to record and weigh up the good and bad deeds of each human life.[119] The explanation of karma that is given in the *Samaveda* reads like a ledger book.[120] Such narrative devices perhaps made the text seem more relevant to the everyday lives of the two men, as well as acting like the passages in Lord's work that affirm the reasoned basis of Hinduism.

Samaveda

The *Samaveda* is one of the four ancient Vedic works of religion collectively called the Vedas. The teachings of the '4 Beads', as Marshall renders them, is referred to several times in the telling of the BhP.[121] One such passage reads:

> [A]ll the Bead & Poran are but to show, as by examples or like, but all tends only to give us judgement that we may know the reality of things by those representations.[122]

Perhaps such references prompted the free interpretation of the *Samaveda* the pair produced. Marshall described the work as 'the Epittomie or the Sum of the four Beads'.[123] Although this is not strictly correct, the *Samaveda* is mostly made up of the same hymns that appear in the Rig Veda. Madhusudana's choice of the *Samaveda* is likely to have been connected to his Vaishnavism: Krishna says in the tenth chapter of the Gita 'of the Vedas, I am the *Samaveda*'.[124] In the version provided by Madhusudana and Marshall, the *Samaveda* is said to contain the greatest part of Krishna's knowledge.[125] Krishna's statement in the Gita is taken literally, so that, in places, the *Samaveda* itself takes on a speaking role, including admonishing the other Vedas.[126]

The *Samaveda* provides a detailed cosmological account, which includes the creation of the seven realms above and below the Earth.[127] God becomes the cosmic man of the Vedas, but it is made clear that this is done merely as a demonstration of divine power.[128] The Vedic gods feature in both texts. However, Indra (given in the Samaveda and the dialogue as 'Juder' and in the BhP as 'Inder'), Lord of the Heavens, functions more as anti-hero than god, while Vishnu in his *Samaveda* incarnation takes centre stage. Interpretations of later events and works, including *Manusmriti* (*Manava-dharma-shastra*, or the 'laws of Manu') and the eighteen *mahapuranas*, also appear in the text.[129] In the exposition of Manu, the castes are mapped onto the European idea of the estates or classes to become 'Kings Noblemen Gentry poor &ca': a similar approach to that taken by Lord.[130]

Passages of both texts are critical of Brahmins. In the *Samaveda*, the Brahmins themselves are immediately denounced as lacking understanding of the true meaning of the Vedas.[131] While such sentiments might seem strange coming from the Brahmin Madhusudana, they were not unusual. Novetzke has noted the prevalence of such self-criticism as a strategy for Brahmins who performed within a world in which they were in a minority, allowing them to deflect criticism and maintain importance.[132] Notably, despite making similar criticisms, the text of the BhP eventually justifies the importance of giving to Brahmins in some detail, with the spiritual rewards carefully matched to the size of the gift.[133]

'A familiar & free Dialogue'

The BhP itself is structured around a series of frame stories, many of which employ the dialogue form, the overarching structure being the conversation between the righteous King Parikshit, who is doomed to die in seven days, and the wandering sage Shuka.[134] The framing of the text using 'a familiar & free Dialogue' between Marshall and Madhusudana is thus apt. The dialogue form was also common in Christian texts that sought to convince the doubter of the truth of theological positions and had been employed by Marshall's mentor Henry More in his heterodox *Divine Dialogues* of 1668.[135] We should suspect that if the texts were the product of oral exchange, the dialogue was a contrived composition that both parties recognised as a route into the text they were about to present. Nonetheless, the dialogue is revealing about the participants' understanding of the key philosophical points of the translated texts.

In his letter to Coga, published in *Philosophical Transactions*, Marshall writes:

> I have always had a profound Veneration for the Dictates of Nature, and the Universal Traditions of Nations, for hereby are Infinite things to be learned, for the establishing of our Glorious Religion against Atheists, and the more easie propagation of the same amongst Infidels and Heathens.[136]

The statement is conventional: as God is revealed in the created world as well as through direct revelation, it should be possible to discern something of God's message through the observation of nature and the human traditions that arise from it. This 'natural theology' can be traced back to Plato's attempt to demonstrate certain basic truths about God and Aquinas' 'Five Ways' to prove the existence of God. The statement concerning conversion resembles the rationale for the Jesuit strategy of accommodation.

'Natural religion', prefigured in Stoic thought and later central to Deist arguments, was a more controversial position founded on the belief that all religions hold the same essential beliefs, a fact obscured by the accretion of doctrines and ritual practices.[137] Marshall comes closer to this viewpoint in the dialogue, from which any attempt to convert his partner in conversation is notably absent. Here, Marshall argues that miracles serve only to convince believers in each religious tradition. But the principles religions teach are in accordance with reason and the senses, and this is their essential essence. Revelation appears necessary only to enjoin belief in the fundaments of religion, which can equally be attained by reason.[138] Perhaps in recognition of the controversial implications of this idea, Marshall proposes that the discussion should be secret. In return, Madhusudana promises: 'I will be free with you, nor do I desire you should conceale what I say to you further than at present, that the people of my religion here in your factory may not know of it now.'[139]

Marshall opens the dialogue by posing three fundamental questions: first, by whom, how and wherefore the world was made; second, how it continues; and third, what will be the end of it? The questions were familiar within the Christian dialogue tradition. However, they also point to central cosmological themes of the BhP, namely: *sarga*, emanation of the subtle aspects of the world from a primordial matter; *visarga*, emanation of the gross elements and formation of the world by gods; *sthana*, conservation of the universe and the pre-eminence of the Lord within the world by protecting the beings, according to their dharmic status in life; *uti*, cause of the emanation; and *nirodha*, the destruction of the cosmos.[140]

Madhusudana responds to Marshall's first question: 'Before God there was nothing nor can it be answered whence he was, but before the world or any part of it was, God was, and he was alone.' The answer mirrors the opening of both the BhP and the *Devi Bhagavata Purana*. The world, Madhusudana continues, came into being as the result of the expansion from a 'seed' (Sanskrit *bija*, which Marshall renders 'Beedy'[141]) or 'care' (Sanskrit *vicara*, given as 'fickur') in the mind of God. This seed is described as immaterial or incorruptible. Marshall uses such passages to argue in his letter to the Royal Society that Hindus 'acknowledge but one God, who they say is *Burme*, that is immaterial' and that God 'dwelt in a vacuity before the world'.[142]

Madhusudana goes on to explain that God's seed – which is said to be imperishable or 'nirgoon' (Sanskrit, *nirguna*, meaning unqualified), like God – gives rise to the 'gunns' or 'goons' (Sanskrit, *gunas*, the conditions or qualities of nature). The *gunas* are described as changeable or perishable, though arising from the seed in God's mind that is eternal and immaterial, and retaining their connection with it and with God. They are '[a]ccidents of things or qualities viz. hot, cold, hard, soft, black, white, & the like, whereas the other is the substance'.[143] This argument draws on the ancient Samkhya system, in which time is said to be correlative with the functioning of the three *gunas* and is thus part of the manifestation of the material.[144] Samkhya philosophy deals with the relationship between the world of matter (*prakriti*), which is composed from the three *gunas*, and its relationship to spirit (*purusha*). While in classical Samkhya, material evolution takes place automatically, in the BhP, it occurs only as a result of God's will. The later *bhakti* tradition, as expressed in the BhP, thus suggests that God is present in the material as well as the spiritual universe. In the Marshall-Madhusudana dialogue and texts, the term used for God's will or strength is the Sanskrit word *tejas* ('teas'), also meaning fire or heat.[145]

Madhusudana refers directly to the debates between the dualist and non-dualist traditions, noting that all men agree on the existence of God, but as for the world, it is 'uncertain whether it be or not, being capable of not being in the minds of men'.[146] Hindu non-dualist thinkers hold that the world of the senses is an illusion and that when the veil of illusion is lifted, the spirit merges with God in oneness. The argument of the Vaishnaiva 'personalists', expressed in both the BhP and the *Vishnu Purana*, hold that the phenomenal world is only an illusion in the sense that it is temporary. When the veil of illusion is lifted, one does not cease to exist as an individual, but rediscovers a permanent identity and loving relationship with the divine.[147] This concept is expressed at several points in the Marshall-Madhusudana BhP. In one passage on reincarnation, it is explained that 'all the soules in the world are tyed to a string or rope', being hurried between one birth and the next and all the time acting from self-interest and lacking understanding. Those that gain understanding, through the selfless love of God, will receive a 'permanent habitation' in which to abide with God.[148] When enlightenment is achieved, the body will remain incorruptible on earth while the soul ascends to join God.[149]

The relationship between the body and the soul had also been a subject of debate in Greek antiquity: Plato favoured the approach that the philosopher should attempt as far as possible to rid himself of concern for the demands of the body, which he regarded as opposed to the life of the soul. The argument is most clearly expressed in *Phaedo*, the text in which he argues for the transmigration of the soul.[150] In his *Meditations*, Descartes had made a sharp distinction between spirit and matter, which are considered categorically distinct, although matter may be animated by spirit.[151] On the other side of the divide in the ancient world were the materialists Epicurus and Lucretius, whose arguments were revived in the seventeenth century by vitalist or atomist philosophers including Gassendi, who argued that the physical world was composed of an infinite or finite number of atoms moving in a void, each of which was endowed with spiritual energy and potential motion. The void was an extension without matter.[152] The Cambridge Platonists – influenced by Hermetic ideas and the work of earlier neo-Platonists including Ficino and Robert Fludd – were important in finding accommodations between these two positions.[153]

Marshall presses Madhusudana on the subtle distinction between spirit and matter that is offered: 'Still I find no real difference as to Essens betwixt God & this Seed & its Goones.' This leads them to the concept of the world as an 'expansion', 'extension', or 'emanation' from the original seed in God's mind.[154] Madhusudana explains that as the seed of the world was expanded from within God, space, time, and matter also expanded.[155] Here he apparently refers to the emanationist doctrine of creation (*Vyutha Vada*): one of the elements of the BhP that is thought to have derived from the *Pancaratra* conception of the universe and to have been gradually harmonised with the Vedic tradition through its inclusion in the *Vishnu Purana* and the BhP itself.[156]

In Western thought, the concept of creation *ex deo* had been developed by Plotinus and his followers, although it was rejected by Augustine in favour of creation *ex nihilo*. Milton creates an emanationist vision in *Paradise Lost*, in which God sends forth his spirit, filling space: not, in Milton's version, permitting a vacuum.[157] The Cambridge Platonists shared this view of God filling the universe: a view that challenged the assumption of both Aristotle and Descartes that matter alone is capable of extension and of motion. More listed the essential attributes of God as 'Self-penetration, Self-Motion, Self-contraction and Dilatation, and Indivisibility'[158] and defined God as 'an extended substance far more subtil than body, that pervades the whole matter of the universe'.[159] Cudworth similarly, and controversially, argued for an extended substance other than body that was self-active, was capable of penetrating matter, and possessed an internal energy.[160]

In England, the philosophical significance of these arguments concerning the material nature of God was soon forgotten: a later editor of More's *Divine Dialogues* simply commented that it contains some 'peculiar sentiments about some abstruse metaphysical questions relating to the immensity and eternity of God', adding that these sentiments seem to have been shared by Isaac Newton, among others.[161] In fact, the philosophical reconstruction of the concept of space through the decoupling of extension and matter that the Cambridge Platonists performed was an important contribution, thought to have influenced Newton's arguments in the *Principia mathematica*.[162] For Marshall, given his convictions concerning natural religion, finding support for these concepts in another religious and philosophical tradition would have served as an important validation of the beliefs of his mentors.

The next part of the dialogue takes us to the concept of *lila*, often translated as God's play, or by Marshall, as 'God's sport'. Madhusudana explains that God made the world 'For a sport to himself'. In the BhP, Vishnu creates the world out of joyous love of mankind and to increase mankind's love for him.[163] Marshall protests that for God to take pleasure in worldly things surely indicates some imperfection. Madhusudana's reply is that while God is not transported, he is always merry. Before creating the world, he foresaw the merriment that it could contain and 'then he caused the world to be made, or to act as extra or without him'.[164] Madhusudana goes on to explain that from the expanded seed that became the world, 'God left himselfe only infinite wisdom or knowledge and pleasure'. This passage demonstrates the connection between God's love for man, omnipotence and omniscience, and mankind's free will.

Marshall does not commit himself to the argument. However, it might have recalled parts of his mentors' work. The radical notion of time in the BhP is closely connected to this idea of infinite possibilities, since God, seeing all human time laid out at once, can descend into it at any given point.[165] Similarly, Henry More argues that God is able to see the unfolding of time all at once: 'The duration of God is like a rock, of natural things like that of a river.' Furthermore, God is able to see: 'all possible successive evolutions that are'.[166] The emphasis on human free will and the native goodness of man were among the elements of the Cambridge Platonists' thought that shocked Calvinist commentators.[167] These concepts were connected to the key arguments for natural religion and toleration, as well as being based in the Platonic doctrine of innate ideas.[168]

The importance of God's love for mankind and a focus on following worldly happiness to a limited and reasonable degree as a means to achieve holiness were also discussed by the Cambridge Platonists. More stresses the importance of God's love[169] and of 'freedom and mirth' while not losing sight of sobriety and falling into the trap of the 'enthusiast', who mistakes his 'tumultuous fancy' for the true message of God.[170]

The issue of God's love for man is connected to the question of how man may love God. The comparison between the longing for communion with God and an earthly lover has been made many times. However, the scenes in the BhP in which Krishna makes love to the *gopis* make some of the most overt connections between religious devotion and human passion. Here, renunciation occurs spontaneously out of desire for God rather than through withdrawal from the world.[171] In the Madhusudana-Marshall texts, the physical nature of the interaction is both made clear and justified: Govind does not sin by his intercourse with the *gopis* as God resides in all places at all times.[172] There is a clear distinction in the BhP between desire in the sense of covetousness and desire prompted by love of God or delight in the world he created.[173]

The early modern English thinker who took the idea of loving communion with God furthest was Damaris Masham, daughter of Ralph Cudworth and close friend of John Locke. Masham's *Discourse Concerning the Love of God* stresses the importance of devotion to God and affection for other human beings that, while guided by reason, are nonetheless joyous and passionate.[174] While Masham's text was far less explicit on the nature of loving relations with God, her text provides some parallels to the BhP in its defence of worldly desire, in the sense that it is an expression of longing for communication with God.[175] As will be discussed below, Masham was among the English thinkers who came to know of the Madhusudana-Marshall BhP.

The potential for a call for loving communion with God and fellow creatures to be misinterpreted as an excuse for licentiousness and chaos must have been well understood by Marshall and his contemporaries, who had witnessed the uninhibited sexual behaviour practised by the Ranters in the name of spiritual freedom during the British Civil Wars. Similarly, Madhusudana would have been aware of the Sahajiya sects, who attempted to re-enact the communion of Krishna and the *gopis* through sexual rituals.[176] This explains the careful attention that is given to the philosophical meaning of the sections of the text dealing with Krishna's sensual adventures.

The final part of the dialogue deals with the formation of the world, or *visarga*. Madhusudana explains that once the seed had expanded and given rise to the three *gunas*, they then resolved themselves into the shape of a woman.[177] From this primordial woman was created all things, beginning with the three primary deities, 'Burma, Bismah, Manhadeeb' (Brahma, Vishnu, and Shiva). The Vedic gods and minor deities follow. The question of creation is connected to the issue of the age of the universe, another question that would have seemed relevant to Marshall, given the contemporary natural-philosophical interest in the issue in Europe.[178]

Marshall tried in various parts of his manuscripts to calculate the amount of time elapsed between the events he records and the present day.[179] In his notes, he concluded that 417,000 years had elapsed between the birth of Krishna and 1670.[180] Like his contemporary Philip Angel,[181] Marshall never attempted to make the length of the *yugas* conform to the Christian idea of the world being 4,000 years old,[182] but makes it clear that Indian chronologies make the world far older. The Flood is said to have taken place 4,732 years ago, at the beginning of the *Kali yuga* ('Colijoog').[183] Marshall seems to have readily recognised the relevance of the calculations related by Madhusudana to astronomy, referring in one passage to the revolution of the ecliptic. The Madhusudana-Marshall BhP gives an even greater figure for the total length of one cycle of the world's existence than is usual: amounting to 18,662,400,000,000,000,000,000 years in total.[184]

The third part of Marshall's question is 'what and how will be the end' of the world. In the *Samaveda*, Shiva is created as the destroyer of all things.[185] In the dialogue, God is said to encapsulate all three functions: creation, continuation, and destruction, within the original seed that becomes the universe. Madhusudana explains that God at once stands outside of and embodies time.[186] Towards the end of the manuscript, it is explained that at the end of a *kalpa*, the universe will be imbibed by Brahma, before being created again. Madhusudana concludes: 'So you see how the world is and while tis thus bound there will be no end of it'; the only escape from *kala* or endless time is to seek refuge with God through the achievement of true understanding.[187]

There is no date at the end of the last book of the BhP, and it is not clear whether the two men had intended to continue. Perhaps Marshall, like King Parikshit, died shortly before the end of the story.[188] However, as the next section will demonstrate, the texts the two men produced were circulated in late-seventeenth-century England, and Marshall was not alone in finding the ideas Madhusudana explained to him relevant to contemporary philosophical debates.

Interpretations

Reading 'the Poran' in England

Follow John Marshall's death, his brother Ralph seems to have made some efforts to carry out the terms of his will, and although his bequest for a marker at the head of the Balasore river was never observed, his manuscripts were sent to England.[189] A letter enclosed with the *Samaveda* shows that the manuscript was sent to Covel by one W. Salmon. In his undated letter, Salmon explains that the manuscript had been 'detained all this while for my Lord Duke's inspection'.[190] The Streatham postmark suggests that the Duke in question was the influential politician William Russell, first duke of Bedford (1616–1700). The manuscripts of the BhP and the 'notes and observations' also reached Covel, remaining in his collection until 1716, when he sold them with much of the rest of his collection to Robert Harley, Earl of Oxford (1661–1724).[191]

Covel clearly took Marshall's works seriously, given his efforts to recover the *Samaveda*. He read the BhP with particular attention, as suggested by his notes at the front of Harley 4256, which compares the work with Thomas Hyde's 1701 *Historia religionis veterum Persarum*. He notes that the religion of the Brahmins, like that of the ancient Persians, considers there to be one eternal God, in contrast to sects such as the Manichaeans that consider good and evil, or God and the Devil, to be co-eternal principles.[192] He notes the existence of a first couple like Adam and Eve and the common belief in the Flood, compares the *devas* or 'dewtas' with the angels in both the ancient Persian religion and Christianity, and even tries to trace connections between the names of the gods and angels given by Marshall and Hyde.[193]

Whether Hyde himself saw Marshall's papers is uncertain; despite several contacts within the EIC, his information about Indian religions was mainly derived from Persian sources. John Covel corresponded with Newton, his contemporary at Cambridge during the 1680s on political matters.[194] Although Newton does not mention seeing Marshall's work, he does refer to making use of Covel's manuscript collections.[195] Whether or not they were aware of Marshall's manuscripts, both Hyde and Isaac Newton shared Covel's interest in the etymology of the word 'Brahma'. In his major work on Zoroastrianism, Hyde claims on this basis that 'Indian idolatry' (although he questions this terminology) was connected to Abrahamic religion.[196] In his *Chronology of Ancient Kingdoms*, Isaac Newton follows Hyde in claiming that Zoroastrianism contained Abrahamic elements derived from Brahmanism as well as its Chaldean heritage.[197]

There is more secure evidence that Covel discussed the BhP with certain acquaintances, including Damaris Masham and John Locke. Locke wrote to Covel in 1700, encouraging him to print the 'Poran' in Awnsham Churchill's forthcoming book of voyages. Locke and Masham, who are mentioned in the letter, shared Covel's interest in the work. In his letter to Covel, Locke adds his hope that 'I shall not need I think to presse you in an affair wherein it is soe visible you will doe a great kindness to the world'.[198] Nevertheless, Covel refused to send the manuscript to the press before he had spent more time with it.

In his letters to Locke, Covel argues that 'the Banyans religion is older much then some of our Learned Authors would make it', based on the cosmography presented in the BhP. He speculates that the myths it contains provide the basis of Greek fables, noting the presence of the story of the Flood and the four ages (*yugas*), which he compares to the aeons or Four Ages of Ovid. He argues that while the connection between the doctrine of reincarnation in the Hindu religion and the teachings of the Pythagoreans had been noted by other scholars, 'I am apt to believe that Pythagoras learn'd from them, and not they from him'. Covel points out that these opinions will be controversial among many of his contemporaries.[199]

Locke sent one further letter encouraging Covel to print the collection, 'solliciting as I doe for the publick good', and agreeing with Covel's views concerning the antiquity of the work and the probability of the origination of the Pythagorean doctrines in India. Despite Locke's concern that 'if you publish it not in your life time it may possibly be quite lost, tis certain it will be lost to the publique at least for some time',[200] Covel apparently refused to relinquish the manuscript. Locke's concern to acquire the manuscript, not for his own interest but expressly for the public good, suggests that he believed it contained evidence of natural law. Unfortunately, it is not clear whether Locke ever succeeded in seeing the manuscript.

The 'Poran' is also mentioned in some passages of the correspondence of the librarian Humfrey Wanley (1672–1726) with John Covel in 1715–1716, while negotiating the purchase of his manuscripts for Robert Harley. According to Wanley's correspondence with an intermediary, Harley was particularly keen to see the work.[201] However, in his correspondence with Covel, who was demanding a high price, he wrote, 'The *Poran*, from what I have read of it in English (for the whole or greatest part of it is in print) I perceive to be full of Dreams and Fables, yet I am willing it should come in among the Herd'.[202]

Unless he was referring to one of the versions of Philip Angel's text that had found its way into the European press, Wanley was probably dissembling in his claim that the 'Poran' had been printed, as there is no evidence that Covel ever relinquished Marshall's manuscript to the press. Nonetheless, Wanley's comment implies that, by the early eighteenth century, some version of the tales of Vishnu's avatars had circulated within seventeenth-century England and that not all thinkers had credited them with the importance that Covel attached to them.

After their incorporation into Harley's collection and later the British Museum, Marshall's manuscripts apparently attracted little attention, aside from some copies made by the Oxford Orientalist John Haddon Hindley (1765–1827).[203] By the time Wanley acquired Marshall's manuscripts, their moment had passed. Platonic ideas were on the wane in England by the end of the seventeenth century.[204] Newton's *Principia* surpassed the Cambridge Platonists in terms of metaphysics and cosmology. Furthermore, as the poor reception of Thomas Hyde's work demonstrated, academic fashion had turned for the moment away from the study of Eastern languages and religions.

If few in England recognised the significance of the 'Poran', the text and its message of religious toleration did circulate among some key thinkers during a time of transition. Its existence might have helped English scholars take another small step on the path from a simplistic view of Indian religions as 'idolatry' or a perverted version of Christianity towards a more complex view of the interactions of religions and languages over time.

Jesuit revivals

While the interest in comparative religion in England waned, French scholarship on the subject was being awakened, on the basis of some projects comparable to Marshall and Madhusudana's and the earlier writings of Thomas Stephens. The strategy of accommodation attracted interest in the last years of the seventeenth century among a group of French Jesuits including Pierre Martin (1665–1716). Martin arrived in Balasore via Surat in 1699 after an unhappy spell in Persia. He applied himself to learning Bengali, and after five months claims to have entered a 'Brahmin university', properly attired.[205] Martin argues that all new missionaries should be sent to the Madurai mission to learn, as he had, the languages and customs of Indians, to 'read and transcribe the books that the venerable Father Robert Nobili and our other Fathers composed', and to resume the practices that they had begun.[206]

The renewed enthusiasm for accommodation may explain the mystery of the *Ezourvedam*. This anonymous text, probably authored during the seventeenth century, reached France by 1760, when a copy was presented to Voltaire. Believing it to be a Hindu text, the philosopher used the work as the basis of his writings about Indian religions. Later scholars debated the status of the text, some regarding it as a version of an original Veda (normally the Yajur Veda) and others as a commentary. However, since the late eighteenth century, most scholars have agreed that the text is in fact a Christian 'pious fraud' disguised as a work of Hinduism.

Recent scholars have focused on the discussion of where and by whom the text was produced, suggesting that Robert Nobili, Pierre Martin, or one or more of several other contemporary Jesuits or converts produced the work in either Bengal or Pondicherry, where versions of the *Ezourvedam* and four other 'vedas' were discovered in the eighteenth century.[207] It was in Pondicherry, in 1769, that the Tamil convert to Christianity Maridas Poullé produced a French translation of the BhP on the basis of an early modern Tamil translation.[208] Poullé worked with a number of French intellectuals, including the astronomer and cartographer Jean-Baptiste Joseph Gentil[209] and the translator of the Zoroastrian texts Anquetil-Duperron. Poullé's text has also been suggested as a possible source for the *Ezourvedam*.[210]

Despite Marshall's claims in the letter to Coga concerning the propagation of the gospel, there is no overtly Christian message in the texts he produced with Madhusudana. Nevertheless, there are undeniable similarities between the *Ezourvedam* and the Marshall-Madhusudana texts. The *Ezourvedam* also opens with a dialogue about the nature of the world; the role of the *gunas* in creation and destruction is discussed. The genealogy of the gods is given, beginning with Aditi, mother of the gods, and her rival Diti in both texts. In contrast to the BhP, however, in the *Ezourvedam* the worship of Vishnu's avatars is explicitly denounced as a fall from true religion into idolatrous fantasies.[211] While God the creator starts off alone, he creates the world not by expanding it from within himself, but as a separate entity. We might conclude that the author of the *Ezourvedam* and Marshall both took selectively from Hinduism's teachings about the creation of the world to support an explanation for creation that supported their own cosmological and theological beliefs.

While Europeans often tried to use their investigations into other faiths to support their own belief, to advance their own viewpoints, or even as tools of coercion, the balance of power did not as a whole lie in their favour in the late seventeenth century. Europeans were also used as informants and vehicles for transmitting particular viewpoints. In our case, the texts were clearly selected by Madhusudana rather than Marshall. It is thus as easy to locate the texts produced by Madhusudana and Marshall within the process of identity making based on creative reworking and performances of Puranic texts described by Rosalind O'Hanlon as it is to see them as precursors to the later practice of Orientalism or Protestant inheritors of the Jesuit practice of 'dressing up' Christian beliefs in the costumes of other faiths.[212]

Conclusions

Marshall and Madhusudana's conversation revolves around the interlinked questions of the nature of God and the creation and the human capacity for free will, doing good, and understanding God's will. Seen from the personal perspective of the two men, their meeting allowed them to find commonalities between a strand of Bengali Vaishnavism on one hand and English Protestantism on the other. While Marshall suggested proselytising as an ultimate motive in his correspondence with his English contemporaries, it is not evident within the texts that the two men produced. Marshall's interest was perhaps directly motivated more by a desire to find a basis in natural religion for his own metaphysical views. Both Marshall and Madhusudana believed strongly in God's love for mankind, the innate goodness of mankind, and the possibility of reasoned enjoyment of earthly pleasures. These shared principles perhaps helped them transcend their particular faiths to discuss the metaphysical questions of the age of the universe, its origins, and the concept of motion and its relation to matter.[213] As the English readers of the 'Poran' realised, some of the points on which their ideas coincided highlighted the legacy of much earlier philosophical exchanges, between ancient Greek and Indian culture.

The urge to find common ground was prompted by practical as well as theoretical considerations. The two men spoke at a time of transition and uncertainty in Bengal; political and religious authority was in flux. The English factory at Kasimbazar was under continual threat from illness and attack, and, as the case of the murdered *poddar* shows, its Hindu brokers often bore the brunt of violent struggles for power both within and outside its walls. Meanwhile, the status of Brahminical religion in Bengal was challenged not only from below by the popular movements focusing on goddess worship, Baul mysticism, and conversion to Catholicism, but also from above by more conservative Mughal rulers and an influx of Protestant settlers. Finding common ground on which to base the relations of trust necessary to survive side by side provided a powerful incentive for religious toleration within the early EIC settlements.

The early arrangements for religious toleration made in the settlements of the European trading companies are likely to have had some influence over the gradual movement towards greater toleration within Europe itself, as merchants established relations of patronage with scholars and were encouraged to undertake investigations into other religions. Arguments for toleration made by early modern European philosophers frequently draw on contemporary accounts of religious practices elsewhere in the world. In the case of the Marshall-Madhusudana BhP, Locke's insistence that it be printed, not merely as a curiosity but in the public interest, evinces his belief that evidence of natural law could be discovered in the practices of other religions.

The impact of the enforced religious toleration that EIC servants experienced in Asia has rarely been considered in intellectual histories of the rise of toleration in Europe despite the direct involvement of philosophers including Grotius and Locke with the EIC and VOC. Nonetheless, its effect was to promote inquiries into different religions that eventually formed an important basis for both studies of comparative religion and philosophical arguments in favour of freedom of conscience.

A nineteenth-century scholar remarked of Marshall's papers that '[i]f Marshall had published his researches in 1680, they would have inaugurated an era of European knowledge of India, being in advance of anything which appeared before 1800'.[214] In the sense that this seems to have been the first attempt to render a Hindu text in English language without the addition of overt Christian polemic, we might agree that the translations can be regarded as a forerunner of the work that was later carried out by the Royal Asiatic Society and others. The interest evident in Marshall's investigations, as well as those of several of his contemporaries, in the religions of those around them, does suggest deeper and earlier roots to Orientalist scholarship. However, in many ways, Marshall's endeavour was closer to the sort of syncretic religious texts that were produced by Jesuits from the sixteenth century onwards than to what has conventionally been assumed to be a more 'Protestant' attachment to the recovery of an original text. Furthermore, in terms of the choice of the text, its mode of delivery, and interpretation, Madhusudana was perhaps more influential than Marshall, as suggested by his assumption of the role as instructor in the dialogue.

4
Botanical and Medical Networks: Madras through the Collections of Two EIC Surgeons

Introduction

This chapter takes a journey around Madras and connected places, as the settlement rose to prominence in the late seventeenth and early eighteenth century. On the way, we will look into some of the public and private spaces in which medicine was prepared and practised and medicinal plants were grown, collected, labelled, and sold or presented as gifts to patrons, local, and international. These include the courts of the rulers of Golconda and the nawabs of Arcot; the army camps of the EIC and the Mughal and Maratha armies; the hospital; the bazaars and apothecaries' shops; the physic gardens that dotted the city; and the ships that carried drugs and ideas to and from the town. Based on this survey of the medical landscape and political geography, I will argue that the practice of medicine was crucial to diplomatic relations in the region. I will show how networks of doctors operated in the negotiation of the EIC's settlement in Madras with the new rulers of Arcot after the disastrous war of the 1680s launched by the EIC against the Mughal Empire. Several volumes of herbarium specimens, along with printed and manuscript descriptions relating to them, collected by two EIC surgeons, provide a glimpse into the range of *materia medica* that were known and deployed by the doctors and their Tamil- and Telugu-speaking counterparts and collaborators.

The evidence for this reconstruction is drawn largely from the correspondence of the two EIC surgeons, Samuel Browne and Edward Bulkley, with the London apothecary and fellow of the Royal Society James Petiver[1] and the collections of dried plant materials they sent to London. Samuel Browne was appointed to the position of surgeon of Madras in 1688, after serving as a ship's surgeon.[2] Petiver became acquainted with Browne through the clergyman and collector Richard Sambach in 1689, and the two corresponded until Browne's death in 1698.[3] Edward Bulkley, formerly the surgeon at Pettipoli, who first joined Browne at Fort St George in 1692, then replaced him in 1697, remaining in the position until 1709 when he resigned and became 'land customer' and member of council and Justice of the Peace until his death in 1713.[4] Their collections of plant and animal specimens remain in London's Natural History Museum (NHM) and their correspondence among the Sloane manuscripts in the British Library, after circulating through the repositories of both the EIC and the Royal Society.[5] However, it is clear that these existing materials represent a fraction of those that each surgeon originally sent to London, let alone of their total collections.

The choice of Madras as the node around which plants, medicines, and information circulated is in part to stress that many of the activities associated with the scientific revolution: the assiduous collection and detailed study of natural objects, the amassing of libraries and 'repositories' of curiosities and books of dried plants, the exchange of information through networks of scholarly correspondence, and the formulation of theories about the natural world were not confined to European capitals but also took place in colonial settlements and outposts, each embedded within a web of local and international connections.[6] Raj has stressed that the collectors relied on by metropolitan scientists were not 'space probes'; in other words, they each had their own priorities and agendas and should be regarded in the light of their immediate economic and social contexts.[7] While the process by which instructions for collecting and observing were composed and despatched by organisations like the Royal Society has attracted detailed attention in recent years,[8] the social situations and knowledge networks of those who responded to these inquiries have received less attention. A close focus on the everyday lives and interests of those in the colonial settlements can throw light on three related questions in the literature: the (im)balances of Asian and European agency in colonial collections, the interaction between agents of the rival European companies in Asia and the interplay between science and colonialism on the ground.

Richard Grove has argued that the study of the networks through which botanical knowledge was distributed demonstrates the shared roots of Indian, Middle Eastern and European medicine and how European science was 'transformed by indigenous technical knowledge' in this period.[9] Raj's reorientation of modern science also highlights the ongoing role of intercultural encounters in the making of modern science. Other authors have made important qualifications to these arguments: Pratik Chakrabarti notes that 'there is a historiographical cul-de-sac in highlighting the creative formulation of scientific and medical knowledge in the British colonies in the eighteenth century'.[10] He warns that to over-stress the creative nature of such encounters risks 'under-playing the equations of real power within the colonial situation'.[11] The histories of medical practice in the colonies, he argues, can only be understood within the colonial context, in which 'commodities became materials of medicine by undergoing erosions of their histories'.[12]

Both positions are useful for examining the materials at our disposal. As I will argue, botanical texts composed in South Asia and printed in European languages can provide information about both Asian medical practices and botany and their influence on European ideas, if properly contextualised. Furthermore, examining the knowledge networks through which they were circulated can demonstrate the process through which the diffusion – and appropriation and reconfiguration – of medical-botanical knowledge took place. In our case, the collections made by Browne and Bulkley provide a small window into the practice of medicine in seventeenth-century Tamil Nadu. Meanwhile, looking at the networks they were embedded in, locally and internationally, shows how the knowledge was disseminated and decontextualised. I will show how these EIC surgeons, in close cooperation with local experts, collected and labelled the items that eventually ended up among the wares of London apothecaries and in the Sloane Herbarium. Sloane's was the collection that formed the basis of the British Museum, later being transferred to the NHM.[13]

David Arnold has called for historians to see European medical ventures overseas as 'more than just a series of independent national narratives; to view them instead in a comparative, transnational perspective', examining the continuity as well as the differences between Portuguese, Dutch and English physicians in the East Indies between 1500 and 1750.[14] Nevertheless, in the literature on the medical and botanical interaction between Europe and Asia, there is something of a lacuna between early works like those of Garcia da Orta and Hendrik Adriaan van Rheede tot Draakestein, and the later work of British botanists including William Roxburgh, Benjamin Heyne, and Joseph Dalton Hooker. In the course of the centuries spanned by these investigators of the Asian environment, European medical thought and systems of botanical classification shifted dramatically. Although authors including Richard Drayton have redefined botanical scholarship during this period as 'the colonisation of Europe by extra-European interests',[15] the study of this process has generally been centred on the intellectual activities of colonial capitals rather than in the settlements themselves. The interaction of colonial agents on the ground both with their Asian counterparts and with one another, and how they contributed to the alterations in scientific thought in the European capitals has been largely overlooked. Here I will demonstrate that the Company's surgeons acted as both medical and political go-betweens, obtaining concessions for the Company settlements in return for medical services to the Mughal courts and camps. I will also demonstrate that, as collectors, the English surgeons were embedded in networks of competing and collaborating agents of other European companies as well as representatives of private and missionary interests. These agents, through their reports and collections, played an active role in altering European conceptions of the natural world.

Cook's recent work has been valuable in terms of drawing out the connections between trade, empire, and botanical/medical investigations in the VOC. One particularly interesting feature of this work is the focus on 'objectivity' – in terms of the literal examination and appraisal of objects as well as in its opposition to subjectivity – in both trade and science.[16] This has bearing on one point that I would like to emphasise – that printed pharmacopoeia were often copies (sometimes even tracings) of *hortus sicci*, or dried gardens, which in turn were often dried impressions of the ways plants were laid out in real gardens. In turn, gardens mirror the real environments around them, as processed by those who tend them. Equally, books inspired gardeners when laying out their plants. At every stage in this continuum between plant and book, the object had a value that was both scientific and commercial: whether as a garden to supply drugs to the hospital, cutting the cost of imports; as a product to be transmitted through international networks to an apothecary; or as a book to grace the library of a curious scholar or to serve as the basis of bio-prospecting elsewhere in the world. Here, I will demonstrate how the practices of collecting and transplanting plants, gardening, and practising medicine as well as publishing natural histories were directly linked to the colonial exploitation of resources as well as to the forging of political alliances.

Medicine and politics

Established in 1639, the EIC's settlement at Madras quickly became the focal point of the Company's operations on the Coromandel Coast. By 1695, Samuel Baron described it as 'the most considerable to the

English nation of all their settlements in India whether ... in reference to the trade to and from Europe, or the Commerce from one part of India to the other'.[17] The migration of Indian traders resulted in a population of around 80,000 by the early eighteenth century.[18] The early attempts to establish trades to China and Japan, to re-establish a presence in Maritime Southeast Asia, and to gain a foothold in Bengal were all directed from Fort St George. With the decline of Masulipatnam after c. 1700, the town also became important to regional trade conducted by both Asians and Europeans, including members of the medical profession.[19] The fortified settlement of Madras was the centre of military power from which the Company fought the territorial conflicts known as the Carnatic and Mysore Wars later in the eighteenth century.[20]

Nevertheless, during the late seventeenth and early eighteenth centuries, Madras was in a difficult position. Since 1657, the settlers had been subjects of the rulers of Golconda, the Qutb Shahi dynasty, which had regained independence from the Mughal Empire after a war of succession.[21] Despite nominally accepting Mughal overlordship in 1677, the same year Abul Hasan of Golconda negotiated a military alliance with the Maratha leader Shivaji, based on the annexation of land in Bijapur and Karnataka, until the final defeat of Shivaji's son Sambhaji in 1687–1689. The EIC had thrown in their lot with these enemies of the Mughal Empire. However, the war proved disastrous, and the English parliament granted a charter to the rival 'new' EIC in 1698, which fought and petitioned the Emperor Aurangzeb against the old Company until a final merger of the two took place in 1702–1709. The Madras factory was therefore under great pressure to form a new alliance with Zulfiqar Khan and Daud Khan, leaders of a powerful faction of the Mughal nobility, who had been sent by Aurangzeb to defeat the Marathas and had consolidated Arcot as a centre by the early 1690s. The network of surgeons played an important part in the diplomatic negotiations between Madras and these new nawabs of the Carnatic.

The involvement of surgeons in diplomacy was a practice that was common within the Mughal Empire.[22] Skill in healing had secured Garcia da Orta his rent of Bombay, and had been important to the VOC's negotiations in Ming China.[23] As a letter from the Surat factory to London dated 1666 shows, the EIC considered it essential to 'the honour of the nation' that Company surgeons should be skilled in the cure of Indian 'Governors and Noblemen', who, they wrote, 'having such civilities done them, in gratitude become very serviceable in ... furthering your business'.[24] The performance of a cure sealed many well-known concessions, including the *farman* granted by the Emperor Farrukhsiyar to the embassy led by John Surman in 1715 after a cure effected by the doctor William Hamilton.[25]

In 1693, Samuel Browne was sent to Arcot to treat a wound of a Mughal general named Qasim Khan.[26] Khan must have been satisfied with his treatment, for he appointed the surgeon *havildar* (a term which could have a military meaning of commandant of a fort or refer to a landholding position just below that of zamindar)[27] of six towns adjoining Madras. Whether Browne took up his Mughal administrative post is unclear – the EIC were apparently unhappy about it, partly because of the uncertain position of Qasim, whom Browne had claimed was poised to succeed as Nawab. The Governor's eventual order to Browne to choose between their service and that of the Mughal Empire[28] was perhaps also influenced by the defection the previous year of another doctor, Richard Blackwall, the surgeon at Fort St David who was said to have 'by his profession, access to the Mughal's camp'.[29] Blackwall had taken on the government of Porto Novo under the Mughals and supported attacks on the Company factories. Nevertheless, Browne did make contact with the Armenian physician to the nawabs of Arcot, through whom he was able to procure a grant confirming the Company's settlements in Madras and Cuddalore, in exchange for gifts of 'curiosities', including a lantern and an ongoing supply of glasses for it.[30]

Browne's involvement in local politics and his stormy relationships with some of his patients occasionally caused the Company some headaches. On one occasion, the Governor was forced to make a personal appeal to Zulfiqar Khan to write off the doctor's debts after he absconded with large sums paid to him by one Aggala Khan to acquire medicines.[31] In 1697, Browne was fined for robbing, attacking, and – to add insult to injury – pulling the beard of a Mughal customs official. Nevertheless, shortly afterwards, he was mediating in a dispute with the very same official, who had cut off the roads between Fort St George and nearby San Thome after a dispute concerning land rents.[32]

Browne used his patrons in the Mughal establishment and the Company hierarchy to build up a lucrative business supplying drugs to the camps of the Mughal generals. Browne's contacts in the Mughal army were also useful for the settlement's Council: he could inform them about goings-on and, as he did in 1695, recover errant servants who had defected to the Mughal camp.[33] The tours of duty that Browne undertook with the Mughal army also allowed him to collect plants. One of the books of grasses he sent to Petiver is recorded as having been collected *en route* to Tirupati in 1696. This was shortly after the Mughal conquest of the region in 1694.[34] Once again, therefore, Browne's trip was probably intended as an exercise in diplomacy. As this example demonstrates, collecting was not an activity undertaken during leisurely hours, but part of the everyday practice of military medicine. Thus, the same plants that became dried specimens in herbaria would also have been tucked into the doctor's bag for immediate deployment.

Despite the accommodation reached in 1693, relations of the English settlers with the rulers of Arcot remained tense. Selim Khan attacked Cuddalore during 1698 and Zulfiqar Khan's deputy Daud Khan repeatedly threatened Madras, blockading it for an extended period in 1702. Again, Company officials turned to the network of surgeons with access to the Mughal hierarchy: on this occasion, they appealed to the self-made Italian medic Niccolao Manucci, then living in Madras, to negotiate with Daud Khan.[35] The need for political acuity as well as medical skill might partly explain the Company's decision to recruit Bulkley, who, they wrote, was 'largely fit to serve us by his large experience of India . . . and as fit for prescribing physic as for manuall operation [surgery]'.[36] In 1707, the year of Aurangzeb's death and a time of political unrest, Bulkley was sent to Arcot on a mission that combined medical and diplomatic aims. While there, he also collected several volumes of plants and information about their medicinal virtues.[37]

Despite its growing wealth and size, Madras was in a precarious position in this period. Poised between Mughal expansionism, the rebellious Marathas, and the unstable kingdom of Golconda, it was regularly threatened with extinction. The network of contacts that could be built up between physicians, who had the advantage of close personal access to those at the centre of power, was an important way to exchange information and gifts. It was therefore crucial for the surgeons to acquire a good reputation: this would have entailed becoming acquainted with the protocols of Mughal and princely court life and medicine and being able to employ the plants they gathered on their travels in their cures. Knowledge of plants, medicine, and the methods of employing them was thus important to establishing the EIC's position in India in terms of their relationship with the Mughal hierarchy.

Colleagues, collaborators, and competitors

As noted above, Browne collected his volumes of plants during tours of duty with either the EIC or Mughal armies. It is clear that they were produced in close collaboration with a Tamil physician who was travelling with him and closely involved with the process of selecting and collecting as well as naming the specimens. For most of the plants in the volumes now in the Sloane Herbarium, the Tamil ('Malabar') name has been written on a strip of palm leaf and pasted into the book. The volumes not only have the transliterated Tamil names and medicinal properties entered for each plant, but they are also organised according to the Tamil names given in the index. When 'species' or families of plants are discussed, it is with reference to their local classifications.[38] Browne also had some interaction with a Telugu-speaking doctor: he has left a space for the Telugu ('Gentue') name in all of the books, and some of these names have been filled in in their transliterated forms.

In 1698, when Browne decided to give up his botanical work, he wrote: 'I have also sent to [Edward Bulkley] one of the Malabar Doctors who is well skilled in the nature of Indian Plants from whome he or his people may transcribe their virtues.'[39] In cooperation with this physician, Bulkley embarked on the process of transliterating the Tamil names of plants and gathering accounts of their medicinal properties. He wrote to Petiver in 1700 promising accounts of both Tamil and Telugu names as well as of the properties of plants,[40] and in 1701, he reported, 'I have lately contracted a friendship with the principall of the Gentue Doctors who promises to be very communicative and give a large account of known plants of these parts'.[41] He sent the results of this collaboration to Petiver in 1703.[42]

Bulkley also relied on local people with medical knowledge to make the collections of plants and fossils he sent to Petiver and others. In a peevish letter in which he complains that Petiver has not sent him requested plants and books, he tells his correspondent: 'it is a greater charge & trouble to make collections than you suppose, there are very few that understand it and they will have extraordinary pay to goe 40 or 50 miles and be a moneth absent from their families and businesse'.[43] In the same letter, Bulkley reports that he had employed a man to travel from Madras to Bengal, making plant collections on the way, but that his collecting activities were being impeded by the conflict between the Mughals and Marathas and the fear of meeting tigers in the woods. Nevertheless, some collections made by this man do survive in the Sloane Herbarium.[44] Bulkley also reports sending his servants to make collections in Achin (Aceh in Sumatra) and Pegu.[45]

Both surgeons constantly differentiate between 'Malabar' (Tamil) names and uses for plants and *materia medica* and their 'Gentue' (Telugu) equivalents. Both are distinguished from the medicine of the 'Moor's camp'. In other words, they viewed the practice of *unani* in the Mughal camps, dominated by North Indian elites, as distinct from the regional and local medical practices they experienced in Madras. The distinction that is made between Tamil and Telugu medicine has to do with the difference in status between the speakers of these two languages at the time. Telugu had become the language of high culture in southern India during the medieval period, and by the seventeenth century its status rivalled that of Sanskrit.[46] The dominance of the Telugu language stemmed from the rule of Telugu-speaking Nayakas over much of Tamil Nadu from the early sixteenth century until the fall of Madurai in the 1730s.[47] Telugu Brahmins were also important in the service of Muslim rulers. A distinctive strand of medicine among Telugu speakers had grown up in the region, with a strong tradition of translation and commentary on classic Sanskrit works. The school received patronage from the Qutb Shahi rulers of Golconda and developed a dialogue with *unani* medicine.[48] The distinction that the surgeons make between 'Gentue' and 'Malabar' medicine is therefore likely to represent a difference between the medicine practised by high-caste Telugu-speaking doctors, rooted in translations from Sanskrit ayurvedic works and dialogue with *unani*, and a more localised Tamil strand of *siddha*.[49] The same distinction was made by Roxburgh in his discussion of the medicinal plants of the Coromandel Coast in the late eighteenth century.[50]

The dominance of Telugu speakers in both administrative and mercantile roles persisted in Madras under European rule:[51] proposals for the town council put forward in 1688 include two 'Gentues' as well as Portuguese, Armenian and English members.[52] The difference in status is clear from the ways in which the two doctors discuss their informants: while Bulkley talks of an equal friendship with the Telugu-speaking doctor, the Tamil-speaking physician is sent to him by Browne like a servant. It is notable, however, that neither doctor is named anywhere in the correspondence of the two surgeons with their contacts in London. Therefore, although the works can be regarded as hybrid and co-produced, the process of obscuring Indian agency was under way as soon as the surgeons sent the results of these collaborations to London.

Studies of colonial scholarship have increasingly emphasised the co-constitution of knowledge through the reliance of European scholars on 'native informants'.[53] However, the term 'informants' can have the effect of homogenising specific relationships in which the balances of power, the incentives for exchanging information, and therefore the type of information that was exchanged all varied. Therefore, it may be best to view those people who provided the doctors with information as they themselves did: as colleagues, collaborators, servants, friends, or rivals in business. More recent studies characterise the history of ideas as embodied in lived experience and collective action, with exchanges emerging from the practicalities of life.[54] Exchanges of knowledge between Indian and European practitioners in Madras took place among the daily business of the hospital, duties in the military camps of the British and the Mughals, and the need to find effective cures. The Indian physicians who worked with Browne and Bulkley borrowed from European methods when it suited them, just as the medical practices the two doctors report in their letters bear the imprint of South Asian medical thought.[55] In the next sections, I will look closely at some of the spaces in which collaborations took place in Madras, before examining in more detail the collections that resulted from them.

Hospital and Bazaar

When they were not on tours of duty with the EIC or Mughal armies, the surgeons were based at the Madras hospital. Originally founded in 1664,[56] the hospital was relocated in 1688[57] to a large building described by Charles Lockyer:

> The Hospital joins the New-House by the Water-Gate to the Northward, is a long building and has a piazza with a paved Court before it: at one end of the Court is the Plaister-Room, and at the other end an Apothecaries Shop, where the Medicines are prepared after the Prescriptions of the ingenious Dr. B[ulkle]y.[58]

In terms of the staffing of the hospital, regional medical conventions appear to have had some influence on Bulkley; he recommended the appointment of a dubash (interpreter and general manager) and a conicoply (accountant or registrar) as well as four 'coolies' (workmen).[59] Although the EIC's hospitals' intake would probably have been restricted to European patients, it appears to have been common for Europeans to turn to Asian doctors and vice versa in the early Company settlements: Fryer noted that in Surat, 'the Brahmin comes every day and feels every man's pulse in the factory, and is often made use of for a powder for argues, which is as infallible as the Peruvian Bark'.[60] The exchanges that took place in the informal medical marketplaces of Madras also often spilled over boundaries of nationality and race. For example, the minutes of the early mayor's court in Madras record a case Manucci brought against a patient Cojee Bauba – probably an Armenian – for non-payment of fees,[61] while Browne gossips about a

Muslim inhabitant of Madras said to have a store of a now rare type of medicinal myrobalans.[62]

Like their Dutch counterparts in the VOC, the EIC Directors relied on the use of local drugs to circumvent the problems of expense and spoiling associated with sending drugs from Europe to its settlements abroad.[63] The apothecaries' shop in the hospital was an English equivalent to the 'medical shop' in the castle at Batavia, and it played a crucial role in provisioning the ships' surgeons who passed through Madras.[64] For Browne and Bulkley, it was also a place of experimentation where they used the knowledge they gained from their surroundings to prepare and market simple and compound drugs. Browne refers to experimenting with creating new drugs based on his observations of existing uses. For example, he notes that 'Care-vail maraum' (*maraum* meaning tree), a type of acacia, is locally used for its bark, but that he has also experimented with distilling an oil from the fruit to use in treating fluxes.[65] Selling these preparations was clearly profitable for the doctors: in 1698, Browne wrote to Petiver that he was retiring from the pursuit of botany to focus on his private 'physick practise' as well as the office of assay master that he had recently assumed.[66]

Among the doctor's duties were to procure drugs from the bazaars to furnish the hospital as well as for their own activities as apothecaries.[67] Browne mentions the drugs available in the bazaar in several of his letters and appears knowledgeable about their origins. The origins that are mentioned give us a taste of the global and regional trade networks through which the drugs available around Madras in this period were supplied. For example, Browne writes that tutia – an argillaceous ore of zinc, the powder of which is an ophthalmic – similar to that available in England can be procured locally, but that its source is Persia.[68] Bulkley was similarly able to inform Petiver that the fragrant black sandal or agallochum available in the Madras bazaar was acquired from Cape Comorin.[69]

Both Browne and Bulkley clearly spent time observing the work of their Indian counterparts, adopting the practices they saw into their own routines, and reporting them to their correspondents in Europe. An example of how Browne observed the use of the drugs that were available in the bazaar appears in a letter to Petiver that discusses the two types of *asafoetida* sold there, giving details of their application to palsy with detailed instructions on preparation.[70] Another interesting example relates to plant products used to clean water including the seeds of the Moringa tree.[71] Some of these beans were demonstrated at a meeting of the Royal Society in 1696 by Dr Havers,[72] a fellow of the

Society and another of Bulkley's correspondents, where it was decided that they worked by sticking to the dirt and dragging it down to the bottom of the container. After the meeting, Petiver wrote to Bulkley about the properties of the seeds, and the surgeon replied, giving the names of the plant in 'Malabar' (Tamil) and 'Gentue' (Telugu), and noting their preparation and use in medicine, including against dysentery, as well as their antiseptic properties.[73]

While the drugs that were used within the hospital were drawn from the international trade in *materia medica* that coalesced in the bazaar, the practices that took place within it also bore the mark of the diversity of medical traditions that met in Madras. A dramatic incident that occurred in 1693 reveals the hybridity of practice. Former Governor James Wheeler had come to Browne asking for ground pearl to treat a stomach complaint after an initial treatment with antimony and potassium nitrate failed. The doctor gave him 'as much as would lie on a rupee', along with a bitter drink. After Wheeler's sudden death, Browne concluded at first that it was the result of some arsenic, intended for a cancer patient, which had been left in a pestle and mortar. After this suspicion proved false based on a post-mortem and circumstantial evidence and a dog to which pearl was administered remained healthy, the surgeon was cleared of all suspicion.[74]

The use of arsenic in the treatment of cancer, which Browne refers to, was first described by Ibn Sina (Avicenna), whose work was well known in both Europe and Asia.[75] However, the perceived danger associated with its use meant that it was seldom used in Europe in this period. The use of ground pearl was mentioned by Garcia da Orta as a medicinal treatment in the East Indies and had apparently spread among the European surgeons in the region, although Bulkley's later attempts to persuade Petiver of its merits imply it had not become widely used in Europe.[76] The animal and self-testing that Browne carried out and the post-mortem by Bulkley reflect contemporary European practice.[77] It is evident that close collaborative identification and naming of specimens, direct observation of cures practised by their Indian counterparts and conversations in the bazaar and at the harbour were crucial to the understanding that Browne and Bulkley gained of the uses of the *materia medica* they amassed. Like the *Hortus Malabaricus*, therefore, the volumes of dried and labelled plants that remain in the NHM should be regarded as collaborative works.

One of the aims that I set out at the beginning of this chapter was to demonstrate that the activities associated with the scientific revolution did not occur only in European capitals but also in colonial settings.

Experimentation is perhaps the most emblematic of these activities. As the example of the Moringa tree seeds demonstrates, the colonial metropolis and its institutions were often 'peripheral' in terms of access to both materials and information. The hospital provides an example of an experimental space, but far from being a solely European institution transferred to alien soil; both its staff and the practices of medicine carried on within it demonstrate hybridity of practice, drawing on Mughal conventions of layout and staffing and regional medical ideas. Furthermore, in terms of the networks that joined the experimental spaces of Madras to those of other settlements and supplied the items to the bazaars which formed the basis of the compound medicines mixed by the surgeons in the apothecaries' shop, they were dictated more by the regional trades independent of colonial control than by the priorities of the EIC itself.

The actor-network theory of Latour has been modified in recent years by a number of studies that demonstrate that scientific knowledge is created not only in the laboratory, but also 'in the field and the forest, on ships at sea and on mountaintops, in the course of exploration and trade'.[78] Other recent studies, particularly of medicine in South Asia, have challenged the traditional assumption that in the face of colonial medicine, science and technology, other systems were limited to 'response or resistance'.[79] The medical practices of the two surgeons as well as their social networks and political agendas were often not determined by colonial priorities or by resistance to them; rather, the two surgeons were often drawn into networks in which they were go-betweens or observers. These networks proved as important to the contents and interpretation of the collections they made as well as their links to correspondents in London. In other word, although the two surgeons sometimes made use of Petiver's instructions for collecting, their collections were not dictated primarily by the demands of the metropolitan collector, but by their immediate networks of trade and politics.[80]

Gardens and libraries

As well as gathering drugs in the forest and purchasing them in bazaars, the inhabitants of Madras cultivated them in their gardens. The map of Madras published by Herman Moll in 1720 but made a little earlier shows no less than ten gardens (Figure 4.1).[81] These include plantain gardens, 'coco gardens' (probably referring to coconuts rather than cacao), and the private gardens of various individuals. As several authors

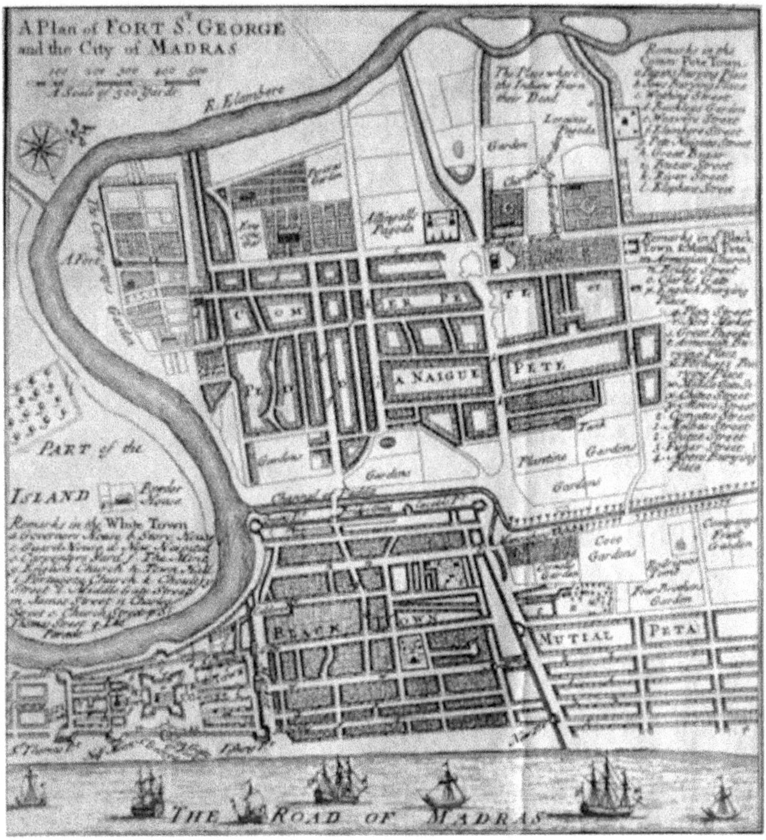

Figure 4.1 Herman Moll, 'A Plan of Fort St. George and the City of Madras', 1726

have pointed out, gardens designed to introduce new crops and to provide *materia medica* for local hospitals had long existed in both Hindu and Muslim traditions and had spread to Europe in the form of the 'physic gardens' and 'acclimatisation gardens' that emerged in Italy, Portugal, later Holland, and finally England and France and their colonies.[82] In Mughal towns of the period, 'householder gardens' were common, along with royal gardens and tomb gardens, and the leasing of gardens could provide civic revenue.[83] The Company's gardens were thus only part of several long and overlapping traditions. As commercial and experimental spaces, the gardens also revealed, in their beds and borders, the networks that Madras was embedded within, as ships brought seeds and plants from other Company settlements, territories of rival European powers, and places of regional trade and of religious pilgrimage.

Mughal gardens were symbolic of the legitimacy of territorial claims: 'microcosms of spatial order within a more tumultuous and uncertain landscape', an idea that would not, of course, have been lost on many contemporary English readers of Shakespeare or Marvell.[84] The consolidation of Arcot included the development of the city's gardens, including exotic fruits and vegetables as well as tanks and tree-lined walks.[85] Gardens were also spaces where armies might rest and cures be performed and diplomatic negotiations took place. Mughal visitors to Fort St George were liable to demand to inspect the Company's gardens as well as its armies, as Daud Khan did on a 1701 visit to the settlement, and thus they were a crucial source of prestige.[86] Perhaps prompted by this visit, the Governor had by the early eighteenth century transformed the Madras garden from a barren space to a pleasure ground complete with 'Bowling-Green, spacious Walks, Teal pond, and Curiosities preserv'd in several divisions'.[87] In 1708, Governor Thomas Pitt received a *husb-al-hukum* confirming the cities' privileges from the *diwan*, Zia-ud-din Khan, in the garden's 'Great Walk'.[88]

Much of the daily work of tending the Company's gardens fell to slaves, and a few who became particularly skilled were ordered to receive minimal pay or other 'encouragements', so important was the pursuit considered by the governors of the early settlements.[89] Some of the correspondence between Browne and Petiver mentions the Company's gardener, referred to only as Randal, who was making a considerable profit from his specimens.[90] Randal, who apparently sent at least six volumes of collections to England, might himself have been a slave. This would explain the single name always used for him and his absence from all lists of European inhabitants and Company employees drawn up in the period.

The surgeons used their space in the Company gardens to experiment with local plants and introducing crops from around the world. For example, Browne describes the 'Country [i.e. Asian] Plants I have collected into a Square of the Company's gardens'.[91] Both Browne and Bulkley also raised plants they received from their networks of correspondents overseas: Browne describes growing China root, a popular medicinal substance,[92] rhubarb, cinnamon trees from Ceylon, and wild agallo, benjamin, and camphire (henna) from Manilla.[93] Edward Bulkley received plants from as far off as West Africa: with one letter, he enclosed a drawing of a small sort of gourd 'gathered on the Coast of Guinea & presented to me by name of the everlasting apple'.[94] He cut the gourd open and extracted a seed, which he planted in his garden, where the plant flourished. Bulkley also despatched seeds to be planted in other EIC settlements: for example, he sent rhubarb roots to Bencoulen (in Sumatra) to see whether they would grow better there than they had in his garden.[95]

The Company surgeons also attempted to introduce European and American plants into their gardens: Bulkley made several requests for Petiver to send him seeds for plants yielding food or medicine.[96] Some introduced crops had a clear commercial value for the Company. By the late seventeenth century, several factories had become involved in the cultivation and vending of tobacco,[97] a plant which in 1693 Petiver still listed as completely unknown in London. Meanwhile, the Company was also renting licences to grow and sell it, along with other mild stimulants or intoxicants including betel and cannabis. An interest in what might be called 'narco-botany' appears in the specimens sent by Browne to Petiver, which include both these specimens.[98] The opium poppy, which later became so important in the trade to China, was also discussed by Petiver and Browne.[99]

Investigations into the origins of plants and the possibilities for transplanting them were thus therefore useful to the Company's strategy of participating in the 'country' trade in medicinal and narcotic drugs as well as food crops at the lowest possible cost.[100] It was therefore useful for them to encourage collections to be distributed among the circles of gardeners spanning Europe and Asia. For the surgeons themselves, the distribution of seeds meant that their friends abroad could provide them with additional comparisons and identifications of the plants they themselves were growing and raise the profile of species they were marketing to private collectors. In return for the seeds they sent, they expected to receive books, seeds or some other form of patronage.[101]

Like the process of comparison of plants from around the world carried out in botanical gardens, the use of printed books was crucial not only to the identification of individual plants but also to the emerging understanding of the distribution of plants across the world. From their correspondence, it is clear that the surgeons owned some key works of European natural philosophy. Examples include Prosper Alpinus' 1591 *De medicina Aegyptiorum*, one of the earliest European studies of non-Western medicine, John Ray's *History of Plants*, Robert Hooke's *Micrographia*, Parkinson's well-known herbal,[102] Garcia da Orta's *Colloquies* and Christoval Acosta's *Tractado*.[103]

Perhaps the most crucial to the surgeons' work were the 12 volumes of the *Hortus Malabaricus*, the work of the VOC surgeon Rheede tot

Draakestein[104] and his collaborators, which Petiver sent to Browne who eventually sold them on to Bulkley. The *Hortus Malabaricus* is constantly referred to in the identification of plants in the volumes Browne and Bulkley sent to James Petiver. The surgeons and their apothecary correspondent even embarked on a project to produce an English summary of the *HM* with additions to describe the plants that grew around Madras. Although this work was soon abandoned, the *Hortus Malabaricus* did prove an important comparative tool. The comparisons between different parts of the world enabled by the libraries the EIC surgeons built up also allowed them to fit the plants that they observed locally into a larger map of the distribution of flora and fauna across the world.

Hortus sicci, herbaria, and herbals

Hortus sicci are dried impressions of the plants that also grew in gardens and were sold as drugs in apothecaries' shops. These collections travelled through interconnected networks. The EIC surgeons of Madras kept the *hortus sicci* that they received from correspondents including Petiver in London and the Jesuit Georg Joseph Kamel (or George Camelli as he appears in the surgeons' correspondence) in the Philippines in return for their own similar collections, using them to identify the plants they received in their medical practice. As noted above, several volumes of *hortus sicci* sent by the two surgeons survive in the Sloane Herbarium at the NHM, along with specimens among the collections of seeds and 'vegetable substances'. Unlike the *Hortus Malabaricus*, however, the volumes of dried plants were never transformed into a printed book, although several illustrations of specimens sent by both surgeons are scattered throughout Petiver's published collections along with the information they sent him.[105] Better-known botanists including John Ray also referred to the surgeons' collections in their work.[106] Some of the drawings and paintings Bulkley made of birds copied from the works of local artists (Figure 4.2) were also included in both Petiver and Ray's work printed works.[107] The two surgeons both also supplied Petiver's rival, Leonard Plukenet, with specimens and are mentioned several times in his *Mantissa*.

Among many other specimens, the collections Browne sent to England during this period included seven volumes of dried and labelled plants with their Tamil names written on bark in the original script and in transliterated form. Many pages of the volume also contain accounts of their medicinal virtues written next to the dried plant.[108] The contents of these volumes were transcribed and published by Petiver in the

Figure 4.2 'Icones Avium Maderaspatanarum', sent from Bulkley to Petiver, BL Add MS 5266, courtesy of the British Library

Royal Society's journal, *Philosophical Transactions*, with comments and cross-references to other contemporary works of botany, particularly the *Hortus Malabaricus*.[109]

The correspondence and collections of the Madras surgeons reveal the types of medical problems that they, their colleagues, and counterparts were encountering. Unsurprisingly, fever emerges as a major problem: thirty-four of the 226 plants mentioned in the first five books of Browne's collected volumes are febrifuges. The Spanish-controlled 'Jesuit's bark' remained difficult to obtain in this period,[110] and therefore it was necessary to find locally available substitutes. However, Bulkley also requested Petiver to 'procure the best method of treating fluxes and feavours', admitting 'they are often too hard for me & allso for the natives'.[111] Smallpox is mentioned several times: often the plants described as useful in treating it are the same as those used against fevers.[112] Treatments for 'bloody flux' (dysentery),[113] gonorrhoea,[114] leprosy,[115] and poison[116] are mentioned several times. In general, the collections sent by Browne do not reflect a particular preoccupation with military medicine or 'tropical' complaints. They also include many 'simples' (medicinal herbs) used in treating asthma, colds, and consumptions, as well as more serious conditions such as epilepsy and diabetes. The collections refer to children's medicine[117] and remedies used by women during childbirth, post-partum recovery, or menstruation.[118] References to the treatment of mental problems such as melancholy, hysteria, and even hypochondria also appear.[119] This suggests that rather than being directed by the priorities of naval and military medicine, the collections were a product of the everyday medicine practised among the inhabitants of the town.

The collaborative nature of their production described above means that Browne's volumes provide a small window into the everyday practice of *siddha* in seventeenth-century Tamil Nadu. Just over one-third (115) of the 317 plants, Browne mentions in the seven volumes are the same as those which occur in the *Hortus Malabaricus*.[120] Several plants are reported to be used in a similar way in late-seventeenth-century Tamil Nadu as they were a little earlier in Kerala. For example, a decoction of the leaves of the plant whose Tamil name Browne gives as 'Mogulamaraum' and cross-references to 'Elengi' (*Mimusops elengi*) in the *Hortus Malabaricus* was employed to treat toothache in both cases.[121] This entry is also cross-referenced to several other works of botany, including Paul Hermann's *Musaeum Zeylonicum* and John Ray's *Historia Plantarum*.[122] Similar descriptions of magical or ritual uses of plants also occur across the two collections, although there are also cases in which magical uses are reported from Kerala but not Tamil Nadu and vice versa. One example is the plant Browne notes is called 'Vusha coddee' in Tamil and which is identified by Manilal as *Ipomoea quamoclit* and was used as a love potion in both cases.[123] There is evidence for the shared use of some plants across Asia: for example, Petiver notes that he has received reports of the use of 'Ville-Vittree' or 'Indian quince tree', from Bengal, Sri Lanka and Batavia, in all cases being used in the treatment of fluxes and against diarrhoea.[124]

Despite the basic similarities in the content of the *Hortus Malabaricus* and the pharmacopoeia collected by the Madras surgeons, however, most references to the same plants differ in terms of the preparation, the part of the plant used, and the diseases that it is used to treat. This suggests that while the knowledge of many plants and their associations was shared by medical practitioners across South Asia, there was also considerable regional variation in terms of the specific uses of *materia medica*,[125] which were often different from those described in those works of Indian medicine thought of as canonical. In the same vein, the substances that are described in the collections of the surgeons and their collaborators should not be held to be representative of a static or bounded school of *siddha* medicine in Tamil Nadu in this period.[126] Rather, they provide a partial snapshot of one of the strands of medical knowledge that existed in the cosmopolitan environment of seventeenth-century Madras.

Globalising local knowledge

As I have argued in the previous sections, the collections that the surgeons made, when studied in the context of their correspondence and the Company records, can be used to construct a picture of the local medical landscape of the globalised space of Madras, albeit one seen through the lens of the specific interests and priorities of the two surgeons. In this final section, I will examine how the knowledge that was gained in this way became decontextualised as it circulated through international networks. This process did not 'neutralise' the objects, as in the case of medicines, it is necessary to communicate not only objects themselves but also instructions about how to prepare, administer and store drugs. These understandings about a substance are thus retained even when the reasons for them are lost.

As noted in each of the sections above, the surgeons received books, dried specimens, seeds and botanical information from a range of international correspondents. As Madras was a usual calling point for all ships heading further east or returning home, the surgeons were able to distribute instructions and materials for collecting – some of which they received from Petiver[127] – and to receive the collections on the return journeys. Bulkley distributed instructions for making collections to the Company surgeons in Bencoulen[128] and Vizagapatam[129] and refers to acquiring collections from Cochin-China from James Cunningham: who was a surgeon in Chusan[130] and in the short-lived settlement of Pulo Condore, where he was imprisoned.[131] These men were also embedded within further networks: for example Cunningham also corresponded with and sent collections to Petiver and the EIC Treasurer and keen experimental gardener Charles du Bois,[132] and had acquired collections from Henry Smith in Malacca.[133] The go-betweens in these networks of doctors were ships' surgeons, who had their own networks of collectors.

As well as the other EIC settlements, contacts included the agents of other European companies – with whom their relationships ranged from friendly exchange to rivalry. Among the Dutch company surgeons, Dr Oldenhand at the VOC settlement at the Cape of Good Hope was known as an avid collector and botanist.[134] The plant collections of celebrated collectors were fought for within the East Indies. For example, in 1702, Bulkley's correspondence reveals a scramble for the remaining collections of Wilhem ten Rhijne and Georg Eberhard Rumpf (Rumphius) in Batavia.[135] One of the most important of the international relationships of the Madras surgeons was with Kamel: a lay father who made the first in-depth studies of the wildlife of the Philippines. Bulkley exchanged frequent gifts of plants and seeds with Kamel and discussed the local uses of the plants common to their respective regions.[136] Bulkley also acted as a conduit in the Jesuit's correspondence with James Petiver, with whom he published articles in the *Philosophical Transactions*.[137] Kamel sent specimens and drawings to Petiver and Ray.[138] Kamel, as well as Browne's patients among the Portuguese of San Thome, also appears to have provided a means for the Madras surgeons to acquire further contacts among Jesuit missionaries: in 1696, Browne wrote concerning obtaining collections of aloes: 'I have written by way of the French and Portuguese padres to their brethren in the vast woods about three months journey to the northwards of Bengal towards Pegu.'[139]

The discussions that the surgeons took part in concerning particular plants or remedies with Jesuit missionaries and the Dutch company's surgeons and botanists[140] demonstrate the closeness of the scholarly investigations taking place among the different European traders and missionaries in Asia at this time. The close connections between the Jesuit Spanish botanist Kamel and the Protestant surgeons remind us that the allied interests of science and trade were sometimes more important in such friendships than the dictates of faith. At the same time, the scramble for the manuscripts or collections of Rumphius and Ten Rhijne reminds us that the acquisition of natural knowledge was a crucial part of the competition between the European trading companies to acquire and exploit the wealth of the Indies.[141]

Browne and Bulkley each sent a huge amount of plant materials to various correspondents in Europe: that which survives in the NHM represents only a tiny sample. For example, Bulkley refers in one letter to having sent 20 volumes of specimens to Petiver.[142] Among the contacts that the surgeons maintained in England were several apothecaries including his brother-in-law, who ran a shop in Bread Street, and Mr Porter, a druggist in Cornhill Street.[143] The circle of botanists who received collections from the East Indies formed a close, though not always friendly,[144] group of experimenters and gardeners who spanned the EIC, the Royal Society, and the Society of Apothecaries.

An example of how collections were distributed is provided by a meeting of the Royal Society in 1698, when Vice-President John Hoskins told a meeting that 'some natural things were in the hands of the East India Company lately sent them from Fort St George'.[145] A committee was deputed to go to the EIC and request that these be submitted to the Royal Society for their consideration. The Court of Directors agreed, giving over to the Society 'a barber's shop full of seeds'.[146] These seeds were examined and catalogued.[147] The following spring, those thought to be fit for sowing were distributed to several correspondents, including Samuel Doody, the Curator of the Chelsea Physic Garden, the Bishop of London, the Duchess of Beaufort, Mary Somerset,[148] the gardening enthusiast Charles du Bois,[149] and Jacob Bobart of the Oxford Botanical Gardens, 'to be by them raised, and an account of the success and specimens of such use sent to the Society'.[150] The following winter, Hans Sloane showed the Royal Society the fruits from some of the seeds they had sent to the Duchess of Beaufort, noting that 'there were some fruits such as were never seen in England as China oranges, Guavas brought so ripe they had their flavour'.[151] Apothecaries also participated in the testing of new drugs from abroad and with growing specimens, particularly in the Apothecaries' Garden in Chelsea.[152] Many more plants, medicines, and specimens sent from Asia were bought by private collectors and for sale in apothecaries' shops as well as coffee houses, taverns, and bookstalls in early modern London.

Although Bulkley and Browne passed on many of their collections to Petiver, du Bois, Plukenet, and others, they were by no means simply acting as conduits of information to London. Contacts in Europe fulfilled a range of functions for the Company surgeons: a powerful member of the EIC's board such as Charles du Bois might be leaned on for patronage in return for indulging his interests in gardening; in one instance, Bulkley sent du Bois five volumes of dried plants, four with their Tamil names, requesting in return du Bois' intervention to procure a further contribution towards his living expenses and the use of a palanquin.[153] Apothecaries such as Petiver and Plukenet were expected to popularise the medicines they were sent and to sell them, remitting the profits back to the their suppliers in the East Indies: for example, Bulkley wrote to Petiver about a root that he named *Radix Traumaticus Indicus*, 'your selfe & some other friends may doe kindnesse to others and allso to me in prescribing it'[154], and later, 'I desire yourself and some other friends to make it publick to the world'.[155] Finally, botanists and medical men such as Ray and Sloane could inform the surgeons about the potential medicinal value of their local plants as well as providing them with drugs, specimens, and books from Europe and the Americas.

The overall effect of these international exchanges of plant material was that the trickle of European descriptions and collections of non-European plants and medicines that had begun with the sixteenth-century work of men like Garcia da Orta and Prosper Alpinus had become a flood by the beginning of the eighteenth century, as not only the emissaries of the rival trading companies but an increasing number of missionaries, such as those of the Danish settlement at Tranquebar,[156] joined these networks of collectors competing for the commercial and diplomatic wealth that could be amassed by the accumulation of specimens, while their European correspondents struggled to order the mass of exotic specimens with which they were bombarded by correspondents hoping to make a profit or to exchange specimens for favours.

These international collecting networks and the flood of specimens from all directions received in colonial outposts as well as European capitals eventually began to alter the nature of thought about plants and disease. This process of change can be seen taking place in the statements of the surgeons about the materials at their disposal. In some cases, they exhibit traces of the older view that the plants of every place were unique, being suited to the environment and the constitutions of its inhabitants. For example, Bulkley comments when recording his experiments with drugs from the Americas including *Psychotria ipecacuanha*, which he had tried using against fevers, although with little success, commenting on a paper by a Frenchman which had recently praised the drug, that 'the monsieur in his paper calls it infallible, if he were here in India he would often find it otherwise'.[157] The argument that certain drugs were only effective in particular environments had been central to the attempt by some sixteenth-century humanists to remove exotic plants and herbs from European pharmacopoeia on the grounds that plants cultivated under a foreign sun would be damaging to the constitutions of people raised under colder skies.[158] However, despite the ongoing concern of many European commentators about the high cost of foreign drugs, the battle over exotic imports can by this stage be said to have been won by the apothecaries, who made huge profits from the substances that the medical establishment condemned. The triumph of this attitude is also evident in the next passage of the same letter, which goes on to compare ipecacuanha with favourable results to a local drug that he sent to Petiver, instructing him 'if you find it usefull pray endeavour: to gain it reputation in the world, it may become a good mercantile drugg'.[159]

Bulkley and Browne also recognised that the same or similar plants could be found in different areas of the world and tended to produce similar effects. For example, in one letter, Bulkley notes that the effects of the sassafras tree from Pegu are similar to those of Virginia.[160] Sending a collection of myrobalans species to Petiver, Browne comments that they are the same species as those that may be received from Arabia and Persia, but in general smaller, something that he attributed to the inferior quality of the soil in the coastal regions of India.[161] In some cases, the surgeons made mistaken analogies between different plants. For example, Browne believed that he had found the tree that was the source of 'Jesuit's bark' (cinchona) growing locally, and sent a specimen of it to Petiver.[162] In fact, as Browne himself noted elsewhere, the tree was the source of *Strychnos nux-vomica*.[163] Such exchanges prompted Petiver to read a paper before the Royal Society in 1699, in which he argued that herbs with similar physical properties normally possess comparable medicinal virtues. He assigned the familiar herbs of England to families along with 'many more no less noble herbs I have received from the East and West Indies'.[164]

The European trajectory away from regarding plants in the context of their local environments and towards slotting them into classification systems such as those devised by Ray and Carl Linnaeus also meant that local knowledge became progressively less visible, as plants were increasingly regarded as common to 'families' across the world. Therefore, while the knowledge that their informants imparted to them was crucial to the surgeons in their daily practice, it seemed less important to substantiate their claims to authority with the kinds of highly visible authorial roles that the Indian contributors to the *Hortus Malabaricus* maintained throughout the production of the herbal. This process can be seen to have continued in the later work of colonial botanists. For example, it is clear that William Roxburgh relied heavily on local knowledge for the production of his later herbal of the Madras region, but the contributors are still less visible than they are in the work of Browne and Bulkley.[165] Although the Telugu names are given in transliterated form in his work, there is no indication of the source of this information. The tendency by the nineteenth century, in the wake of the so-called 'chemical revolution' to separate the study of botany from the practical search for medicines, which increasingly relied on a smaller number of ingredients, intensified this tendency to obscure local knowledge.[166] This process was not, of course, value-free, but part of the final solution to the drain of European wealth to purchase foreign substances and its informational dependency as concerns their use. In some sense, we might regard it as the scientific establishment regaining control from the disorderly ranks of apothecaries and empirics and their suppliers, who had been the chief promoters of exotic medicines.

Conclusions

In his letter of October 1698 to Charles du Bois requesting his correspondent's favour and patronage, Edward Bulkley sent five volumes of plants, four with 'a catalogue of their names in Malabar', two parcels of seeds, a gourd from Pegu, two pickled fruits with their names in Tamil and Telugu, and a litany of woes. His collections were being impeded by the 'trouble the Moor has given us of late' on the pretext of the acts of piracy lately committed against their shipping. These, he admitted, 'have been divers', but he blamed the agents of the 'new' EIC for the 'mischief we feele and fear', adding, 'I pray God direct you to some speedy redress of the grievances your trade labours under, which are truely very great'.[167] This letter gives a sense of the conditions under which the collections I have discussed were made. Medicinal plants were not gathered and exchanged in the leisured hours of gentlemen, but in woods and fields over which lay the shadows of marauding armies and wild beasts. They were intended for immediate use in medicine and sale for profit, whether locally or internationally, and to win favour within the shifting and competing hierarchies of the old and new EICs and the regional constellations of power, whether in the form of increased funds from the Company treasury or Mughal office.

Bulkley wrote at a time of transition in both England and India. The Glorious Revolution had just swept King James II from power, spelling trouble for many of his former allies within the Company's ranks. The Mughal Empire looked to be at its height of its powers after the death of Sambhaji in 1689 and in 1702, Aurangzeb ordered the suspension of trade with all European companies and a survey of Madras was commissioned with the intention of occupation.[168] Meanwhile, the conflict between the two companies continued until the final merger in 1709.

In the following decades, the situation in England gradually became somewhat less precarious, and the introduction of 'Dutch finance' ended the fiscal crisis. While theories of Mughal 'decline' and the significance of European powers in India during the eighteenth century remain contentious,[169] it is clear at least that by the time Bulkley died in 1713, being buried like a Mughal nobleman at the end of his garden,[170] the United Company was more securely established at Madras, as expressed in its now immaculate gardens. The networks of doctors had been crucial diplomatic actors in a critical period during which many believed that Madras was fated to be eclipsed altogether. The settlement in this period in fact survived more through the performance of healing – of both the literal bodies of rulers and the figurative body politic – than through the infliction of wounds on the battlefield. In fact, it was the new relationship with the rulers of Arcot, which these doctors were important in establishing, which would eventually allow the Company to consolidate its base at Calcutta.[171]

As Raj has shown in his investigations of Calcutta as a 'brokered space', the Company as well as metropolitan scientists became gradually better able to regulate the work and retain the allegiances of their go-betweens by the nineteenth century.[172] However, as the example of Paul Jodrell, physician to the Nawab of Arcot, demonstrates, having English physicians in the employment of Mughal and princely powers continued to be important both to the diplomacy of the EIC and in providing information to the scientific community in England.[173] The servants of the Company in the colonies also continued to place more importance on the local names and meanings of plants than their metropolitan counterparts, as exemplified by William Jones' alternative system of plant classification according to the Sanskrit names of plants.[174]

Colonial botanical gardens have received attention in recent years as nodes in the imperial exchange of knowledge and wealth-generating crops. However, their history is normally begun with the founding of

Kew in the 1760s and Calcutta three decades later. As the example of Madras shows, gardens were important in Company settlements as sites of prestige and profit, experiment and exchange from a much earlier stage. Similarly, while the beginnings of English involvement with Indian botany are normally located in the work of William Roxburgh,[175] the older collections, although not transformed in their entirety into printed works, were extremely significant in their volume and their comprehensive nature. Because of the greater embeddedness of the surgeons in regional politics and local medical traditions compared with many of those who followed them, the collections are often more revealing about the environments within which they were collected than later works like Roxburgh's.

I have argued here that the collections the surgeons made might, with caution, be considered revealing about the practice of medicine among sections of the Tamil-speaking population of Madras in the late seventeenth century. Features such as the presence of the original labels and dried specimens with details of the time and place they were gathered offer further potential for interdisciplinary study to reveal more accurate information on this subject. Such studies have potential to provide a historical dimension to current ethno-botanical research or to reconstructions of historical environments.

There are, of course, many important details missing from this attempt to reconstruct a medical landscape of seventeenth-century Madras. Unlike the contributors to the *Hortus Malabaricus*, we never learn so much as the names of the Tamil- and Telugu-speaking doctors who were so crucial in collecting and revealing the medicinal uses of the specimens the surgeons sent to London. Furthermore, we do not know the language in which they communicated: could Browne and Bulkley, like Manucci, speak one or more Indian languages, or were their counterparts already skilled enough in English to explain the details of the uses of the medicinal plants? Or were the exchanges mediated through another language such as Portuguese? It is possible to see this process of the obscuring of the creative role of informants in colonial records as progressive over time by comparing the authorial role of Itty Achuden, Ranga Bhat, Vinayaka Pandit, and Apu Bhat in the *Hortus Malabaricus* with the unnamed artists and collectors who contributed to the works of Browne and Bulkley and later Roxburgh and Wallace. On the other hand, agency on the ground must be assessed as far as possible on a case-by-case basis according to the specific relationship of informant and informed; in our example, while some specialists were able to charge large sums for their collections, others made little profit from their knowledge.

The specimens that reached England and became preserved in the Sloane Herbarium were the same as those sold in apothecaries' shops, along with instructions for their use ultimately derived from South Asian sources. In this way, we might agree that Europe was in fact transformed from below by the medical knowledge of Asia, often contrary to the intentions of the metropolitan medical hierarchy. Contrary to the arguments of Mary Louise Pratt and Londa Schiebinger[176] that imperial collecting practices stripped local meanings from objects, in the case of drugs, it was essential that at least some of the meaning was retained in order to enable it to be administered correctly for the right complaints. Nevertheless, the information was subject to what has recently been called 'epistemological slippage';[177] it became dissociated from its original social and cultural moorings.

The surgeons' collections reflect the hybrid environment of early modern Madras. The compilations which Browne and Bulkley sent Petiver must therefore be regarded as complexes of both Asian and European thought: they are neither simple reflections of the *siddha* medicine of Tamil Nadu nor works which simply draw Asian medicinal plants into a European pharmacopoeia. Instead, they demonstrate the tensions of the different allegiances and influences that the two surgeons were susceptible to. In many cases, these were contradictory impulses: while on one hand they felt the need to claim the superiority of the remedies they discovered locally, on the other they were drawn by the wider networks in which they were enmeshed to make comparisons between these plants and those they saw depicted in the herbals from abroad they received from around the world. On one hand, they participated in local medicinal practices and regarded local drugs as best suited to treat local ills: on the other, they wanted to market their own findings internationally. The collections of these two surgeons, who were key players in the transformation of politics and botany in the region, straddling local and international concerns, are thus in many ways the perfect instrument through which to view Madras as it was transformed from a trading post subservient to the interests of regional powers to a major player in British colonial expansion.

5
Bio-Prospecting and Experimentation: Producing and Using *An Historical Relation of Ceylon*

Introduction

In mid-1680, Captain Robert Knox (1641–1720) arrived in London after almost 20 years spent in the central Kandyan kingdom of Ceylon (Sri Lanka), an experience that he began to record during his long voyage home as a passenger.[1] The first person to greet the bearded and be-whiskered traveller was a 'drugster': a peddler of the exotic remedies which were growing ever more popular in late-seventeenth-century London.[2] He had come aboard as soon as the ship docked, eager to buy the produce and the recipes of those aboard. By chance, the drugster, or apothecary, recognised Knox and reunited him with his brother-in-law and sister. Knox would soon meet several other people interested in the knowledge he had brought back from Ceylon. After being called into the EIC's Court of Directors to give an account of his travels,[3] he was taken aside by Jeremy Sambrooke, a member of the Royal Society.[4] Through either Sambrooke or his own brother James,[5] Knox was introduced to the polymath Robert Hooke. By the following year, these new contacts and the contribution of his cousin, the minister and historian John Strype, had helped him to compose his notes into *An Historical Relation of the Island of Ceylon*. The book was printed by the Royal Society's printer, Richard Chiswell, in August 1681, financed by subscribers from within the EIC. It became hugely popular during the author's lifetime and has remained a standard source for the island's history ever since.

In Chapter 1, I looked at the composition of Baron's *Description of the Kingdom of Tonqueen* in the context of the author's relationship with both the EIC and members of the Royal Society as well as his connections with the society he describes. This chapter examines the final stages in the production and use of a natural history. After briefly discussing the arrangement of Knox' text with the assistance of scholars in London, I will examine its translation into other European languages, being added to and re-orientated in the process. I will then demonstrate that the text circulated in the possession of travellers including Knox himself, and show that the process of comparing the written account with experiences led to annotations and borrowings that served as the basis for further writings. Using EIC records and Knox' own unpublished works reveals how the *Relation* was used as the basis for bio-prospecting for naturally occurring drugs and food sources and in efforts at agricultural transplantation spanning the Indian and Atlantic Oceans. Even political lessons learned in Kandy found application elsewhere during Knox' travels. The aims of this chapter are to understand the cumulative, collaborative, and comparative processes involved in producing natural histories like the *Relation*; to demonstrate that their circulation extended outside European capitals; and to examine the practical uses to which they were put, in order to throw light upon the connections between description and experimentation in the scientific revolution.

Directing travel

Recent scholarship has examined the connections between travel narratives, geographical accounts, or 'natural histories' and European expansion and the contribution of both to emerging forms of scholarship.[6] The methodising of travel through the interactions of scholars and merchants, sailors, local informants, and colonial and Creole officials has been traced from the Iberian world into the writings of the later Renaissance of Northern Europe.[7] Following on from the writings of men such as Francis Bacon and Samuel Hartlib, the Royal Society, from its inception in the early 1660s, was involved in producing instructions for travellers and guidelines for the authors and compilers of natural histories.[8] The EIC's own use of writing, including travel narratives and directed enquiries, as a source of information and thus power has also begun to be explored in recent studies.[9] The production of Knox' *Relation* should be seen as part of an ongoing collaboration between the EIC Directors and members of the Royal Society to lay claim to the wealth promised by the knowledge of the East by producing natural histories and travel accounts making use of such guidelines. At the same time, as argued in Chapter 1, each such work was the product of the specific circumstances of its author and was designed to win patronage by revealing some pieces of information and concealing others.

Bacon's essay 'On Travel' advises that the traveller should carry with him 'some card or book describing the country where he travelleth; which will be a good key to his inquiry'.[10] This highlights a point which seems rather obvious but that has been somewhat overlooked in many of the recent discussions of travel literature: such accounts were not intended primarily for the entertainment and information of European elites. Instead, they were designed to be carried on journeys, with the traveller using them as starting points for his own impressions and scribbling comparisons or corrections in the margins. Knox' work provides an especially interesting example of this process of comparison and annotation because he took an interleaved copy of his own work with him on his later journeys, using the extra pages, as well as unpublished journals, to draw comparisons between his knowledge of Ceylon and the observations he made on his later travels. Here, I will explore how the comparisons Knox made were used by the EIC Directors in their search for new settlements capable of producing coveted Asian crops for food and medicine. The connected processes of observation and experimentation were guided at every stage by Knox' ongoing relationship with members of the Royal Society.

Robert Knox' career

Robert Knox was captured along with his father and other members of crew of the EIC ship *Anne* in early 1660 by agents of King Rajasingha II of Kandy (r. 1634–1686) after arriving in Trincomalee in Ceylon and remained in the central kingdom[11] for 19 years, becoming fluent in Sinhala and well acquainted with local ways of life through his employments as farmer, moneylender, and pedlar. Despite never entering the employment of the King, Knox was evidently well informed about events at court. Knox says that he began to compose the *Relation* on his return journey to England in 1680, after his escape via the Dutch fort at Arippu.[12]

The *Relation* contains a detailed description of the geography, politics, wildlife, agriculture, religion, languages, laws, learning, medicine, and domestic lives of the inhabitants of Kandy as well as an account of the author's own capture and way of life and the circumstances of other Europeans in the kingdom. It quickly became very popular and was soon translated into German, Dutch, and French.[13] John Locke drew on the work in his *Second Treatise of Government* (c. 1689).[14] The *Relation* is also said to have provided one inspiration for Daniel Defoe's heroes in *Robinson Crusoe* (1719) and *Captain Singleton* (1720). Along with the Dutchman Philippus Baldaeus' description of the mainly Tamil-speaking north of the island, published in 1672,[15] and the Portuguese Captain João Ribeiro's description, which also focuses on the mainly Sinhala-speaking part of the island,[16] Knox' account continues to form the basis for accounts of the island in the mid-seventeenth century.[17] The work has also been studied for its literary style,[18] its description of the caste system and religion of Ceylon,[19] the Sinhala vocabulary that it incorporates[20] and, most recently, the natural history it contains.[21] Sarojini Jayawickrama's literary study provides a re-reading of Knox' text within the context of the customs, institutions, and representational practices of the Kandyan kingdom.[22]

Knox also left several other manuscripts, which, added to the records of the EIC and the Royal Society, allow a fairly comprehensive reconstruction of his life after leaving Ceylon.[23] Those found so far are his autobiography, the interleaved copy of the *Relation* containing significant additions to the original manuscript and intended for publication as a second edition,[24] several letters to his cousin John Strype[25] and one to Lady Worcester, daughter of the EIC Governor Sir Josiah Child,[26] a truncated account of his voyage to Tonkin in 1681[27] and his will.[28] References to Knox appear in both the Journal Books and Council Books of the Royal Society and in the Court Books, Letter Books, Original Correspondence, and Factory Records of the EIC as well as in the diary of Robert Hooke.[29]

In January 1681, Hooke recorded the first of several meetings with Knox, who was accompanied this time by his cousin James Bonnell.[30] Around the same time, Hooke showed a leaf of the talipot palm (*Corypha umbraculifera*) that Knox had brought back from Ceylon to the Royal Society.[31] In August the same year, Hooke notes that he had given Knox 'queries for the Indies'.[32] At around the same time, Knox was presented with the interleaved copy of the *Relation*.[33] Knox carried these documents with him when he departed the following month on the *Tonqueen Merchant* via the Cape Verde archipelago off West Africa to Tonkin,[34] where, as mentioned in Chapter 1, he encountered Samuel Baron.[35] After this, Knox visited Bantam and Batavia, returning via Batavia in early 1682.[36] In November 1683, the Council voted to give Knox a present in return for a long list of items that he had given to the Society's repository.[37] As well as collecting these items during his second voyage,

of Knox' map of Ceylon, clearly lifted from Philippus Baldaeus' 1672 work *Malabar en Coromandel*.[61] The map retains some Dutch labels, one marking the end of the cinnamon fields at Matara.[62] Baldaeus' work was published by Johannes Janssonius van Waesberge, whose plates Moses Pitt had acquired as part of his *English Atlas* project.[63] This suggests that Hooke – who also had significant input into the *English Atlas* – might have had access to the original plates as well as consulting Baldaeus' published text in the Royal Society's library.

The draft second edition of the *Relation* gives a clearer indication of how the process of using guidelines and queries directed to foreign correspondents and drawing on other works is likely to have functioned in the composition of the original text. For example, Hooke seems to have shown Knox an illustration of the coconut tree from the *Hortus Malabaricus*, before presenting him with detailed questions about the tree's stages of growth and uses.[64] A second passage enlarging on the process of tapping a coconut tree for 'toddy' refers again to the picture of the tree. These questions are addressed in the interleaved sheets that follow.[65] Hooke's hand appears in the preamble and the first two chapters of the interleaved copy making editorial corrections and explanations, expanding a section concerning the medicinal use of leeches, and speculating about why it tends to rain more in mountainous places.[66] At the end of the text, Hooke has added a list of queries for Knox to address. These include requests for detail about medicines and manufacture in Ceylon, and for descriptions of particular plants and topographical features in various locations including Tonkin and Mauritius.[67] Hooke then made small corrections and elaborations to the text that Knox produced in response. The cooperation between Hooke and Knox to produce the *Relation* thus functioned in a similar way to Bowrey and Hyde's collaboration in making the *Dictionary English and Malayo, Malayo and English*.

A similar process of questioning, editing, and arrangement of travel texts or natural histories produced by members of the EIC with the assistance of scholars is evident in several other texts of the period. In 1681, the year of publication of the *Relation*, it is also possible to see the guidelines for the composition of natural histories being put into practice in the production of two other texts: the EIC surgeon John Fryer's account of his experiences in Persia and India,[68] and Moses Pitt's *English Atlas*.[69] The *English Atlas* was explicitly intended as a composite work, as demonstrated by the public advertisement for any gentlemen with 'any curiosities of any country whatsoever' to bring them to be incorporated in the atlas, if 'approved of and judg'd fit to be Printed by those Learned men, whose Judgements are consulted'.[70] Such projects were part of the movement towards multi-authored reference works that began to be envisaged with Bacon's *Instauratio magna* (1620) and culminated in the encyclopaedias of the Enlightenment period.[71]

While the Royal Society and other scholars were routinely consulted about the content of natural histories, the EIC Court of Directors often financed them by collecting subscriptions from among their members. Knox' dedicatory epistle praises the EIC for bringing 'not only the Wealth but the Knowledge of the Indies ... home to us'.[72] The Company, whose status was often uncertain during its first century of existence, was keen to be seen to benefit the nation by sponsoring the publication of useful information. French and Dutch works of the period exhibit a similar concern to claim that their work contributed to the national good.[73]

However, the circulation of information around European capitals and their settlements in the East Indies also raises the question of the tension between the urge to publicise claims to the knowledge and wealth of the Indies and the need to prevent certain information from falling into the hands of European rivals by enforcing secrecy. Therefore, most of the writing generated within the Company, although often copied and circulated, was not published: in fact, most was closely guarded from European rivals, who tried equally hard to gain access to it. For example, in a letter to England, the President at Surat notes that the adventurer and writer Jean-Baptiste Tavernier, whom he describes as a 'Dutchified Frenchman', had been given passage to Persia several times on EIC ships and had therefore been trusted by the Persian factory to carry their letters to Surat. However, Tavernier had then visited the 'Dutch house' in Persia, where the factors, overcoming him with 'drink or persuasion', had stolen the contents, replacing them with alternative letters written in the English style and returning the packet to Tavernier.[74] Cartographic information was often acquired by the English from the Dutch by a similar combination of theft and covert copying.[75]

The process of assimilation of the work of Baldaeus and others into Knox' work was followed by the translation, circulation and re-appropriation of the published *Historical Relation of Ceylon*. The translations reorient it through the additions of new prefaces, dedications, and illustrations. One particularly interesting feature of the Dutch version is a new illustration that shows the text of the *Relation* written on an ola, or book made from the strips of dried talipot leaf, a technology of writing that Knox describes. The text is being presented to the King of Kandy himself (Figure 5.1).[76] This illustration serves as a reminder

Figure 5.1 An imagined scene in which the *Historical Relation of Ceylon* is presented to Rajasingha II of Kandy from Simon de Vries trans, t'Eyland Ceylon in syn binnenste (Utrecht: by Wilhelm Broedelet, 1692), courtesy of the British Library

that travel texts in European languages also circulated outside Europe: a point that will be discussed in more detail below. In this case, it also raises an intriguing question: was Rajasingha II aware of Knox' work? The Kandyan king was undoubtedly well informed about the Europeans in his kingdom and had sponsored the translation of European works into Sinhala.[77] Knox' letters to his fellow captives in Ceylon, which, like other correspondence, are likely to have been screened by the Kandyan court,[78] explicitly mention the publication of the work. Whether the text was known in the Kandyan kingdom is likely to remain speculation. However, Knox' depiction of the kingdom of their enemy was certainly scoured and used by the Dutch.

The geographical information provided by the *Relation*, as well as Knox' oral comments on a map he was shown in Colombo in 1679, was incorporated into Dutch maps.[79] The map that accompanies Simon de Vries' translation of the *Relation* is a copy of Knox' own map but also contains more information taken from other Dutch sources.[80] The other illustrations are free interpretations of Knox' originals, some with details added and others combined into composite scenes.[81] A few changes in the orthography of transliterated Sinhala words in both the map and the list of vocabulary are also made in de Vries' edition.[82] The French translation of Knox' work contains an accurate copy of the original map with labels translated into French as well as relatively faithful reproductions of the original illustrations. It also includes a new frontispiece depicting the capture of Knox and the rest of the *Anne*'s crew.[83] Knox' map seems likely also to have been used in a French map attributed to 'Sieur de l'Isle' of which several copies survive, including that which appears in the French translation of João Ribeiro's account of the island printed in 1701.[84]

As shown by the examples of the Dutch and French versions of Knox' map, early modern translations often involved appropriating and correcting information, which was then fed into further projects to produce useful knowledge. In Ceylon and elsewhere, the Dutch, the French and the English trading companies were in competition for territory and allies, and accurate cartographical information was essential for those who wished to reach little-known places such as the remote and well-defended kingdom of Kandy. The incentives for closely guarding information were therefore balanced against the importance of staking claims over knowledge through publication when considering whether or not a manuscript should be published and what information it should include or omit.

Diplomatic information could also be sensitive. The *Relation* gives a the fairly extensive, and often unflattering coverage of Dutch relations with Rajasingha II, including the claim that the conflict between the Dutch and Rajasingha II after the expulsion of the Portuguese from Columbo in 1656 was due to Dutch treachery. Knox also depicts the VOC as unable to match the Kandyan King militarily and being forced to resort to flattery. In contrast, the *Relation* makes little mention of English attempts to negotiate with Kandy, which had included correspondence between Rajasingha II and Charles II. For the EIC, Ceylon was a coveted site for trade, particularly in cinnamon and pepper, and they had made several attempts to treat with the Kandyan King to establish an alliance against the Dutch. These included a meeting between Edward Winter and Antonio d'Almeyda, a servant of Rajasingha II and letters that combined inquiries about Knox and the other prisoners with proposals to set up a settlement and pepper plantation on the island and inquiries about the possibility of opening up trade in cinnamon or acquiring cinnamon plants for transplantation to St Helena. These attempts had been obstructed by the Dutch[85] and forestalled by the rebellion of the Kandyan King's son in 1664. Knox seems to have hoped that the *Relation,* with its description of the remaining English captives, could even have served as a potential pretext for further interventions. In a letter copied in his Autobiography, he assures his fellow captives that their circumstances have been 'published in print', although he also notes that there seems to be little current hope of intervention on their behalf and suggests that they resign themselves to remaining in Kandy.[86]

Competition over medicinal drugs and recipes was especially fierce: as the drugster who boarded Knox' returning ship recognised, there were considerable profits to be had from the knowledge that sailors and merchants acquired overseas.[87] The next sections will explore how the imprecise descriptions of medicinal plants and agriculture given in the *Relation* were built upon in Knox' unpublished manuscripts to form the basis of practical experiments in bio-prospecting and transplantation.

The desirableness and facility of this undertaking: Using the *Relation*

Bio-prospecting

As noted above, travel texts were meant to be read and used not only – or even primarily – in European capitals, but also in settlements abroad and during travel. When travel books were taken on voyages, comparison between the written descriptions and first-hand experience took place, sometimes generating further written accounts, as was the case with Baron's corrections to Tavernier's work discussed in Chapter 1.

In another example, Allen Catchpole's discussion of the possibility of establishing a factory at Pulo Condore makes extensive comparisons with a work of Dampier in his appraisal of the fruit and timber yielded by the local trees as well as referring to Guy Tachard's account of a voyage to Siam and an apparently unpublished description by Henry Smith.[88]

This process of comparing geographical descriptions and verifying and refining the information they provide is also evident in the use of the *Relation*. Members of the Royal Society and their social circles had other contacts with personal experience of Ceylon, whose accounts they used to supplement the information in Knox' *Relation*. Paul Hermann, Professor of Botany in Leiden and a former employee of the VOC in Ceylon, is first mentioned in the Royal Society records in 1680[89] as preparing a catalogue of the plants of the island, which was received on its eventual publication in 1698.[90] Hermann also exchanged specimens with Hans Sloane,[91] and his draft illustrations for a second edition of his catalogue eventually came into the hands of James Petiver.[92] Hermann was another informant who provided multiple types of information from the East Indies to his European correspondents. For example, in 1683, Hermann sent his *Vocabularium Selanense* to Thomas Hyde, who compared his transliterations of Sinhala words with Knox' versions and in collaboration with Thomas Bowrey had a copper plate engraved of the Sinhala alphabet and some words of vocabulary.[93] Hyde used his comparisons between the vocabularies to make some connections between Sinhala and Indian languages and further compared Hermann's vocabulary to a Persian account, noting that both indicated that the meaning of the word *sinha* was lion.[94] The Royal Society was also in contact with a physician named Strachan, who had lived on the island for 17 years and who published articles in the *Philosophical Transactions* concerning the use and transportation of elephants in Ceylon and the religions, plants, and wildlife of the island, in several cases drawing on Knox' work.[95]

As noted above, Knox was carrying the interleaved version of his own text with him and making annotations on at least two of his subsequent journeys: the first to Tonkin via Cape Verde and Java and the second to Madagascar and St Helena.[96] Knox referred back to the text on both voyages, directed by the process of questioning by members of the Royal Society discussed above and the practical instructions of the EIC Directors. In both cases, Knox' expertise on Ceylon, along with the written text, was used to make comparisons between the island and other prospective sites for colonial settlement, trade, or transplantation.

'Bio-prospecting' describes the process of examining and describing plants and identifying them, with the aim of either using them *in situ* or transporting them to other locations: whether botanical gardens in Europe or colonial plantations.[97] The acquisition of plants that could yield drugs and food crops had a high potential monetary value and has been identified as a major factor in Iberian political power in the sixteenth century,[98] as well as in European colonial expansion in the New World.[99] There is some debate about how far European investigations of plants that could provide food and drugs in the East and West Indies were comparable. While both involved drawing exchanges with native 'informants' – involving varying degrees of consent and cooperation[100] – some scholars have argued that, in contrast to the largely oral exchanges that took place in the West Indies, the European contribution consisted largely of compiling or rearranging South Asian knowledge that had been previously systemised and recorded in textual form.[101] I suggested in Chapter 4 that European reports about nature in the East should be regarded as complexes of Asian and European ideas, mediated through the specific experiences of those involved in gathering and transmitting the information and packaged for their intended audiences. In the same way, Knox' reports about plants useful for food, drugs, and other purposes are best viewed in the light of both his experiences in Ceylon and his relationships with European scholars.

Knox' *Relation* contains no direct reference to written medical texts or specialists,[102] although a ritual that would have been performed by a priest as part of a cure is twice mentioned.[103] Furthermore, there is no reference to any influence of systems of medicine that may have resulted from the influence of Muslim traders in Kandy, or to any imported medicines. Probably because its author lived mostly in rural areas rather than in the city or court, the *Relation* concentrates its short section on medicinal plants on household medicine using simples. Knox states that 'the woods and trees are their *Apothecaries* Shops, where with Herbs, Leaves they make all their Physic and Plaisters, with which they sometimes will do notable cures'.[104] Rather than listing these medicines, 'of which there are hundreds', Knox simply notes the cure of a broken arm with unspecified herbs and his own cure of a sore throat by chewing the bark of the 'Amaranga tree'.[105] In the interleaved version, Knox expounds on the preparation of the simple from this bark 'as I was instructed by one of the C[o]untry men'. The survey of edible plants, whether wild or cultivated on a small scale for food, is similarly perfunctory: he notes the existence of two types of aloes, gives the

Sinhala names for four other vegetables, and notes that several European herbs have already been transplanted, from 'which I perceive all other *European* plants would grow here'.[106]

Despite the lack of detailed descriptions of edible and medicinal herbs and plants in the *Relation*, comparing the original text, the interleaved copy, and the truncated journal of Knox' voyage to Tonkin in 1681[107] reveals that Knox was involved in bio-prospecting for naturally occurring plants yielding drugs and food by comparing the plants growing in Cape Verde with those he was aware of in Ceylon. The plant in which the process of bio-prospecting and transplantation can be most clearly traced across these three texts is the 'jack', a name used both for the tree and its fruit. In the *Relation*, the following description is given:

> There is another Fruit, which we call *Jacks*; the Inhabitants when they are young call them *Polos*, before they be full ripe *Cose;* and when ripe, *Warracha* or *Vellas;* But with this difference, the Warracha is hard, but the *Vella* as soft as pap, both looking alike to the eye no difference; but they are distinct Trees.[108]

The fruit is described here as a foodstuff: 'they are a great help to the People, and a great part of their food', being compared to a turnip or cabbage, with the thick white juice being used to catch birds.

In his account of the 'I[s]le of May' (Maio, Cape Verde), Knox also describes the type of jack yielding 'warracola', this time revealing a medicinal use for the fruit, although omitting the description of the tree itself. He notes:

> The leafe resembles a Cabbage both in thickness & colour; onely not so large. It is ful[l] of white thick milk, or juice, if you break the leaves which is soon done, being very brittle. The Chingulais put to many medicinal uses: & here in this I[s]land of May they grow in great plenty, just above the sand on the sea shore.

Knox argues that the plants were probably indigenous rather than having been transplanted, as the inhabitants – who he correctly assumed were mainly descended from slaves transported from around the Gulf of Guinea during the period of Portuguese rule – appeared to be unaware of many of their medical uses. He goes on to link the type of plants that are likely to prosper to the relative positions of the two islands, noting that 'the difference of Longitude doth not much chang or alter the nature of the Climate as the Latitude doth. For notwithstanding Zeilon Lyeth

Degr. log ½to the Eastward of this I[s]land, yet here I saw several Plants growing wild, which grow in the same manner upon Zeilon.' As well as describing two other plants, 'Bintombracole'[109] and 'Endraatta',[110] growing on both Ceylon and May, he also discusses the neighbouring island of St Iago (Santiago), where he notes that plants common throughout the East Indies such as oranges, limes, coconut, grapes, plantains, and watermelons grow, as well as briefly discussing the effect of the soil type on the native flora.[111]

In the interleaved copy of the *Relation*, Knox describes the jack tree in much greater detail. As well as its value as a foodstuff, he notes the strength of its roots and the hardness of its timber, which he compares to that of the English oak and notes that it is used as a building material for houses in Ceylon.[112] Also in the interleaved copy, Knox mentions that the people of Ceylon are not familiar with grafting, perhaps implying that he himself had tried this technique. Finally, Knox notes his own failed attempt to transplant the jack to a plantation in Barbados.[113]

Knox' discussions of the jack tree and its products across these three passages serve several functions. The survey of the Cape Verde islands was part of the EIC's vision of establishing a chain of settlements in the Atlantic and Indian oceans like St Helena, the Dutch factory at the Cape, and the later French settlements in Madagascar and Mauritius.[114] As will be discussed in more detail in the next chapter, these bases were intended to function like the Caribbean islands as 'plantations', supplying ships trading to both the East and West Indies,[115] and acting as points of defence. Knox' account of Cape Verde is clearly geared towards this objective. He includes accounts of the fortifications, buildings, as well as possible products, such as cotton, which he notes the inhabitants produce by spinning it with a stick, again comparing it to the method used in Ceylon. In addition, he gives a brief (and unflattering) account of the inhabitants.[116]

A similar example of a commission to inspect an island for its potential occurs in the Company's instructions to Knox to make an 'exact survey' of Tristan d'Acuna, giving an account of harbours, vegetation, animals, etc., and to give an opinion whether it would be 'advantageous for the Company hereafter to make settlement upon any of those Islands', which 'Capt. Gayer & other inform us would save the lives of many men'.[117] Again, this commission was to be undertaken using as a starting point an earlier text containing advice, in this case the unpublished journal of a mate who had landed at the island the previous year.[118] Like his description of Ceylon, Knox' account of his visits to Madagascar in the draft second edition of the *Relation* would have been intended for the use of other traders. By amassing and re-circulating such materials among their servants, the EIC built up a base of knowledge about the potential of various locations for trade or settlement.

This instance of bio-prospecting on Cape Verde can thus be seen as part of an ongoing speculative survey of several possible island bases for the EIC. The search for food and drugs was coupled with an analysis of the defensive capabilities of the island, its potential for manufacturing cotton or extracting natural resources like ironstone, and its capacity to supply a workforce (whether through independent trade or free or coerced labour). The race to identify, settle, and defend suitable oceanic islands was run against the Dutch and French companies, who were making similar surveys in this period,[119] as well as competing for those islands that had already been settled.[120]

The Royal Society during this period was engaged in a number of investigations about the effects of longitude and latitude, as well as climate, altitude, and soil type on plant and animal life and even disease.[121] It therefore seems likely that Knox' discussion in his later texts of plants that grew in distant places with similar latitudes and of grafting was developed through his contacts with these scholars. The instruments Hooke provided in 1683 were described as a 'picture box' (perhaps one of the portable pinhole cameras invented by Hooke and Boyle), an 'azimuth perspective' (probably an 'azimuth mirror', an instrument placed over the mariner's compass to measure the azimuths), a 'longitude clock' (making use of balance wheels governed by dual springs), and a 'pendulum watch' (a watch employing a conical pendulum).[122] These instruments were intended to facilitate Knox' observations concerning latitude and its significance.

Theories of the effects of different latitude on animal and vegetable life naturally had implications for the types of plants to look for in the process of bio-prospecting, as well as the types of crops that were likely to be transplanted effectively. For this reason, reports such as Knox' were crucial for scholars advising the EIC on their settlements. Thus, as I have argued throughout, the development of scientific theory regarding the natural world was inextricably linked to European expansion and proto-colonialism.[123] Knox' use of Hooke's instruments alongside his text to draw conclusions about the effect of latitude on plant life also provides an illustration of the ways in which natural philosophical focus on experimentation and the natural historical practices of description and collecting intersected.

Transplantation and the transfer of techniques

As well as surveying potential bases for naturally occurring foods and medicines, the EIC was also involved in attempts to transplant both agricultural and manufacturing techniques from both the West and East Indies to its own settlements. These efforts were based on consultations with experts in the use of a particular crop or technique, as well as on books and manuscripts in London. They also involved transporting both people and texts around their settlements, a process I will discuss further in Chapter 6. Consultations in London encompassed a range of people from merchants including Thomas Bowrey and scientists such as Robert Boyle, to an unnamed black man from Mauritius whom Knox was ordered to bring before the Court of Directors in 1690.[124] They drew on evolving theories such as those about the effects of latitude to select suitable crops for transplantation. For example, the Directors noted in a 1683 letter to St Helena:

> We have discoursed with many persons about the growing of wheat upon our island and Wee know that very good doth grow in many parts of the world with similar Latitude and therefore we would have our Governor continue to make trials of the Surrat Wheat and a Bushel of English Wheat herewith sent, untill you shall find by Experience which are the fittest places of the Island for the sowing and growth of it.[125]

Knox was involved with this transfer of expertise and materials from the East and West Indies in several ways. He was instructed to procure indigo and cotton seeds and root ginger from Madagascar.[126] The indigo was to be planted under the supervision of a planter from the West Indies, as the Company had failed in their efforts to entice expert growers to relocate from Surat.[127] The Company also instructed Knox when in Madagascar to 'secure likewise any other Plants Roots seeds or fruit that you think may grow at St Helena and deliver them to our Govt & Councill, to be propagated for the benefit of our Island', adding, 'Be very curious and carefull in this as a matter of great publique concernment wherein you may oblige us & be a benefit to your Country.'[128] As well as the journal of a Captain Bass, on the basis of which the Company urged that salt works should be established on the island, Knox also carried a rule book from Barbados, which was to be consulted by the Governors for advice in managing the Company slaves.[129]

Knox was intended to contribute to the Company's attempts at the transplantation of crops based on his own descriptions of agriculture in Ceylon in the *Relation*. The description of agricultural techniques in the original text is more extensive than that of medicinal plants and focuses in particular on rice cultivation. Rice production in seventeenth-century Kandy was organised on a village level, with paddy lands worked communally and a portion of their produce set aside for the King's storehouses.[130] Knox describes the creation of these paddy fields by the management of water. He notes and names five different types of rice requiring various amounts of time to ripen,[131] and explains that the type to be sown will be selected depending on how long the necessary amount of water is likely to be available and the approach of the time of year when the fields are due to revert to grazing land for cattle.[132] Finally, he notes that there is a type of rice that can be grown without always standing in water, which is cultivated on high lands that cannot be flooded. Knox goes on to describe the preparation of the fields using light ploughs drawn by oxen and the cooperation of the villagers in reaping the harvest, and the method of treading out of the grain using oxen.[133] Again, the interleaved version of the text expands on this section, giving more detailed instructions and, in one section, a diagram of a tool used to channel water.[134]

On the basis of the passage relating to the breed of rice that ripens without standing in water,[135] in addition to the oral advice of Knox and 'several others', the Directors formulated a plan to sow this type of rice on the high lands of St Helena. In 1684, they first sent to their agents in Surat to acquire 'three or four baggs of fresh Paddy, of that particular Kind…wee intending it for seed at St Helena'.[136] The letter goes on to give instructions to store the rice in a cool cabin during the journey and to transfer it between bags at least once a fortnight in order to 'preserve its prolific quality'. The letter to Surat stresses the potential value of the crop, which is described as 'no meane concerne to this Kingdom'.[137] Knox and a former fellow captive in Ceylon were then instructed to oversee the sowing of the rice, the Directors in London informing their Council in St Helena:

> [T]here is a peculiar sort of Rice that groweth best on high and dry Lands, the seed and cultivation whereof he knoweth very well. He and Ralph Knight that is on board his ship having wrought many years upon it with their own hands in Ceylon.[138]

Knox was also instructed to demonstrate to the inhabitants of St Helena the uses of coconut trees, which, instructions from London to 'cherish and increase your Coco-nut trees all that possibly you can, they being of much more use than you do at present apprehend', imply were growing on the island.[139] He was intended to demonstrate to them how to draw the oil from the tree: a process described in the interleaved version of the *Relation* as well as the *Autobiography*.[140] A final lesson drawn from the *Relation* relates to the production of iron. The Directors note that Knox believes that ironstone is plentiful in St Helena and that 'in the Country where he was captive, every poor man... made his own Iron for the use of his family'. Knox therefore was instructed to demonstrate the technique on St Helena in the hope that the stone could eventually be transported to the Company settlements in Asia and manufactured there.[141]

Rice was clearly an important foodstuff on St Helena in the early period, and its transportation and storage seem to have been problematic: references occur in the early records to shortages and the poor quality of the stores.[142] An earlier proposal to send two slaves experienced in cultivating rice paddies had already been dropped after advice that the crop was unsuitable for the high ground of St Helena:[143] hence the specific type of rice that Knox was intended to supply.

Whether Knox ever put into practice the intended experiments in rice growing, the production of iron, or the growing of cotton, indigo, ginger, and other plants is unclear.[144] In fact, Knox' most important contribution to the development to the island was probably in his role as a slave trader: he arrived at St Helena with at least two consignments of slaves from Madagascar. As will be discussed in more detail in the next chapter, slaves were often valued not only for their labour but also for their expertise, and were often transported along with the crops they were skilled in growing.

Knox' involvement with the attempted transplantation of rice is only one example among several indications that he was involved in transplantation around the settlements he visited. For example, the list of items brought to the Royal Society in 1683 includes 'the root of a tea tree, which the Captain designing to bring home growing had planted, and kept in a pot of earth aboard the ship, but which, by the way was gnawed and killed by the rats'.[145] Whether or not it was eventually carried out, Knox' commission was one of a long series of agricultural experiments on St Helena. Scholars in London continued to play an advisory role: during the eighteenth and nineteenth centuries, such well-known scholars as Joseph Banks, Daniel Solander, William Roxburgh, and Joseph Dalton Hooker all investigated the island's plant species.[146] Transplantation was often also carried out *ad hoc* by travellers and migrants, as with Knox' attempt to sow the jack in Barbados.

Travel writing and natural histories like Knox' continued to form the basis of such experiments into the eighteenth century. For example, Joseph Banks' recommendations about the transplantation of tea to certain areas of India were based on the writings of Knox' contemporary James Cunningham about the types of tea growing at similar latitudes in China.[147] Banks also owned a copy of Knox' *Relation*, which he annotated at the back with identifications of several of the plants Knox mentions.[148]

The *Relation* as spiritual journey

As noted above, the *Relation* is a work of two parts: natural history and personal account. Although the text should be regarded as a composite work with considerable input from European scholars, this does not negate the presence in the text of Knox' personal outlook or the effects of ideas received during the nineteen years he spent in Ceylon or in his subsequent travels. Like Baron and many of the other authors discussed in this book, Knox's personal experiences rendered him both a mediator between cultures and a hybrid product of their intersection.

As Katherine Frank has stressed, Knox continued to annotate and add to the *Relation* and *Autobiography* throughout his life, even after the prospect of a second edition being published had faded.[149] Knox's sincere piety, inherited from his mother Abigail Bonnel, has been noted in several accounts of his life.[150] As Knox himself wrote, he regarded his life, with its many changes of fortune, as having been guided throughout by divine providence.[151] Reflecting back on his life was therefore a means to better understand the workings of God, who he believed had repeatedly intervened to ensure his escape from danger and captivity, as well as enabling him to find a Bible during his period of captivity in Ceylon.[152]

Drawing on ideas of natural religion similar to those Marshall expresses in his letter to Coga (Chapter 2), Knox makes several favourable comparisons of the morals and charitable works of the Buddhist inhabitants of Ceylon with those of European Christians.[153] While the description of Rajasingha II in the *Relation* has often been considered to be a prototype for the European stereotype of the 'Asiatic despot', often based on Locke's use of the work, many of the negative comments about the monarch seem to have been the result of later interpolations in Knox' text.[154] Overall, Knox' account of Rajasingha II does not suggest that he believes 'despotism' to be a necessary characteristic of Kandy's government, nor does he believe it to be without divine legitimacy. Knox also invests another non-Christian monarch, King Rybassa, with the divine right of a sovereign in his description of their edgy encounter in Madagascar in 1684, when Knox was again briefly held captive.[155] Despite describing the King and his men as 'inraged barbarians', Knox relates that the King did not execute him because, 'the hands of Kings as well as the hearts are in the hand of God & he could not strike without his blessed will'.[156]

If Knox believed in God's direct intervention in the human world, he also credited malevolent or demonic forces as having a tangible impact on the world. In the section of the *Relation* that discusses demons, Knox draws on indigenous traditions that consider Sri Lanka to be the abode of demons, remarking that their cry is heard whenever the King is about to execute someone without reason.[157] However, it is notable that the King's government itself is not presented as demonic, as was the case in some earlier European accounts of Asian rulers; rather, the cries of demons are heard at the times when the King has overstepped the boundaries of his natural and legitimate rights as sovereign, thus disturbing the harmony of ruler and kingdom. Like the contemporary Sinhalese he encountered, Knox believed that demons could cause illness. Knox's theology also pervaded his understanding of the natural world. As will be discussed further in Chapter 6, Knox and many of his co-religionists firmly believed that human wrong-doing could lead to environmental devastation.

While the impact of his religious faith should be considered, Knox' writings should also be examined with reference to his personal interactions with members of the societies he visited.[158] As Raj notes, attention to these relationships is the missing element in many discussions of the interactions between European 'sedentary men of science' and the travellers who answered their inquiries. Knox lived in Ceylon for 19 years, during which time he adopted the local language, dress, and way of life,[159] as well as bringing up a child. In fact, it was his return to Europe rather than this level of assimilation that was unusual: his other crew members remaining behind and marrying locally.[160] Knox' identification of himself as a member of the society of Kandy is evident in his language in the description of Cape Verde, in which at one stage he describes two plants, 'Allacola' and 'Attangcola', 'which *wee* have on Zeilon' (emphasis added).[161]

This perspective also colours his account of his subsequent travels. For example, the inhabitants of Cape Verde are dismissed as lacking in 'ingenuity' from Knox' perspective as a resident of Kandy. Rather than inquiring into their own medicines and industries (or taking into account their enforced migration), he assesses them based on their knowledge of the particular plants he recognises from Ceylon and from the viewpoint of a transitory visitor rather than a resident. The customs of Madagascar similarly receive a less than complimentary treatment in Knox' writings.[162] While Knox retained a strong strain of Puritan thinking in his ideas about the workings of divine providence in the world, this identity was mingled with others, including that of a long-term resident of Ceylon. It was apparently in reference to his hybrid identity that his crew, deserting from St Helena with the *Tonqueen Merchant* in 1685, referred to him as 'an old Portuguese rogue'.[163]

Conclusions

Investigating the production and circulation of Knox' *Relation* suggests several conclusions about the place of travel texts and natural histories in both the process of European expansion and the development of ideas about the effects of latitude on plants and people. The involvement of Strype and Hooke in the first and second editions shows how the process of questioning and comparison with and borrowing from other accounts was used to form composite 'natural histories' with a number of applications. The cooperation between the EIC and the Royal Society in producing this text should be viewed as part of an ongoing collaboration between the two entities to collect and lay claim to the material, intellectual and spiritual wealth promised by amassing useful knowledge about the world. Such claims were important in establishing the authority of both institutions, as well being part of the competition with the French, Dutch and other European nations to lay claim to the knowledge as well as the wealth of the East and West Indies on behalf of the English nation. Other examples from the same period include the published works of John Fryer and William Dampier, Baron's manuscript *Description of the Kingdom of Tonqueen*, as well as the unpublished letters, responses to questions, and descriptive accounts that fill the archives of both organisations. By providing merchants and travellers with queries, guidelines on making observations, and instruments such as those Hooke gave to Knox, scholars were able to amass information that returned to aid the Company's agents in the form of advice about appropriate crops or medicinal plants likely to flourish at certain latitudes or under particular environmental conditions, as well as in the forms of guidebooks or 'natural histories', maps, and dictionaries.

Locating Knox' *Relation* within a number of interrelated published and unpublished manuscripts produced by collaboration between merchants and scholars in the late seventeenth century also allows for a clearer picture to develop of how such texts were used to explore the possibilities for transferring non-European practices and techniques to European settlements such as St Helena or identifying other possible areas for settlement such as the Cape Verde islands. For example, Knox' involvement in bio-prospecting based on the *Relation* is only recognisable if his unpublished second edition and journal of his voyage to Tonkin are also consulted. Similarly, it is necessary to trace the movement of ideas from the *Relation* into the letters of the EIC to note the contribution of the work to the attempts at transplanting crops and agricultural techniques, and transporting human experts around the globe. Many merchants including Knox moved across areas that are now often studied separately, such as the Indian and Atlantic Oceans, taking their ideas and prejudices as well as their papers and specimens with them. They also occupied roles such as author and slave trader that are often studied within quite different conceptual frameworks. Following their journeys across these spaces and roles can contribute to a greater awareness of how ideas and techniques were transferred between different areas of the world in the early modern period.

6
Transportation and Transplantation: Slave Knowledge and Company Plantations

Introduction

In the Oxford Museum of the History of Science hangs an oil painting (Figure 6.1), which contains a rare depiction of slavery in the context of the early East India Company. The young man of African origin, whose iron collar suggests that he is enslaved, is holding a map of the Persian Gulf. The other figure in the painting is John Chardin (1643–1713), a French Huguenot who became wealthy designing jewellery and trading diamonds in Persia and India. He immigrated to England in 1681 and quickly became a leading investor in the EIC and a member of the Royal Society. He produced several volumes of descriptive accounts of Persia and India, assisted by the natural philosopher John Evelyn. Chardin was also instrumental in brokering the 1688 agreement between the EIC and the Armenians for trading collaboration.[1] Much like other characters discussed in this book, Chardin made his name within both the EIC and the Royal Society by trading on his knowledge of foreign lands and customs. The focus of this chapter, however, is not on Chardin, but on the inhabitants of EIC settlements represented by his companion.

Previous chapters of this book have discussed the movement of knowledge that is contained in texts, maps, pictures, and objects, from dried plants to compasses. This chapter focuses on embodied knowledge. It examines how members of the early EIC used the coerced and forced movement of people to transfer skills and techniques between settlements and how those people resisted, often using the same forms of knowledge that made them useful to the Company. People of several different nationalities, free as well as enslaved, were used to transmit knowledge in the world of the early EIC. For example, many of Chardin's co-religionists were sent to EIC settlements along with grapevines, as the French were considered to be particularly expert in winemaking.

Figure 6.1 Sir John Chardin and an unnamed slave, oil on canvas, dated 1711, courtesy of Oxford Museum of the History of Science

The connection that is evident in this example between the transportation of people and the transplantation of crops occurs frequently in the debates of EIC members and natural philosophers of the period. As with other means of transmitting knowledge, however, the EIC Directors in London were often unable to control the movement of people between Company settlements. Although they have received little attention in most previous studies,[2] slaves formed considerable percentages of the populations of all the EIC settlements and were vital to their success, not only by providing manpower, but also in terms of their skills and knowledge. The movement of slaves between the early EIC settlements set important precedents for the later transportation of convicts and coolies within the British Empire.

The movement of slaves between the EIC settlements was part of much larger trends in global migration. The movement of an African diaspora across the Indian Ocean, the Mediterranean and finally the Atlantic was a process that occurred over centuries. These migrants disseminated knowledge, particularly of plants and their cultivation.[3] The indigenous slave trade within the Indian Ocean included various forms of bondage.[4] European participants in the early modern trade intensified and altered these forms of bondage with slave raiding and long-distance transportation.[5] A recent estimate puts the total numbers of people enslaved and transported across the Indian Ocean by Europeans between 1500 and 1850 at a minimum of 530,000.[6] Both the *Estado da India* and unlicensed Portuguese traders were major participants in slave trading from the East African coast to the Indian colonies and within Asia from the sixteenth century onwards.[7] By the seventeenth, they were joined by the Dutch,[8] English, and other European traders in buying, capturing, and employing slaves in their settlements. The French became major players in the Indian Ocean slave trade after the acquisition of Mozambique in the eighteenth century.[9]

Occluding slavery

References to slavery are rare in the records of the early EIC. Of the many maps of the early settlements, only two show evidence of the presence of slaves. One is in the City Museum of Mumbai; estimated to date from the early eighteenth century, it shows 'Madagascar Town', a relatively large area beyond what is now Back Bay, close to the burial or cremation grounds of Christians, Muslims, and Hindus.[10] The second map (Figure 6.2) shows the 'Coffreys' barracks' (quarters for soldiers of African origin) in mid-eighteenth-century Madras.[11] Both indicate significant number of African slaves in the EIC's Indian settlements. A casual reference to the export of 200 slaves from Fort St George in 1686[12] and the Dutch sources that indicate 665 slaves being taken to Aceh by English traders the same year also indicate that the early EIC settlements in India profited from the regional demand for slaves.[13]

Despite such indications, identifying references to slavery amid the EIC's records can be difficult. This is partly due to terminology. 'Servant' is often used interchangeably with 'slave', as in some passages that refer to 'buying black servants'.[14] The term 'black' is used for both Africans and Asians, and while the terms 'coffrey', coffree, or 'caffre' (from the Arabic *kafīr*) and 'negro' usually indicate African origin, their

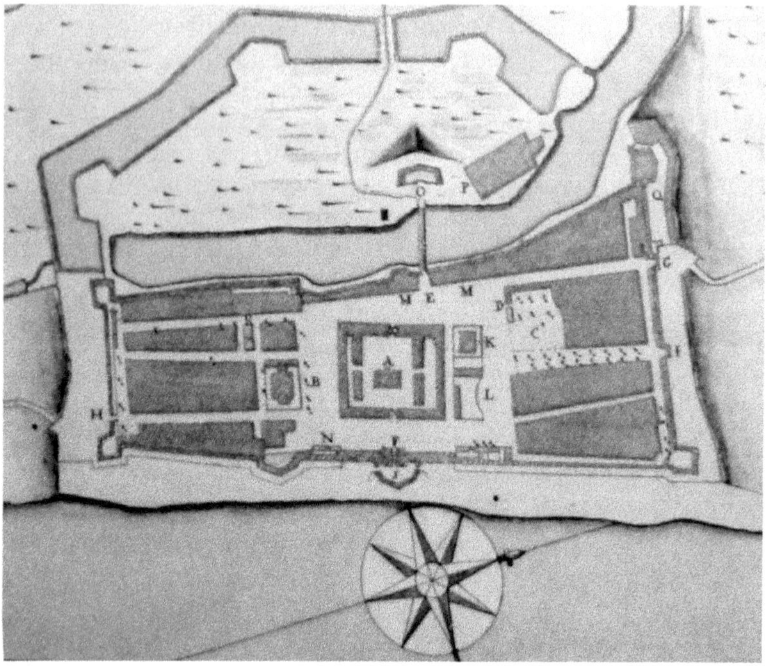

Figure 6.2 'Coffreys' barracks' shown just outside Madras Fort (marked 'P') in detail from 'A plan of Fort St George and the Bounds of Madraspatnam', surveyed and drawn by FL Conradi, 1755, courtesy of the National Archives of India

applications varied. Though often implying slavery, these terms were also used for the Siddis of Janjira and the few free black people in Company settlements. The names 'Siddi' and 'Habshi' can give some indications of origins, but again they were often used indiscriminately.[15]

The practice of giving slaves Christian or classical names also makes reconstructing their origins problematic. Only occasionally does a name – like 'Batavia Jack'[16] – imply a place of origin, and this may only indicate a previous location rather than a birthplace. Terms such as 'coolie', 'lascar', 'topass', and 'peon' usually refer to those who willingly entered EIC employment and received some remuneration, but, as I will show, the status of these workers often overlapped with slavery.[17]

At the beginning of the period, slavery was not necessarily associated with race. In one case as late as 1719, a Portuguese man was enslaved on St Helena, although he received preferential treatment, becoming overseer of the slaves until it was rumoured that he planned to raise them to rebellion.[18] The status of a number of people within the early settlements was less than free; for example, the 'hospital boys' sent to people the factories or factors sent in disgrace to the west coast of Sumatra had little control over their movements. This may have made the condition of slavery seem less worthy of particular comment to contemporaries.

The invisibility of slaves in most of the EIC records also results from a deliberate attempt to present the Company as enlightened. Statements condemning slavery were issued by the Directors, particularly when, as happened in 1688, English participation in the slave trade attracted condemnation from Asian rulers. The Mughal Emperor Aurangzeb was said to be particularly averse to the trade and had expressed his displeasure at Dutch exports of slaves from Masulipatnam.[19] On this occasion, the Council of Fort St George noted the 'great complaints and troubles from the Country Government for the loss of their children & servants sperited and stolen from them'.[20] They strictly enjoined that no inhabitants of the town, 'Christian or other do directly or indirectly buy or transport Slaves from this place to any adjacent port' on pain of a fine of 50 rupees. Those who already owed slaves, who 'by reason of their sickness or want of opportunity to transport them still remain by them', were permitted to keep them provided they delivered a list to the choultry and return any that were proven to have been 'stolen'.[21] The list of slave names registered by the Justices of the Choultry has unfortunately not survived, nor is it recorded whether the fine or the gory promise of the loss of an ear in the pillories for a second offence of 'people stealing' was ever imposed, although some people accused of this offence were imprisoned in 1693.[22]

'Slave stealing', however, was regarded as distinct from the purchase or ownership of slaves. People who had been captured in war, convicted of a crime, sold themselves or been sold by their families or masters remained fair game. These ideas accorded with contemporary interpretations of Roman and Islamic law and with codes of many societies around the Indian Ocean.[23] Early instructions to the EIC factories in the Gulf of Guinea stressed that none of the inhabitants should be pressed into service but should instead 'voluntarily' enter Company service.[24] Arriving at Madras from the Gulf of Guinea in 1663 without slaves, one captain excused himself on the grounds that 'the blacks were unwilling and hee had noe order to force them'.[25] Similarly, in a letter to Fort St George of 1676, the Directors address the 'Purchase of Natives to be sent to St. Helena'. They note that:

> We doe not approve of your sending any persons to St Helena against their wills. One of them you sent theither makes a great complaint and wee have ordered his liberty to return againe if he desires it, for

> we know not what effect it may have if Complaints should be made to the [Mughal] Emperor [Aurangzeb] that wee send away the natives besides it is against our owne inclinations to buy any blacks and transport them away from their Wives and Children without their own consents.[26]

The idea of the 'consent' of the slave to being purchased also occurs in the next letter, which approves of 'buying 100 black-servants if you can procure such as are willing to serve' but orders that the slaves be instructed in Christianity and that those who embraced it should be freed after three years. Despite these passages, the officially sanctioned trade in slaves between EIC settlements did not decline; in fact, the number of expeditions to Madagascar to acquire slaves for the factories actually increased in the years after these injunctions (Table 6.1). While ideas about acceptable forms of slavery within the early EIC were not hugely different from local norms, the process of transportation and the particular situation of the factories meant that the experience of slavery in EIC settlements differed from indigenous patterns of debt bondage.

Histories of Indian Ocean slavery traditionally began from the premise that in contrast to the plantation slavery of the Atlantic World, most slavery was domestic and therefore not economically significant.[27] In the EIC settlements, there were two categories of slaves: those who were bought privately by the inhabitants of the settlements and those purchased by the factors on behalf of the EIC, known as 'Company slaves'. The label 'domestic' could perhaps be applied to the slaves of the free inhabitants of EIC settlements.[28] Yet, in settlements such as Bencoulen and St Helena, these slaves were also employed alongside the 'free planters' in activities such as farming and planting. The roles of Company slaves were by no means domestic but – as I will demonstrate below – were crucial to the subsistence, profit and defence of the settlements. The EIC was certainly concerned to maximise slaves' productivity: letters to St Helena from London expressed dismay that slaves of free inhabitants performed more work than Company slaves.[29]

Although the European companies' use of slaves certainly borrowed from and overlapped with indigenous forms of slavery,[30] there were some differences. The gender bias towards girls or women evident in the records of the Mughals and their contemporaries[31] was not present among 'Company' slaves; later EIC orders often specified that a cargo of slaves should be made up from two-thirds men and one-third women.[32]

Company slaves lived in designated areas, often in self-built housing,[33] and were encouraged to marry among themselves, since the EIC believed that this limited unrest.[34] The structure of slavery in the English factories had a closer affinity with the Dutch factories they sought to emulate than to indigenous forms of Indian Ocean bondage.

Slave origins

The EIC, like its European rivals, both bought or captured slaves independently and participated in indigenous slave-trading networks. The earliest reference to EIC participation in the slave trade dates from 1621 and concerns prize slaves taken from Portuguese ships at Surat.[35] By the last half of the seventeenth century, the Company was profiting from the sale of slaves through registration payments and export customs as well as from transportation and sales.[36] EIC slave-trading expeditions encompassed Cape Verde and the Gulf of Guinea and extended up the East African coast from Madagascar. In South Asia, they included Bengal and the Coromandel Coast. Southeast Asian regional markets provided another source of slaves. Transportation was crucial: Madagascan and Bengali slaves worked in Sumatran factories that also profited from the export of Malay slaves.

Table 6.1 shows officially sanctioned movements of slaves between 1660 and 1720. It is based on a survey of selected records of the major factories: given the vastness, incompleteness and diffusion of the records, this is a task that can be performed only imperfectly. It excludes private and secret commerce in slaves and all cases where the origins, destinations, or any indication of the numbers of slaves being transferred are omitted – including many cases in which captains are told to acquire 'several' or 'as many as possible' – as well as all cases in which the number of slaves being transported is less than ten. Therefore, it represents only a fraction of the real trade.[37] However, it provides a sketch of slave origins and transfers between settlements.

Although the figures are low compared with the Atlantic trade, we should keep in mind that there were very few Europeans in the EIC settlements in the period. For example, just 24 European EIC employees, 16 freemen, and 5 unmarried women were recorded at Fort St George in 1679.[38] Slaves therefore formed a significant proportion of the populations of all the early factories. The increase indicated here in the numbers of slaves being transported between the 1660s and the 1720s continued thereafter, peaking between the 1730s and 1750s and beginning to decline by the 1770s.[39]

Table 6.1 Slaves transported to EIC settlements, 1660–1720

Year[a]	From	To	Number
1660	Guinea	Madras (for Pulo Run)	10[1]
1661	Fort St George	Bantam	30[2]
1661	Guinea	Madras (for Bantam)	15[3]
1662	Madras	St Helena	10[4]
1670	St Iago[b]	Bantam	20[5]
1671	St Iago	Bantam	20[6]
1684	Madagascar	Barbados	80[7]
1684	Madagascar	St Helena	180[8]
1686	Madagascar	Priaman[c]	105*[9]
1686	Madras	St Helena or Bencoulen	130[10]
1686	Madras	Aceh and Quedah[d]	200[11]
1687	Madras	St Helena	20*[12]
1687	Madagascar	St Helena	30*[13]
1687	Madagascar	Bencoulen	100[14]
1687	Madras	Bencoulen	100[15]
1687	St Helena	Bencoulen	14*[16]
1688	Madras	St Helena	50[17]
1697	Cuddalore	Anjengo	10[18]

Notes

1 Foster, *The English Factories, 1661–1664*, 43 – 'ten lusty black, men and women' ordered to be taken from Guinea to Pulo Run (Court to FSG, 1660, Letter book II, fol. 330, received 1661).

2 Foster, *The English Factories, 1661–1664*, 51; these slaves include the ten mentioned above, destined ultimately for Pulo Run and seven African and thirteen Indian slaves.

3 Foster, *The English Factories, 1661–1664*, 157 (Court to FSG, 31 August 1661).

4 Foster, *The English Factories, 1661–1664*, 271 (Court to FSG, 27 October 1662), ten Indian Hindu slaves (six men, four women).

5 IOR E/3/87, [n.f.] p. 419, London to Bantam,18 January 1670; E. B. Sainsbury, *A Calendar of the Court Minutes of the East India Company, 1671–1673* (Oxford: Clarendon Press, 1932, 4, 6) (Court of Committees, 13 January 1671, and Instructions to Captain Anthony Earning, 18 January 1671).

6 Sainsbury, *A Calendar, 1671–1673*, 4, 6 (Court of Committees, 13 January 1671 (Court book, vol. xxvii, p. 178), and Company to Anthony Earning, 18 January 1671 (Letter book, vol. iv, p. 420)).

7 SHA, *Consultations*, II, fol. 24, 9 February 1684.

8 IOR E/3/90, fols. 182–183, 4 April 1684.

9 IOR, E/3/91, fol. 27 (instructions to Captain Unkettle).

10 IOR, G/19/4, fol. 32v, 21 June 1686 (100 slaves), and fol. 62v, 4 October 1686 (30 slaves).

11 IOR G/19/4, fol. 68r, 11 October 1686.

12 IOR G/19/4, fol. 94v, 17 January 1687 (ten slaves to be sent on every 'Europe ship').

13 BL IOR E/3/91, fol. 186 (373), 3 August 1687 (letter to St Helena), and fol. 204, 12 October 1687 (Instructions to Captain William Dorrow). Instructions given to Dorrow and others to leave 15 slaves on the island.

14 BL IOR E/3/91 fols. 175r–175v, London to Bencoolen, 3 August 1687.

15 IOR G/35/2 fol. 85, Fort St George to Bencoulen, 8 September 1687 (100 slaves).

16 E/3/91, fol. 189, London to St Helena, 31 August 1687.

17 IOR, G/19/5, fol. 1, 20 February 1688.

18 Letters from Fort St George 1697, VII (Madras Government Press, 1919), 30, 18 December 1697, and 31, 27 December 1697.

Table 6.1 (Continued)

Year[a]	From	To	Number
1701	Madagascar	Surat	100[19]
1702	Madagascar	Madras	200[20]
1705	Nais	Bencoulen	112[21]
1708	Calcutta	Bencoulen	20*[22]
1709	Calcutta	Bencoulen	20*[23]
1713	Madagascar or Delagoa[e]	Bencoulen	200[24]
1717	Madagascar	St Helena	60[25]
1717	Madagascar	Bencoulen	366[26]

* Indicates lower-bound estimate

[19] MSA, *Surat 'Inwards' Letter Book*, XXXIX, fols. 303–305, 1 August 1701.
[20] TNSA, *Public Consultations*, XXX, 134–138, 10 January 1702.
[21] IOR G/35/7, York Fort General (summary), fol. 10, 1 January 1705.
[22] NAI, *Home Misc.*, Vol. 3, fols. 181–189, 26 January 1708.
[23] NAI, *Home Misc.*, Vol. 3, fols. 245–252, 18 February 1709.
[24] IOR E/3/98, fols. 127v–129r, London to Bencoulen, 22 January 1713, and fols. 130v–135v, Commission to Alexander Reid, Commander of the *Arabella*, 22 January 1713.
[25] IOR E/3/99, fol. 250v, Enclosure with London to St Helena, 21 March 1717.
[26] IOR G/35/7, [n.f.] p. 127 – York Fort General, 11 September 1714, noting the arrival of 121 men and 46 women on the *Clapham Galley* and the order given to the *Arabella* to procure 200 more slaves from Madagascar, and p. 129, Malborough Fort General, 10 December 1714, arrival of the *Arabella* with 199 slaves – 121 men, 70 women, 5 boys, 1 girl, and 2 infants.
[a] The date of the order rather than the date it was carried out is given.
[b] Santiago, Cape Verde
[c] Sumatra
[d] A state on the west side of the Malay Peninsula
[e] Mozambique

The EIC and its English rivals tried on several occasions to establish settlements in Africa, following the example of the Portuguese, who held fortifications in both West Africa and, until the Omani conquest of the late seventeenth century, along the East African coast. In 1635, the EIC had received a major setback when the rival Courteen Association was chartered to 'settle factories and plant colonies after the Dutch manner'.[40] Madagascar was the first target for the Courteen Association, inspired by enthusiastic accounts of the island's potential drawn from a 1635 voyage.[41] However, the colony in St Augustine, begun in 1644, like Portuguese, Dutch and early French attempts, was swiftly abandoned, relations between the settlers and locals having descended into open warfare by 1646. The Madagascar factory records show that a major objective was the transport of slaves to the Middle East, East Africa, the Comoros Isles and Brazil.[42]

Despite the Courteen Association's difficulties in procuring slaves, due to competition with the Dutch and French as well as local resistance,[43] Madagascar became an important source of slaves for EIC ships bound for the Indian factories, St Helena and Bencoulen, and the American plantations. At least ten officially sanctioned EIC voyages called at the island between 1660 and 1720 and took, at the very lowest estimate, over 1,240 slaves (Table 6.1). The most common destinations were Sumatra and St Helena, but Madagascan slaves were also taken to Indian factories. James Armstrong estimates that a total of 40 English vessels called at the Cape *en route* from Madagascar during the seventeenth century.[44] The island was an infamous meeting place of pirates and private traders who, despite the EIC's initial disapproval, also provided slaves for St Helena. Purchasing slaves from private traders or 'interlopers' and Company members' unlicensed slave trading, though officially frowned on,[45] provided additional sources of slaves for EIC settlements. From 1715, the EIC issued licences to private ships to trade in slaves from Madagascar in return for leaving nine slaves at St Helena for every £500 worth of goods exported from Britain.[46] Items exchanged for slaves in Madagascar included 'curiosities' such as looking glasses, beads of agate, coral and cornelian, brass, tin and iron rings, and alcohol.[47] Most important were weapons: the price of a slave was often computed in guns and powder. This ready availability of weapons fuelled warfare, perpetuating the slave-guns cycle.[48]

Regaining its monopoly in 1657, the EIC made an attempt to gain a foothold in Africa, this time on the west coast. It acquired control of Fort Cormantine (in modern Ghana), previously held by the Guinea Company, for a period of seven years. After this, it was again excluded from the trade by the British authorities, who passed the settlement on to the Royal Adventurers.[49] The Company, however, continued to purchase slaves from the Royal African Company and markets around 'Callebar' (in modern Nigeria), also called the 'Bite (or Bight) of Guinea', and Angola,[50] and EIC ships often stopped to conduct unofficial private trade in the region. Rather ironically, during his voyage out to conduct trials for those accused of being 'interlopers' in India in the 1680s on the authority of the Admiralty, Sir John Wyborne was accused of piracy in West Africa, including the theft of a large amount of ivory and eight slaves. The total value of the cargo was claimed to be 49,000 pounds.[51]

Portuguese fortifications in coastal East Africa provided the *Estado da India* with a source of slaves during the sixteenth and much of the seventeenth centuries. As the seventeenth century Italian adventurer Giovanni Careri wrote, 'there are… [an] abundance of Cafres and Blacks:

for there are Portugueses who keep thirty or forty... These slaves are carried to sell at Goa and all along the Portuguese towns by the Company's ships belonging to Lisbon and India, who buy them at Mombaza [Mombasa], Mozambique and Zofala [Sofala].'[52] Following the Omani conquest of the Swahili coast, slaves continued to be offered for sale in regional markets, notably via the Comoros Islands. On a few occasions, the EIC sent their own ships up the East African coast to trade for ivory, gold, and slaves.[53]

In 1712, Thomas Bowrey drafted a plan to recapture the old Portuguese forts on the East African coast and establish 'plantations' worked by slave labour and producing medicinal drugs for Asian and Middle Eastern markets.[54] Bowrey noted that Madagascan slaves were often taken to Johanna (modern Anjouan, also known as Ndzuwani in the Comoros), where, he wrote, in May, June or July, 300 slaves might be purchased in six or seven weeks or a yearly total of 500–600 people.[55] Bowrey's plan was never put into action, but it does demonstrate an early English interest in colonising East Africa.

The EIC also participated in the Atlantic slave trade. This activity exceeded the bounds of the Company's monopoly, which was restricted to trade east of the Cape of Good Hope, and encroached on the territory of the Royal African Company. During the early eighteenth century, the EIC lost a bid to win a monopoly on the provision of slaves from Madagascar to the American colonies.[56] The Company's slave trading in the Atlantic was therefore carried on with extreme secrecy: two fleeting references indicate that during the 1680s and 1690s, the EIC ordered returning 'Europe ships' to carry slaves from Madagascar.[57] Hans Sloane's references to meeting both 'East Indians' and Madagascans in late seventeenth-century Jamaica imply that the India fleet either called there or met American traders at an intermediate point.[58]

While they were a useful source of silver, sugar and slaves for St Helena, the EIC Directors officially regarded English ships trading in slaves from Madagascar to the American colonies as infringing their own monopoly and launched actions in parliament to prevent the trade after allegations in 1695 that these traders were supplying the pirates who attacked the Mughal fleet in the Red Sea. After a ban in 1698, the trade was resumed until 1716 but then prohibited again after the Company had been denied the right to engage in the slave trade to America themselves.[59]

Despite the complaints made by several Asian rulers, the EIC factories on the Coromandel Coast purchased slaves for local use and transportation to other settlements. During the famine of 1686, the Fort St George factory noted in its consultations that '[b]y reason of the great Scarcity of Grain, Slaves are so very plentiful and cheap that it is thought convenient and ordered that a hundred be bought at as low price as possible'.[60] These slaves were sent to St Helena and Sumatra. As mentioned above, Indian slaves were in high demand in the sultanate of Aceh; Samuel Baron indicated in his advice of 1695 that they had previously been a mainstay of the trade between the sultanate and the EIC's Indian factories.[61] Fort St David in Cuddalore on the Coromandel Coast provided a site for slave markets and supplied the new factory at Anjengo (Tranvancore) from 1697.[62] Sporadic conflict in Bengal from the 1680s onwards provided another source of coerced labour, and by the early eighteenth century, Fort William (in what became Calcutta) was the new source for Anjengo's slaves.[63]

Southeast Asia was an exporter as well as an importer of slaves. Various traditional forms of bondage were common across the region, but chattel slavery was only introduced with the arrival of Europeans.[64] Slaves were central to the urban economies of Aceh, Malacca, and Makassar during the seventeenth century, where increasing commercialisation meant that slaves were traded in open markets and transported beyond their localities. The Bencoulen factory had developed a reputation as a slave market by 1686 when the local ruler demanded a customs payment for each slave sold or exported. The customs to be received from the export of slaves at Sillebar were cited as one motivation to establish a factory there.[65] The appearance of slaves from the small island of Nais in the Bencoulen factory by the early eighteenth century reflects their emergence as a regional community, emerging from a previously closed system.[66]

Destinations

Slaves were employed in and transferred between all the EIC settlements. However, they were especially important to the settlements that were intended to serve primarily as 'plantations' rather than trading posts. These included Bencoulen and its subordinate factories in Sumatra, which were intended to produce pepper and sugarcane for export. Another settlement often referred to as a 'plantation' was St Helena. As discussed in the previous chapter, the island was not only intended to grow sufficient food to provision passing ships, but was the target of a number of experiments in growing Asian and American crops that were intended – although often failed – to generate profits.

In many ways, these two locations symbolised the extremes of European ideas about the tropical environment. From the time of its discovery on 21 May 1502,[67] St Helena provided a place for European ships returning from the East Indies to gather provisions and rest their sick crews. After the Portuguese abandoned the island, it was taken over by the EIC in 1659 and, aside from a short period of Dutch occupation in January–May 1673, remained a Company possession until it was transferred to the Crown in 1836.[68] Visitors from Sir Thomas Cavendish and Jan Huyghen van Linschoten onwards described St Helena as an 'earthly paradise', abounding with fruits, fish, and fowl,[69] a vision that also appears in Johan Nieuhoff's map of the island (Figure 6.3). Francis Godwin's 1638 allegorical meditation on the new discoveries made possible by the telescope describes the assent of the imaginary traveller Gonsales from St Helena, 'the onely paradise, I thinke, that the earth yeeldeth', to a lunar utopia peopled by Christian Martians who combined the civilisation of the Chinese with the innocence of Native Americans.[70] Even as late as the 1690s, Dampier eulogised the fresh fruit and healing herbs of the island, which cured sick seamen.[71]

Figure 6.3 Johan Nieuhoff, map of St Helena, courtesy of Leiden University Library[72]

In contrast to the perceived healthiness and fecundity of St Helena, Bencoulen, which was suffering from high rates of malaria, symbolised for the European settlers the horrors of disease and degeneracy considered to lurk in the tropics. On hearing of the decision made by the Madras factory's embassy to settle at Bencoulen rather than Pariaman, the EIC Directors complained that 'our people were made very sickly, and many dyed at Bencolen and Indrapura'.[73] These concerns were confirmed by the early consultations of the factory. One passages reports that the garrison had only five Englishmen and some Portuguese soldiers on duty, the others being sick and 'the sick neglected for want of people to tend them: all or most of the Cooleys & black workmen are sick & dead... the sick cryes for remedies, but no[ne are] to be found'.[74]

The complaints of the factors soon entered into contemporary consciousness in England. One natural philosopher wrote of Bencoulen: 'grief and venery are the most dangerous passions... for here the Constant heat [and] the unconstancy of the air causeth such a dissipation of the spirits that men know themselves to be weak & are subject to faintings'.[75] By the time Robert Knox arrived in 1690 with a consignment of slaves from Madagascar, he observed that Bencoulen had 'bin since my memory imputed to be the most sickly Country in the known world'.[76]

As well as the foci of Edenic fantasies, islands were sites of experimentation. St Helena, an enclosed space with no indigenous inhabitants, served as an arena in which plants and people could be experimented with before being re-circulated. However, these experiments were often altered in unforeseen ways by the plants, people, and techniques that arrived unplanned on the tides of the Atlantic and Indian Oceans. Attempts to transfer both European and Asian crops and methods of agricultural production to St Helena began with the Portuguese introduction of the fruit trees that early travellers marvelled at as well as the European and herbs and species brought in 1502.

The English fleet sent to 'settle, fortifie, and plant' the island in 1659 was instructed to stop at Cape Verde to pick up specimens of cassava, yam, potatoes, beans, and chickpeas as well as orange and lemon trees for transplantation.[77] This began a long tradition of experiments with the introduction of crops: some, such as New Zealand flax and coffee,[78] enjoyed a greater or less degree of success, while others, such as rice and cinchona (the source of quinine),[79] never became established. Many proved disastrous to the island's indigenous and endemic flora.[80] Both sugar and tobacco, which the English hoped to emulate the Portuguese by producing in their own settlements, were marked failures in St Helena but seem to have had some success in Bencoulen.[81] Coffee trees were planted in St Helena, and later Bencoulen, with some success, although the crop never became profitable in either settlement.[82]

Several attempts to introduce rice, indigo, and cotton to St Helena – including those of Robert Knox discussed in Chapter 5 – also failed.[83] However, yams thrived and provided the main foodstuff for much of the population. Seamen, settlers and slaves also carried out transplantation independently. For example, in 1700, the consultations mention a tree brought from the Cape of Good Hope by a ship's captain that had succeeded in establishing itself in an area where no other vegetation could survive, thus protecting against further soil erosion.[84]

To people their plantations, the Company employed techniques ranging from persuasion via material incentives to coercion, with varying degrees of success. Despite bills posted in London describing the joys of life in St Helena, voluntary settlers were few in the early years of the colony, and the population was only 50 free men, 20 free women, and 6 slaves in 1666.[85] The Company therefore resorted to offering incentives to tempt seamen to stay on the island[86] and, according to John Ovington, tricking young women into migrating in search of wealth husbands.[87] These efforts to attract free settlers met with only partial success, and slaves made up a considerable proportion of the population of St Helena through the period, outnumbering the white inhabitants for a time in the early eighteenth century. Table 6.2 shows the total number of slaves, including those enslaved by the Company and those bound to private inhabitants. The first detailed census of the 'Company' slaves, in 1716, records 18 slaves from 'Guinea' out a total of 108, the rest coming from Madagascar.[88]

Table 6.2 Slaves as a percentage of St Helena's population[1]

Year	Slaves	Total population	Percentage
1659	6	46	8
1665	21	74	28
1716	327	743	44
1717	411	953	43
1718	411	801	51
1720	443	834	53
1721	424	800	53
1722	406	924	44

Note: [1] Derived from census data for these years, given in SHA, *Consultations*.

Because of its reputation for sickness, Bencoulen was still less popular as a destination for settlers, and although incentives were offered for people to move there from St Helena,[89] it was more common for them to be sent there as punishment from either St Helena or one of the factories in India. The first detailed census of the Bencoulen factory dates from 1748 (Table 6.3) and similarly shows that most slaves came from Madagascar, with smaller numbers from India and the Malay world, notably Nais. A similar distribution is probable in the earlier period: in 1685, the slaves were described as 'caffreys' and 'gentues', that is black Africans and Indian Hindus. The former were valued at 100 dollars and the latter at 60 dollars for an adult male, probably due to the perception of African slaves as being able to perform more hard labour.[90] Malagasy slaves were also particularly prized for their reputation for hard work, ingenuity in crafts and military prowess.[91]

As noted above, the transportation of people was not carried out simply to supply labour, but also as a means of transferring skills. Plants and people with skills in growing them were often transported together. Examples include the recruitment of French Huguenot refugee vintners who were transported to St Helena along with their vines, the leader of this group, Stephen Poirier, eventually becoming the Governor of the island in 1697.[92] In 1704, the Bencoulen factors wrote to London to request their own supply of vines and 'French refugee vineroons'.[93] Similarly, the West Indian planter Nathaniel Cox was sent with his sugar canes first to St Helena and then to Bencoulen.[94]

Seventeenth century writers often compared plants and people and applied the language of experimentation and improvement to both: for example, Ovington wrote disapprovingly of St Helena, that 'the minds of the Inhabitants are generally as Uncultivated as the neglected Soil, their Intellects as ordinary as their Qualities, but what is worse, the pravity of their Manners compares them with the rankest Soil, producing nothing but noxious Herbs, untractable to all the Arts of Husbandry and Improvement.'[95] In 1668, the Court requested their representatives at Fort St George to procure eight 'young Gentues or Aracans and their wives to be sent as our servants to remain on our Island of St Helena. Wee being very desirous to make trial of them, supposing that they may bee more useful and ingenious than those people which come from Guinea.' In a similar vein to other letters of the period, this letter urged that no violence should be used, 'but that they may be such as will willingly embrace our service.'[96] In 1682, the Directors in London wrote to their Council in St Helena 'it must be your care to see the Seeds Plants

Table 6.3 Occupations, average ages, and places of origin of Bencoulen slaves, January 1748

	Occupation									Age	Place of origin			
	Bricklayer	**Carpenter**	**Armourer**	**Gardener**	**Bookbinder**	**Cooley**	**Superannuated**	**Blind**	**Total**	**Average age**	**Madagascar**	**Malabar (S. India)**	**Nais (Indonesia)**	**Malay**
Men	7	2	11	3	1	36	21	1	82	40	47	15	19	1
Women						66	19		85	37	80	2	3	0
Boys									32					
Girls									36					
Total	7	2	11	3	1	101	40	1	235					

& Negroes so used and employed that we may be careful that we may at length reap some Benefit of all our Care and Cost'.[97]

Theorising transplantation and transportation

The comparisons that were being made between plants and people at this point were not solely allegorical. In this period, when ideas about human life were being revitalised by the recent discovery of the circulation of the blood and a new attention to anatomy, some interesting speculations were circulating. Comparisons were being made between the animal and vegetable worlds. Both the Dutch anatomist Frederik Ruysch and the English naturalist Nehemiah Grew drew parallels between the veins and arteries and similar structures that were being observed in leaves.[98] Microscopic observation of cells both in the sap of plants and in blood also contributed to some sense of connection between their functions.

There was a general belief that both plants and people were connected to their physical environments, which gave them their particular characteristics. Although, as discussed in the previous chapter, there was an emerging sense that transplanted crops might thrive at similar latitudes to those they were accustomed to, up until the time of Linnaeus' experiments with growing coffee trees in Sweden, it was thought that plants might adapt themselves to radically new environments. That people were conditioned by the environments they lived in was an accepted idea in early modern medicine and was also factored into the philosophical historical work of the jurist Jean Bodin.[99]

What might happen to people when they moved was also a matter of concern. With no fixed idea of race during the seventeenth century, it was assumed that people might change in their physical make-up and temperaments within the course of a generation when moved to a new environment.[100] However, it was considered that sudden exposure to different climates posed a threat. Discussing slave mortality and morbidity in the early eighteenth century, the Governor of St Helena Isaac Pyke wrote of slaves that 'this climate being different from that of other Nativities subjects them to many other Diseases so that of old Blacks one in ten usually dies every year and of those new Blacks, 2 in 15'.[101]

Thought about the environment and the physical health of the inhabitants of the Company's 'plantations' was closely connected to the issue of moral conduct.[102] Knox ascribed the ultimate cause of the 'contagious aire' of Bencoulen to 'the wrath of God, who turneth a fruitfull land into Barrennesse for the wick[ed]ness of them who dwell therein'.[103] In St Helena, the rapid degradation of much of the land as a result of deforestation and overgrazing quickly raised concerns about the fragility of the island,[104] which was linked to the lax morals of the islanders. Speaking at a trial for adultery in 1698, Governor Poirier reflected on the biblical example of the destruction of Sodom and Gomorrah before stating his belief that the same sin had brought on a recent disaster in Jamaica (probably in reference to the earthquake of 1692).[105] His warning that a similar fate could befall St Helena was repeated in warnings issued after a spell of droughts in 1717 that the island was not yet safe from divine displeasure.[106]

The EIC Directors and the members of the councils of both settlements were therefore concerned to 'cultivate' and 'improve' all their servants, a process that was considered vital to the moral and even physical survival of their settlements. However, slaves were subject to the most intense scrutiny, speculation, and experimentation as their condition meant first that, at least theoretically, more control could be exercised over them than over free employees and second that the Company was more concerned about the possibility of their resistance.

Slave education and professions

One of the most obvious ways in which the 'improvement' of the morals of the settlements' inhabitants and particularly its slaves was intended to take place was through religious instruction. As the 1680 tract of Morgan Godwin argued – drawing on the assumed connection between climate and constitution – the failure to instruct slaves in Christianity demonstrated 'the Crime of such degenerate English, who with that air, have imbibed the Barbarity of the Countries they live in'.[107] The minister appointed to St Helena in 1673 was commissioned to teach children, including slaves, to read and write English as well as instructing them in Christianity.[108] A 1670 letter to St Helena from the Court of Directors in England, which Robert Boyle copied into his commonplace book, instructed that any slaves who publicly accepted Christianity were to be freed after seven years.[109] Similar instructions were sent to the EIC settlement in Bantam and, as noted above, to Madras.[110]

The list of slave baptisms at St Mary's Church in Fort St. George suggests that the domestic slaves of the English residents received some religious instruction.[111] It is difficult to tell whether many EIC slaves converted to Islam, as was often the case elsewhere, but Catholicism evidently provided competition. The Fort St George records contained an order in 1658 for the expulsion of a Catholic priest named Pascal for his attempts to convert to Catholicism an African slave named Frank, a drummer in the army, who had previously accepted Protestantism. The same year another slave was imprisoned until he revealed the identity of the Catholic priest who had baptised him.[112]

Slaves also received practical training in masonry, carpentry, and agriculture and as blacksmiths and gardeners.[113] In general, slaves were bought in their teens because they were thought to be more amenable to training.[114] Literacy was expected to enable slaves not only to read the Bible, but also to occupy managerial roles in the plantations.[115] Slaves who had acquired particular skills in one settlement were often transported again to other parts of the world where their expertise was in demand. For example, in 1685, the Council requested that English-speaking slaves with experience in sugar or saltpetre works be sent from Bengal to St Helena.[116] When the factories in Sumatra were in their infancy, English-speaking slaves trained in bearing arms or in crafts such as carpentry and making lime were sent from St Helena to Sumatra to act as overseers for slaves acquired directly from Madagascar.[117]

Slaves' involvement in construction work would not have been purely manual. After losing several of their slaves to smallpox in 1748, the Bencoulen factory complained that without them 'our Fortification[s] and Buildings will of course run to decay…for the greater part of the Slaves we lost were deemed the ablest and most skilled in Fabrick as well as other Imploys'.[118] As this example shows, the construction of the Company's forts, factories and warehouses was largely carried out by slave labour. Some were skilled craftsmen, training in stone masonry, carpentry, and smithing. There seems to have been some hierarchy and system of 'apprenticeships' within these professions: the Bencoulen consultations mention the chief bricklayers and smith receiving (minimal) wages as encouragement, 'being the most ingenious and industrious fellows among them'.[119] The St Helena consultations refer to one young slave apprenticed to the smith and another 'working with free Jack, to learn stone laying'.[120] One slave, 'Old Will', who was said to have learnt carpentry in London, had instructed another slave called Jack Grewer.[121]

Like Frank, mentioned above, many slaves were employed in military capacities.[122] The large building just outside the Madras fort labelled as the 'coffreys' barracks' in the map of 1755 (Figure 6.2) suggests that a considerable force of Madagascan slaves were employed to defend the factory, while being located just outside its walls in case they themselves posed a threat to its inhabitants. Slave militias were often seen by the EIC as an alternative to the dangerous proposition of arming local people, and slaves were often among the first to be sent to new factories.

A letter sent with a consignment of African slaves to the new factory at Pariaman ordered that some be employed to keep guard armed with 'lances, darts and swords'. Here, the Court of Directors expressed the hope that the presence of these slaves would provide 'some kind of balance in case of need ag[ains]t an enemy, they being as much strangers to the Sumatrans as they are to Us'.[123] A mixture of nationalities among slave soldiers was deemed crucial to the security of the Company. When Josiah Child sent an urgent appeal to the Sumatra factories to send soldiers from Makassar and Java to strengthen the garrison at Fort St George, he requested that the troops 'have no mixture of the Portugall caste blood in them' because he already had 'too many of the Portugall or mesier cast[e].'[124]

In 1688, Fort James (Indrapura, Sumatra) wrote to Bencoulen and London demanding 200–400 Madagascan slaves. Otherwise, they wrote, 'yo[u] had better withdraw the factory for unless we are able to force the Mallays neither men nor prows are to be had'.[125] One hundred slaves sent to the new factory in Bencoulen in 1687 were to be trained as soldiers, as sailors, or in related work in the garrison.[126] The Company wrote to St Helena requesting some more experienced slaves who spoke good English or Portuguese to join the company of soldiers.[127] In a 1697 letter, the Court warned the Fort St George factory against moving many slaves from Bencoulen to work on the fortification at Fort St David (in Cuddalore on the Coromandel Coast), 'because it must be our care to keep York Fort so well manned as to preserve it from any attempt of the perfidious Malays'.[128]

In a letter to Sir John Goldsborough of 1693, the Governor of York Fort in Bencoulen noted that he had made soldiers of several of the younger 'coffreys' (African slaves).[129] This policy was taken on by the Indian factories, and the Court wrote approvingly to Fort St David in 1694:

> [Y]ou doe wisely advise to the raising of a Company of Cofferies which would be the truest People and the stoutest Blacks you could trust to in that Place, having no affinity or Relation to the Black People of these countreys nor speaking any of their languages'.[130]

Concern about training and arming slaves is also evident in the Company records. For example, in a letter to Pariaman, the Court of Directors urged their factors to limit the types of weapons that the Madagascan slaves had access to, warning that they should not be trained in the use of firearms. Meanwhile, they suggested that '50 of our old Peons from fort St George' could serve as musketeers and provide 'a good ballance against the Sumatrans, Madagascar blacks, and mutinous Englishmen ...'.[131]

Despite such warnings, slaves were clearly important to the defence of early settlements: by 1734, the garrison of St Helena consisted of 75 slaves 'belonging to the garrison' and 102 'belonging to the planters' with military training as well as 134 white men and a 'planter militia' of 31.[132] In employing slaves as soldiers, the EIC was in fact following the earlier example of other powers in Asia in using slaves in the military. Indian rulers had employed African slaves as guards, soldiers, and advisers since the medieval period.[133] A few rose to hold important positions: the best known being Barbak Shah (Sultan Shahzada), Habesh Khan and Shams-ud-din Abu Nasr Muzaffar Shah (Sidi Badr), who seized power in fifteenth-century Bengal;[134] the Ulugh Khans, commanders of Arab forces in Gujarat in the mid-sixteenth century;[135] and Malik Ambar, ruler of Ahmadnagar from c. 1600 to 1626.[136] Although free Africans came to Asia, particularly before 1500, the slave trade was the origin of many of the Siddis of the small island of Janjira, a formidable naval force, which at the command of the Mughal Emperor Aurangzeb besieged and defeated the English factory in Bombay in 1687–1689. The Dutch in Ceylon (Sri Lanka) used considerable numbers of slave soldiers, a policy that was taken on by the English upon their acquisition of the island.[137]

The steady flow of African slaves into Bijapur and Ahmadnagar was augmented from the seventeenth century by escapees from Portuguese territories, who often found employment in the armies or navies of these Muslim-ruled states.[138] The Portuguese attempted to stop this by signing agreements to return them with rulers such as Nizam Shah and, on the transfer of the island of Bombay, with the English.[139] That some fugitive former Portuguese slaves were nonetheless employed in English territories is evident from a 1698 exchange between one Don Lewis Henriques of Goa and the then chief agent at Bombay, John Gayer. Countering Henriques' complaints about the non-return of Portuguese slaves, Gayer claimed that the Portuguese also kept English runaways. He added that he could not persuade the former slaves to return, 'they desiring rather to be cut in pieces'.[140] Whether many slaves achieved freedom in the military service of the English is uncertain, although an offhand reference in a contemporary story implies that some did.[141] Certainly, none were able to rise to comparable positions of authority as had been achieved earlier in Asia.

The Japanese 'namban' screens produced during the sixteenth and seventeenth centuries, and now scattered throughout global museums, record the arrival of the 'southern barbarians', as the Portuguese were known. These depict Asian, African and European sailors, a picture that would have accurately reflected the crews of any of the European companies in the seventeenth century. The term 'lascars' usually refers to paid, non-European sailors.[142] But in some records, lascars were explicitly equated with slaves. In 1688, the *Royal James'* crew was described as '67 Europeans...with twenty lascar slaves belonging to her'.[143]

The status of lascars could also shift towards slavery during the voyage. In 1688, a ship's surgeon complaining of an abusive captain claimed that 'all the lascars left him at Indrapoora, because he used to almost starve them, he would beat them and abuse them like doggs & threaten to sell them to the Mallays when they came to him for money or Rice'.[144] That slaves were also employed as sailors on this voyage is implied by another complaint from a fellow crewmember.[145] The practice of abandoning penniless lascars in London after they had crewed ships back from the East Indies was only halted after it became embarrassing for the EIC.[146] Other sections of the records describe incidents in which 'coolies' were seized from the Portuguese and forced to serve in English ships before escaping or being recaptured.[147] Robert Drury also noted the capture and enslavement of lascars by an English ship.[148]

Slaves being transferred between different factories would have been expected to work during the voyages, as shown by a muster of the ship *Beaufort*, which included 50 slaves being transported to St Helena.[149] On the ship's arrival at St Helena, slaves replaced ten sick sailors.[150] In 1693, the Fort St George factory wrote that a vessel had been sent on a voyage to acquire pepper on the west coast with '10 of the Company's Coffrees to supply the place of lascars'.[151]

In the case of the ship *Pryaman*, sent to Indrapore in Sumatra after stopping to acquire slaves at Madagascar, the slaves became part of the crew; on their arrival, the factors let them remain on the ship to assist those of the crew still alive. Shortly afterwards, these slaves ran away, perhaps in the vessel's longboat.[152] In 1688, eight slaves were sent to Tonkin to assist with fortifications on the factory, if necessary, otherwise to 'continue as Marriners for the Ship'.[153] Slaves were also involved in the construction, repair, and loading of ships: in 1685, the Company instructed Fort St George to train its slaves in carpentry in preparation for establishing a shipyard.[154] The Pariaman settlement was also intended to host a shipyard staffed by slave labour. Again, the British followed Asian rulers in using slaves as workers in shipyards.[155]

Slaves often acquired several different languages as a result of forced relocations. Portuguese was a lingua franca across the Indian Ocean in this period, and slaves who spoke it were therefore useful interpreters. Sometimes slaves also served as interpreters on return voyages to their places of origin. One passage in the Company records refers to a man from Johanna returning with an English expedition against the French[156] and another to the capture of 'Francisco, a black designed Linguaster for Soccatra', on a voyage to that island.[157] A slave named Chinéâ and known as 'Jambee', possibly in reference to Jambi in Sumatra, was frequently sent as a messenger between the English factories in Taiwan and Amoy and was highly prized for his knowledge of Chinese.[158]

Company directives also expressed concern about the linguistic capacities of slaves. For example, a letter from London to Fort St David about the garrison of African slaves warned:

> [A]s you can buy them from the Portugueez you may add six or eight, but we would have them bought as they come raw from Mosambeek or Mumbas before they can speak the Portugueez language... and the more Native languages they are off, we think it may be the better...[159]

Elsewhere, however, the EIC accepted the usefulness of Portuguese. When, in the early eighteenth century, Governor Collet was planning a school for slaves in Bencoulen, he assumed that Portuguese would be the language of instruction.[160]

As noted in Chapter 4, experimental gardening was considered crucial to all the early Company settlements: plants could prove food, medicine, and valuable items of trade. Gardening was therefore an important profession, and slaves who became proficient were sometimes singled out for preferment: in the Bencoulen consultations of 1695, two slaves listed as gardeners were ordered to receive four dollars per month and a suit of 'perpetuanos' or wool.[161] Similarly, a slave belonging to the Governor of St Helena was described as worth three times that of other slaves because of his skill.[162] The Bencoulen factory was intended to supplement the unreliable supply of pepper from inland by running a pepper plantation staffed with slave labour, as well as the paddy fields needed to feed the settlement.[163] As in the Atlantic world, slaves in the early Company settlements were responsible for producing their own food, working on provisioning plots as well as labouring on the Company pepper plantations in Bencoulen.[164]

As noted above, slaves were often transferred along with plants that they were skilled in growing. However, they were not merely passive subjects in such experiments. While the important impact of food grown by slaves on the wider cuisine of the Americas has received attention in some recent studies,[165] the contribution of slaves to the foodways of the Indian Ocean world has attracted less attention. Nonetheless, there are some indications that slaves were responsible for some major innovations. In one example, William Dampier noted that a type of rice cultivation had been introduced into Aceh by the slaves brought from the Coromandel Coast by English and Dutch traders and had since become widely practised.[166]

Despite all the attempts of the Directors of the EIC to introduce crops to St Helena, that which flourished best appears not to have been introduced by Europeans but to have arrived with African slaves. This was the yam, which became such a ubiquitous dish on the island that the inhabitants were eventually nicknamed 'yamstocks'.[167] Yams are mentioned on the island since the 1670s and had become a regular foodstuff, but the transplantation of one particular variety received particular attention. As noted in Introduction, these were introduced to St Helena by a slave, Maria, who had wrapped them in her clothes for the journey from Madagascar, planting them by a spring upon arriving at the island.[168] As this story demonstrates, women tended to be particularly involved in the cultivation and transplantation of food crops.[169]

Natural knowledge was often allied to medical knowledge. Historians of the Atlantic world have noted the employment of 'black doctors' in American colonies,[170] and consulting the medical knowledge of slaves was also common in the world of the early EIC. The traveller in Francis Godwin's parable is left on St Helena with a slave, Diego, to tend to him,[171] and the use of slaves in this capacity was in fact common. In 1663, the Court wrote to the Council at Fort St George to ask them to purchase 'two Gentue barbers such as are most expert among them in letting blood and send them on our ships to St. Helena'.[172] Whether the request was granted is uncertain. In 1687, a group of English-speaking slaves were sent from the island to Bencoulen, where they were to be employed tending to sick soldiers.[173] Slaves and 'coolies' were also employed within the hospitals of Indian factories.[174]

That medicine was also practised among slaves on St Helena is evident from a set of rules in 1708 collected from instructions from London and the planters with the consent of the Governor and Council. These state '[t]hat no Negroe whatsoever shall prescribe or Administer any Physick or Medicine whatsoever to any Negro or Negroes without the Consent of his or their Mistress... under the penalty of severe correction'.

Slaves were also forbidden to receive medicine from other slaves.[175] However, slaves were still called on to perform medical services. For example, in 1709, a ship captain complained that lodging sick crew members with the inhabitants in the usual way was too expensive and requested that slaves be sent to tend them. The Governor and Council agreed.[176]

The list of occupations of the female slaves of St Helena implies that most followed professions as washerwomen, cooks, midwives, and workers in the dairy.[177] In the 1723 survey of St Helena slaves' occupations, only five women are listed as involved in construction work. In contrast, all the women of working age in the Bencoulen census are listed as 'cooleys' (Table 6.3). Enslaved women's involvement in medicine might have focused on reproductive issues, as indicated by a curious story from 1650s Madras about a 'Gentue negro' from San Thome brought in to perform an abortion for one of the factor's wives.[178]

Resistance and rebellion

Unlike in parts of the Atlantic world, the EIC never faced a large-scale, organised slave rebellion. In 1695, however, slaves on St Helena made a detailed plan to kill their owners, seize stocks of ammunition, use them to besiege the main fort and take over the island before escaping on the next ship.[179] One of the ringleaders boasted that 'I can handle a gun as well as anyone: for I was used to carry one when I was a little boy'. This may reflect previous experience in a European militia or the militarisation of areas involved in slave trade. The plotters were caught and punished with extreme public torture and execution and restrictions on the movements of other slaves.[180]

The same year, 30 slaves, armed with muskets and clubs, escaped from their enclosure in Bencoulen in the early hours of the morning; this was probably a response to the reappointment of an overseer who had killed a female slave. After killing two of the pursuing soldiers, the escapees vanished into the woods. The factory responded with more restrictions on the movements and interactions of the remaining slaves.[181] In Bencoulen, unlike in St Helena, the punishments of slaves who defected were tempered by a concern that those with military training might join local enemies.[182]

As with the knowledge of arms, skill on-board ships occasionally enabled slaves to resist their condition. During the plans for the rebellion of 1695 in St Helena, 'Jacob . . . that runn away from Captain Nynn's ship' was designated 'Captain' for the planned escape on the first vessel to arrive at the island.[183] In 1718, some slaves who had been working on St Helena seized a boat and left the island, aiming for Madagascar. Although the voyage was probably tragically unsuccessful, it was carefully planned: preparations included packing eight gallons of water, yams and extra oars and sails.[184]

Shared linguistic skills also helped slaves resist their condition. An account of a planned uprising on St Helena in 1707 describes four slaves, Mingo, Caesar, Clause, and Tobey, conversing in Portuguese and planning to take the next Portuguese ship to arrive with the aid of 'all the Portuguese blacks'.[185] Organised rebellions have been used in the literature on slavery to show the formation of new bonds between slaves of different ethnic, religious and caste backgrounds. This solidarity is evident in the EIC records. However, it is also clear that the Company's plan of diversity was effective to some extent: in both the 1695 St Helena uprising and another failed insurrection on Sumatra, the plans were made and revealed by slaves of different origins.[186]

Smaller groups or individual slaves also resisted their condition by going into hiding – there are several references to escaped slaves living in the woodland that still covered much of St Helena in the late seventeenth century. The free inhabitants sometimes sheltered them: one man was sentenced to a public lashing when caught doing so.[187] In one case, two escaped slaves resisted attempts to recapture them by establishing themselves in a high hill on the island, which they were able to defend by rolling stones down on anyone who attempted to reach them.[188] In Bencoulen, individual slaves also escaped. In one case, an African slave called Chubdar escaped after murdering an Indian slave after a conflict over stolen sugar canes. Reports suggest that he had continued to live nearby, returning to the African slaves' enclosure at night, where he was secretly fed.[189] In Bencoulen, escaped slaves might, as the Company feared, have occasionally been able to find employment elsewhere. However, local rulers profited from the Company's slave trade and were rewarded for returning escapees.[190]

Skills in gardening and medicine would have been useful to maroon communities, and slaves were often feared for their reputed skills with potions and poisons.[191] Reference to a covert international trade in poisons appears in the records of the trial of a slave, Sattoe, for attempting to murder his master John Boston in 1679. Sattoe had threatened Boston with an axe and was also found in possession of a 'red and white poison' that came from Bantam and was given to him by Rowland, another slave, as well as a pipe of tobacco 'which afterwards he found it very hot in his head and belly'.[192]

Sattoe was punished by having the hand with which he wounded his master cut off, the original sentence of having his head displayed on top of the market house having been commuted, perhaps partly as a result of concern that a larger rebellion was being planned among the slaves.[193] Reports of attempted poisonings in 1687 and 1689 mentioned poison made from a 'weed growing on the island' as well as a mixture of blood, ground glass, and 'earth from dead people's graves'.[194] Like the use of the poison from Bantam, the use of the grave dirt had transnational as well as local implications: it was used in slave oaths in the Americas and probably represented a combination of the various beliefs about death and the power of corpses derived from discussions between the slaves from Madagascar, around the Gulf of Guinea and elsewhere who were brought together in St Helena.[195]

For most of the inhabitants of St Helena, the threat of black magic practised by slaves was as powerful as the danger of poisoning. In 1693, a slave named Jamy was tried by jury, convicted of witchcraft, and burnt at the stake.[196] Two slaves apparently died in terror after another slave, known as 'John Batavia', who was punished for pretending to be acquainted with the devil, had put his hands on their faces and recited some 'hocus-pocus' words.[197] While at the beginning of the period, Company officials tended to take witchcraft accusations seriously – several women being tried and convicted of the offence – by the second decade of the eighteenth century, accusers were more likely to be treated as foolish or insane. When, in 1734, Mr Bates complained that his slave was trying to poison him by witchcraft by burying a phial under his chair and produced a copy of Hamilton's *Voyages* for evidence, Governor Isaac Pyke ridiculed him.[198]

The decline in belief in magic in the EIC settlements paralleled that in contemporary Europe, where witchcraft trials had ended by the 1720s, and the act of 1736 instead criminalised the claim that any person could possess magical powers.[199] Despite his disdain for magical beliefs, however, Pyke found it useful to investigate and make use of slaves' beliefs in his attempts to prevent them from suicide, the more desperate means of resisting the condition of slavery. In 1714, several slaves from Callebar (Calabar, Bight of Biafra) attempted to kill themselves.

Among the St Helena slaves, like their contemporaries in the Americas, it was widely believed that death would enable a return to the homeland. Noting this belief, Pyke had one man who had succeeded in hanging himself from a fig tree hung up by his feet in plain sight of the public highway, writing that the other slaves 'have a notion that as long as he is tyed by his feet he cannot get to his own Country, which wee hope has been a means to hinder others from destroying themselves in the same way'.[200] Pyke's actions also mirrored the 'spiritual terror' practised by slave masters in the Caribbean.[201]

Pyke was a regular correspondent of the Royal Society and avid experimentalist,[202] and it is clear from his description of his act that far from random cruelty, the punishment was intended to maximise the usefulness of the remaining slaves. After this experience with these slaves from Callebar, whom he also depicted as cannibals, Pyke wrote to the Council to request that no more slaves be sent from that area, dismissing them as 'a race of people not to be Esteemed, they being so little use to the rest of mankind'.[203] As Pyke's phrasing here reveals, definitions of race were hardening by the eighteenth century.

Free blacks and gentlemen of colour

While race was becoming a more important category in colonial thought, the EIC struggled to impose racial distinctions between free and unfree on the ground. We do not know whether the Company's early directives to free Christian slaves were ever enforced: probably as in the American colonies, the idea was quickly abandoned. Significantly, although he had a copy of the original letter, Governor Godwin did not copy it into his rulebook of 1708.[204] Nonetheless, masters often freed slaves, sometimes in their wills.[205] Slaves were also occasionally able to earn money, and in 1695, the Bencoulen factors assumed that eventually the Company slaves would purchase their own freedom.[206] Slaves were also occasionally freed for particular services: most notably 'Black Oliver', who was freed for his part in recapturing the island of St Helena from the Dutch in 1673 and became a free planter.[207]

The number of free black people mentioned in the St Helena records gradually increases, and in 1721 they are referred to as 'many', with 18 recorded in the census of 1722.[208] In a meeting of 1723, apprehension was expressed about interactions between free and enslaved black people. The former were said to be 'tampering with' or 'corrupting' the slaves after a planter sold a slave named Abigail to a free black man, Tom Collier.[209] The Company consequently ordered that any freed slaves should be immediately transported to the Indies.[210]

Freed slaves could also be returned to a state of slavery: the family of Black Oliver provides an example of this process. Oliver's son Jack was imprisoned and then enslaved, despite a lack of evidence to support the accusations against him.[211] Oliver's two daughters were also punished for bearing illegitimate children: Mercy (or Marcy) Oliver receiving a public lashing and Martha being deported to Bencoulen. However, Martha apparently retained her freedom when arriving in Bencoulen: the records report that 'a free Black woman with two small children (one but 14 daies old) came ashore here, who was sent from St Helena for several misdemeanours'.[212] It was probably Martha who wrote to Will Oliver, another member of the family, from Bencoulen in 1701.[213]

The father of Martha Oliver's child, born in 1697, was the white planter Gabriel Powell.[214] Another black woman called Elizabeth Lansdowne was whipped for having a bastard child by the late surgeon David Law in 1701.[215] Later approaches appear to have shifted the punishment onto the father of the child. After the birth of the child of a black slave and white soldier in 1716, the soldier had his face blackened and was forced to 'ride the wooden horse' as punishment.[216] As was the case in other colonies, including Virginia, the children of enslaved women were normally regarded as unfree, regardless of the condition of their fathers. In some cases, the father of the children was able to purchase their freedom and that of their mothers. For example, a Portuguese soldier in Triamang was permitted to purchase the freedom of a Company slave by whom he had two small children.[217] In one extraordinary case of 1713, described in a letter from Joseph Collet, Governor of Bencoulen, a high-caste Indian man was voluntarily enslaved so that he could marry a Company slave.[218]

Despite efforts to prevent the blurring of boundaries between their free and enslaved subjects, the Company failed to prevent liaisons between people of different races. An account of Bencoulen in 1812 by the surgeon Benjamin Heyne notes that people of mixed race had become integral to the functioning of the settlement:

> Gentlemen of colour, from yellow to jet black – the descendants of Jews and Christians of all nations, by Malay or Bengal women: some of them…are covenanted servants of the Honourable Company; others are monthly writers at our office in Marlborough and the outer stations. All are gentlemen of considerable circumstance.

Heyne describes about 300 domestic slaves living among the inhabitants of the town and notes that they tended to be treated more humanely by both Indians and people of mixed race than by Europeans.[219]

In St Helena, racial mixing among the inhabitants meant that the Directors and Council eventually had to abandon all attempts to legislate difference, and that by the 1930s, the islanders were being described as a 'race'.[220] Slavery was eventually abolished in St Helena in 1832, and the children of slave parents became free from 1818 onwards.

Conclusions

The EIC's involvement in the Indian Ocean slave trade neither rivalled that of the Portuguese, Dutch, or French nor approached the numbers involved in the Atlantic trade. Allen's recent estimate suggests that the Company accounted for just 2.2–3.4% of slaves traded in the Indian Ocean by Europeans during the seventeenth century and 1.4–1.7% during the eighteenth.[221] Nevertheless, the Company slaves made up a significant proportion of the population of the early EIC's small settlements.

As the repeated demands for more slaves show, they were considered essential to the success of the early factories. Slaves were trained in crafts and agriculture and often relied on for their linguistic and medical skills. Therefore, transporting slaves was one way for the EIC to transfer skills such as making lime, growing indigo, or the knowledge of languages or medicines between settlements. The movement of unfree labour should therefore be considered central to the 'circulation' involved in making knowledge in the early modern period.[222]

The Company considered 'cultivating' slaves through education and training in professions as equally important to the success of the early plantations as tending the plants that were introduced. However, like the plants that ended up on the island, the introduction of ideas and people was not a simple stream of ideas flowing between England, St Helena, and the EIC's Asian settlements, but instead followed multi-directional currents – people moved in all directions from Africa, Asia, and the Americas as well as Europe, and, like the informal introductions of plants, ideas about the moral, environmental, physical, and spiritual 'government' of the settlements came from a range of places and peoples. Those who inhabited the settlements cannot therefore be considered the passive subjects of a controlled experiment: they also took part in medical, agricultural and social experimentation as they moved between several different locations, occupying different roles, learning new skills, and forging new social ties and practices. An ongoing struggle for control over the physical, moral, and spiritual condition of the settlements is evident in both slaves' resistance to their condition and the response of Company officials.

The EIC records demonstrate that deliberate efforts were made to create racial and linguistic diversity among workers and slaves in the early Company settlements. This worked to gradually decrease the dependence on local rulers described in the introduction and previous chapters, as the EIC was able to build and defend fortified settlements using slave labour. The transportation of slaves in this period set the pattern for the movement of people in the later Company period and during the Raj. In the later British Empire, as in the early period, people and plants were often transferred together, including the 'plantation Tamils' brought to Sri Lanka to work the tea, coffee, rubber, and coconut plantations. Bencoulen and St Helena also provided early grounds for experimentation with these other forms of coerced labour. Bengkulu (Bencoulen) was the first site of a penal colony for Indian convicts from 1787, around the same time that its slaves began to be liberated.[223] Chinese 'coolies' were brought to St Helena from 1810.[224]

While there is long-running debate over the similarities and difference between slavery and indentured labour, it is clear that there was a connection between the two systems in terms of the groups that were targeted and the locations of labour. The EIC's long history of involvement in slavery has been overshadowed by the legacy of the anti-slavery patrols mounted by the British navy from the nineteenth century. These are still described mainly in terms of their humanitarian mission. From 1839, St Helena housed the Liberated African Depot, where liberated slaves were cared for until they were ready to move on. However, these former slaves were not returned home but were either retained on the island as workers or sent to the West Indian colonies. Around 10,000 extra labourers were acquired in this way.[225] Thus, St Helena continued to serve as a place where coerced labour could be contained and trained before being re-circulated. Throughout the records of the early EIC, an expressed humanitarian concern for the welfare of slaves coexisted with a pragmatic recognition of the need for their labour. This did not change in the nineteenth century, when the anti-slavery patrols became a convenient way of tapping into the sources of labour that had previously been more successfully exploited by other European powers. The project of suppressing slavery also enabled empire building in East Africa. As the repeated efforts by the early EIC and their rivals to establish factories in both East and West Africa show, this had been a long-term aim. As this chapter has demonstrated, the experimentation on plants and people in the early EIC settlements set important precedents for later English imperialism.

Conclusion

In around 1720, Jai Singh II of Amber (c. 1686–1743), the Hindu ruler of Jaipur in Rajasthan, began to construct a series of large observatories, or *jantar mantar*, in Jaipur, Delhi, Benares, Mathura, and Ujjain (Figure 7.1).[1] In pursuit of the latest natural philosophical knowledge, Jai Singh gathered around him Muslim astronomers, Hindu pundits, and Jesuit scholars.[2] He sent observers to distant islands to make observations and despatched a delegation to Lisbon to acquire the latest scientific information. Like the Royal Society and their circle, Jai Singh and his assistants amassed and generated copies of earlier works of scholarship including the star catalogues of Ulugh Beg,[3] the grandson of the conqueror Timur, who built an observatory in Samarkand. Jai Singh discusses the building plans for this observatory in his *Zij-i Muhammad Shahi.*[4]

Ulugh Beg's star catalogues had also been the focus of scholarly interest in Europe during the mid-seventeenth century. John of Greaves (Gravius) had acquired a copy in Constantinople during the 1650s and translated parts of his work.[5] Prompted by the interest of the Royal Society,[6] Thomas Hyde made a more extensive translation in 1665, later discussing the astronomical tables with the Chinese visitor Shen Fuzong and comparing them with those given by Nasir al-Din al-Tusi.[7] Jai Singh also commissioned Sanskrit translations of the work of Ptolemy and Euclid from earlier Arabic versions and of other works directly from Latin and Portuguese as well as relying on several earlier works of Hindu astronomy and Arabic and Persian works, also including that of al-Tusi. Works such as *Samrat Siddhanta* of Jai Singh's principal pundit Jagannatha demonstrate extensive engagement with these works as well as measurements that are accurate as any that may be obtained with the unaided eye.[8]

Figure 7.1 The Samrat Yantra at Jaipur

In many ways, the story of Jai Singh's observatories and the contemporary European investigations provide an apt ending to this discussion of hybrid knowledge. The aim of this book has been to produce a specific understanding of the character of late seventeenth- and early eighteenth-century scholarship – which has been described as the scientific revolution and the early Enlightenment – in its interaction with the spaces of commerce, government, and encounter constituted by the early EIC settlements, each understood in their local context. I have emphasised in particular the importance of patronage and clientage in establishing and shaping scholarly, political, and trading relations. I suggested in Introduction that the characterisation of the scholarship of the period as more revolutionary or radical than much of what followed must be considered in global context.

When considering the issue of global contacts, this period should be regarded neither as merely the last phase in an era of adventuring and trade, nor as simply a precursor to the era of colonialism that followed. Of course, it cannot and should not be entirely divorced from either: in some senses it was the culmination of a period that began when an Indian navigator guided the first Portuguese ships across the Indian Ocean from Malindi to Calicut at the end of the fifteenth century, while in others it was a time in which the foundations were laid, not only for colonialism, but also for modern institutions and scholarly practices. To argue for the specificity of this period is not to deny its connections to either the past or the future, but to highlight its particular characteristics. As I have argued throughout, the understanding of knowledge as hybrid, or 'motley', assembled from a series of local ways of knowing, is key to the aim of understanding how the commercial and scientific revolutions can be understood in global context.[9] In characterising this period, it is necessary to understand the particular ways in which local knowledge was assembled into a larger body of knowledge. I suggested in Introduction some general characteristics of knowledge was made in the period. They were summarised as openness, collaborative work, collecting or prospecting, classification, and circulation.

As monuments to both Hindu kingship and contemporary science, Jai Singh's observatories provide a demonstration of the extraordinary degree of openness in this period, in which people from around the world combined diverse forms of knowledge in new configurations for varied audiences. One reason that early modern texts such as those of Knox, Baron, and Bowrey continue even today to be regarded as sources of information about the parts of the world they describe is because they are the result of a high degree of engagement between different cultures that was freer, or even 'wilder' – returning to an early sense of 'hybrid' – than the more structured exchanges of information that took place in the colonial period. The claim for greater openness in the earlier period does not imply fairness or equality in exchanges; as Drayton notes, 'the promise that universal exchange might lead to cosmopolitan utopia was flawed in its earliest age'.[10] Confusion, anxiety, and anomie, as well as greed, not only clouded but often shaped the exchanges that took place in the period.[11] Nevertheless, it was a period in which ideas moved across social, cultural, and ethnic boundaries, often because such boundaries were not yet perceived or constructed. The concept of openness may be linked to that of adaptability: this was a time when people moved in all directions, moulding and reshaping their identities. As Raj puts it, perhaps consciously mirroring the rhetoric of the time, the very nature of the men who made knowledge in this period was mutable.[12] The contemporary tales of Samuel Baron, a Eurasian who fashioned an English identity for himself, and the construction of an Asian identity by the Frenchman who called himself George Psalmanazar provide apt illustrations of this mutability.

Jai Singh's project of gathering around himself scholars of different religious and scholarly traditions illustrates the principle of collaborative work, which follows from the idea of openness. The scale and diversity of the new information gathered meant that to process, analyse, and assess the validity of such information, scholars needed to collaborate. Venues for collaborative scholarship included learned societies such as the Académie des Sciences and the Royal Society, religious organisations such as the Jesuits, and the cosmopolitan courts of rulers such as the Mughal Emperors from Akbar onwards, and, in this period, of King Narai of Siam and Rajasingha II of Kandy. Less formal collaborations also took place in hospitals and apothecaries' shops, markets and bazaars, gardens and plantations, aboard ships, in bookshops, coffee houses, and places of worship.

International networks connected sites of knowledge production, following the routes through which trade goods were distributed. The types of information that were inquired after often crossed what we would now recognise as disciplinary boundaries – bio-prospecting, 'ethno-prospecting', and linguistic prospecting were often undertaken simultaneously: for example, Samuel Browne and Edward Bulkley collected the names of plants as much as the specimens or drawings of the plants themselves, Thomas Bowrey collected plants as well as words for Thomas Hyde, and Samuel Baron collected and copied maps, drawings, and objects. The information produced by collectors in the form of travel texts, natural histories, dictionaries, maps, ethnographic drawings, or books of dried plants was then circulated and re-circulated through networks of trade and diplomacy; being, translated, reinterpreted, and used for the ends of several competing powers. Information also circulated through the movements of people, both free and enslaved.

The overlaps between ways of collecting different types of information fed into common strategies for dealing with the collections that were amassed. This was the early modern project of classification. It was becoming increasingly apparent in this period that to comprehend the diversity of the natural and human world required both large designations and minute distinctions. Both plants and languages began to be classed into 'families', then branches, then individual species or languages, and finally subspecies or dialects. In terms of social relations, the idea of classification can be held up against that of openness to suggest how this mutability of human nature, a product of openness to new ideas, eventually came to be restricted by more rigid concepts of race and nation. Just as the system of 'tribes' of plants based on their humoral qualities began in this period to give way to binomial classification, the idea that the characteristics of humankind could be broadly explained by the climates in which they lived – and thus might be mutable as they moved between 'nations' of people, altering their religion, language, and behaviour – gradually began to give way to more fixed ideas of particular races with inherent characteristics.[13]

An understanding of early modern knowledge must be balanced by an understanding of the production of ignorance or 'agnotology', which also resulted from the friction generated by the global contacts of the period.[14] Each of the attributes of the period listed above could be matched with a negative counterpart. Openness was always accompanied by secrecy and the close guarding of valuable information,[15] collaborative work was limited by competition and the functioning of individual patronage, collecting and prospecting were based on a sense of what was useful to observe and acquire and what was not and classification inevitably privileged some forms of connection over others. In many cases, the failure of knowledge to endure or circulate, or the preservation of ignorance, was the result of the violent situations in which contacts occurred.

In several senses, the history of Jai Singh's observatories is a story of the failure of knowledge to travel. Despite their impressive scale, the scholarly and technical expertise that clearly went into their construction, and the high quality of observations they generated, there are some significant and surprising omissions. For example, none of the work produced around Jai Singh mentions the telescope, although Jesuits had apparently made observations with one in Jaipur during the 1730s.[16] It also neglects to engage with the contemporary work of Newton and Kepler. The grand scale of the instruments in the observatories was perhaps a conscious, but in scientific terms mistaken, attempt to improve on the small European and Arabian equipment Jai Singh was presented with, as well as reflecting Indian cosmological principles.[17]

In turn, the scientific academies of Northern Europe remained unaware of the work of Jai Singh for several decades. An article was eventually published by Robert Barker in *Philosophical Transactions* in 1777 with drawings of the instruments at the Jaipur observatory. However, by this time, the Orientalist myth of a dreamy or stagnant modern India forgetful of a great civilisation of the past[18] was apparently so powerful that Barker accepted without question when he was told that the buildings were ancient monuments rather than contemporary architecture, adding 'arts appear to have declined equally with science in the East'.[19] William Hunter eventually made the correct attribution to Jai Singh over 20 years later, in the fifth volume of the *Asiatick Researches*. Hunter translates sections of the *Zij-i Muhammad* relating to the observatories, gives a detailed commentary on the remaining instruments in each observatory, and notes the accuracy of many of the measurements made under Jai Singh. Despite his generally positive assessment of Jai Singh's science, however, he attributes this to his wisdom in having recognised the superiority of Western science, a sentiment he also attributes to his contemporary, Tafazzul Khan, the translator of Newton's work.

Many of the attempts at cross-cultural communication discussed here seem to reveal similar stories of missed connections. Baron's attempts to mediate between the cultures he spanned often ended in narrow escapes from death; the pioneering translations and dialogue produced by John Marshall and Madhusudana Rarhi remained unpublished despite Locke's efforts to make them known; despite Petiver's ambitions, the collections that the surgeons sent from Madras never resulted in an English- and Tamil-language rival to the *Hortus Malabaricus*; the planned second editions of the *Dictionary English-Malayo, Malayo-English* and the *Historical Relation of Ceylon* never reached the press; and the chair in Malay at Oxford was never established during the eighteenth century. Finally, the long-running project to transplant crops to St Helena failed to yield a profit and continues to this day to wreak environmental devastation on the island, while moves to educate and train slaves were always contradicted by the ultimate aim of exploiting their labour and skills for profit.

The attention to record keeping and the dissemination of information by both the EIC and the Royal Society in London was paralleled by the EIC Director's recognition of the importance of maintaining trade and diplomatic secrets, and of the Royal Society's increasing focus on guarding the flow of information in the interest of safeguarding intellectual property. Within both institutions, the information that was debated openly and recorded was balanced by the oral discussions of select committees. Furthermore, as I argued in Chapter 5, the incentives for publishing texts, maps, and pictures, or the ingredients of a medical prescription were always weighed against the possible commercial and strategic benefits of secrecy and the reputational advantages of being known to possess valuable secrets. Deliberate attempts to smooth out contradictions and present work as the result of direct experience, often in the quest to claim sole authorship with the aim of attracting prestige and patronage, often served to obscure collaboration. As discussed in Chapter 4, the informants of the Madras surgeons slip from view as their contributions are integrated into Petiver's project to map the world's flora and fauna, and as revealed in Chapter 2, the multiple sources of the *Dictionary English-Malayo, Malayo-English* are hidden behind its claim to be the sole product of Bowrey's remembered conversations during a voyage home.

The emerging emphasis on 'useful' knowledge in European thought was paralleled by the assumption that certain forms of knowledge could be considered useless. This shifted according to time and context so that, as discussed in Chapter 3, the texts Locke considered invaluable to contemporary philosophical debates in the late seventeenth century could be dismissed as idle fancies by the early eighteenth century. I have noted throughout that the experiments in cultural *métissage* led not only to peaceful exchanges of ideas but also to violent conflict. Many of the revolutionary scholarly and social experiments discussed here floundered because of the fragility of the networks on which they were based and the turbulent and conflict-ridden situations in which they were created. Often, scholarly collaborations were ended by the death of one party or degenerated into rivalries, while merchants' collections were lost or destroyed in the violent expulsion of EIC factors from yet another settlement where they had upset locals with their drunken rudeness or worse.

Early modern encounters often proved formative of national identities in the modern world.[20] As noted in Chapter 1, the reaction of the rulers of Japan, China, and Siam to the disruptive effects of contact was to limit the access of foreigners to political power and markets while restricting the international travel of their own people. Despite continued cosmopolitan exchanges in Tokugawa Japan, these measures were paralleled by the development of official narratives about self-sufficiency and an unwillingness to interfere in the sovereignty of other nations, policies that later became known as *sakoku*.[21] In China, a similar discourse was maintained even while regional contacts remained strong and the empire expanded under the Qing.[22] The rhetoric of non-interference has remained important to the present day, even while China's influence on the global stage continues to grow. In Europe, by contrast, increasing investment in interventionist overseas policies was backed by investment in military technology, a financial system based on the creation of wealth, after the mainstreaming of 'Dutch finance' from the eighteenth century, and internal competition founded in the possession of overseas territories. This was paralleled by an ideological emphasis on trade as the path to civilised society and the value of empire.

As noted in Chapters 1 and 4, the armies and navies of the European trading powers were unable to take on powerful Asian polities such as

Siam, still less the military might of the Mughal Empire in the late seventeenth century. On the other hand, as noted in Chapter 2, strategic alliances such as those the Dutch built up in Southeast Asia proved effective in destroying powerful enemies such as the Sultan of Bantam. It was through the performance of military services for local rulers that the EIC factories continued to build up their influence in India after 1720.[23] This, added to the circulation of people described in Chapter 6, enabled the settlements to construct forts and establish garrisons composed of soldiers who were increasingly removed from local life and inculcated with new social structures internal to the garrison. These detached and well-drilled soldiers, equipped with the technology of bayonet and flintlock, were eventually able to win battles such as Plassey, often considered the turning point in the transition to European empire in India.[24]

Although it is important to demonstrate the legacies of the early Company, we should be wary of the teleological assumption that the practices and ideologies of this period, whether political or cultural, decisively contributed to the later success of the British Empire. The causes of the Company's success in Asia may be considered contingent at their source. In other words, the EIC's ability to assume increasing governmental powers in India was the consequence of the rise and fall of empires that occurred with no regard to Company policy in London. The circulation of knowledge alone does not lead to power, especially when, as I have argued for this period, the sources of patronage for knowledge brokers are multiple. However, combined with the strategic circulation of people and the specific circumstances in Asia at the time, the types of knowledge making discussed here formed the basis of later colonial strategies of information gathering and policy-making. The fact that the Company in this period laid the foundations for the empire that followed should not lead us to disregard the foundational importance of the discourses that emerged in opposition to those of the political power of private corporations and colonial hierarchies. These were equally important to those that arose in the context of the Atlantic world to later struggles against slavery and colonialism.

Hybrid forms of natural or political knowledge did not cease after the early modern period: indeed, they became more important. As many studies have shown, there was never any such thing as a separate 'imperial' science, only a series of local, albeit networked, colonial sciences. Mapping the new territories of empires depended heavily on the records of earlier administrations as well as local techniques and knowledge.[25]

Physicians worldwide continued to borrow from one another's pharmacopoeia, and botanical works continued to be co-produced,[26] while the illustration of European works on botany by Indian artists became a veritable industry.[27] Collaborative projects of collecting words continued as communication because globalised, and linguistic projects continued to be financed by the EIC and institutions for language learning such as Fort William College were begun under its auspices.[28] Political forms also rested on hybrid organisational and theoretical foundations. Throughout the history of the EIC and the Raj, Europeans continued to enter the service of Asian rulers, and Asian bankers, go-betweens, soldiers, and merchants continued to determine the fate of the British Empire.[29] As Ranajit Guha argues, Indian concepts such as *danda*, *dharma*, and *bhakti*, allied to imperial ideas of order, improvement, and obedience, were central to the conceptual framework maintaining the British Empire's hold over its Asian colonies.[30]

It is indeed possible to go further and to argue, following Latour, that the proliferation of hybrids might be considered the prime characteristic of 'modernity'.[31] However, as Latour also argues, this proliferation is managed by drawing thick dividing lines between nature and culture, marking out the respective territories of science and politics. If the early modern was global, so is modernity: in the core senses of connectivity and co-creation. The colonies have been described as 'laboratories of modernity', in the sense that they functioned as spaces where the mapping of territory, the exploitation of resources, and the surveillance and control of subjects could be tested.[32] In the post-colonial world, even where claims for an 'alternative modernity' or the rejection of modernity exist, they tend to re-frame or critique 'universal' precepts of science and governmentality.

There remains, nonetheless, a popular sense that certain sections of the world might be sectioned off from now, such as the terrorist groups whose actions, coordinated on social media, are misleadingly described as 'mediaeval' by commentators as well as regarded by their supporters themselves as a return to an earlier, idealised, period of history. This imbalance, in which travel across space also becomes time travel as different places are claimed to inhabit different eras, in fact reflects scholarly thought since the Enlightenment, in which modernity is claimed to be a purely Western characteristic, born of rationalism. Such views have often been propounded not only by Western champions of modernity, but as part of the 'self-Orientalizing rhetoric' of Indian, Iranian, and Ottoman historians and the postmodern rejection of modernity.[33] Latour himself is no exception: as Pollock points out, his 'non-modernity' is devoid of specific characteristics, becoming as timeless as that of the Orientalists.[34] Nevertheless, his concept of increasing hybridity accompanied by strict conceptual divisions between the concepts of the natural and the social remains useful in understanding how, while the practice of natural knowledge making became ever more diverse, it was presented in ever-more standardised fashion as the product of rational enquiry detached from social and political context.

The idea of a 'great divide' between East and West, founded on a Europe-centred idea of scientific revolution, has been challenged by the claims for a global early modernity, in which characteristics once considered to define Europe alone were shared or forged in the process of cross-cultural encounters. In addition, it has been pointed out that industrialisation might be regarded as an unintended consequence of a response to particular resource needs rather than a planned developmental strategy.[35] The concept of a great divide was founded in the same Enlightenment ideals of Western progress and rationality, paralleled and supported by narratives of decline and superstition elsewhere, that came to explain and justify the expansion of empire by denying modernity to the rest of the world. Thus, by the mid-eighteenth century, attitudes such as those expressed by Hunter concerning Jai Singh's observatories – the claims that their virtues must have derived either from a forgotten past or from acceptance of the ideas of Western science – were gathering force as strategies to justify colonial rule by denigrating the intellectual capacities of the ruled.

As argued in Chapter 1, the habit of creating inferior 'others' to be colonised was not exclusively a European one in the early modern period, but a common feature of the cultures that came into contact and conflict. However, as Europe became more powerful, so did the stereotypes created there to describe others. Meanwhile, lines of patronage that had translated knowledge into practice were broken as older sources of support for scholars disappeared along with the expansion of European empires. All this meant that publicly challenging such stereotypes became increasingly difficult.[36] Nevertheless, colonial expressions of contempt for other forms of knowledge were imposed over a far more complex reality in which global knowledge continued to be built on hybrid exchanges.

What of those who mediated exchanges of knowledge upon which works of scholarship and state power were founded? The main actors in this book have been cosmopolitan go-betweens, whose loyalties were divided between several local and global sources of patronage. While private trade and private means of acquiring knowledge remained important to EIC servants, because of the military and strategic advantages gained over the course of the eighteenth century, the EIC gradually became able to exert more influence over their servants in Asia by providing more stable sources of patronage.[37] While the profession of go-between remained vital after the period I have discussed here, non-European and subaltern agency in mediating exchanges became progressively less evident, so that the brokers who remained visible to historians by the 'second scientific revolution' of the late eighteenth and early nineteenth centuries were predominantly European and almost invariably literate and male.[38]

I remarked on this process of the progressive obscuring of agency in Chapter 4, noting that the influence of Browne and Bulkley's nameless Tamil and Telugu collaborators on the works of botany that were sent to Petiver is harder to discern than the slightly earlier contributions made to the *Hortus Malabaricus* by Itty Achuden, Ranga Bhatt, Vinayaka Pandit, and Apu Bhat while remaining more apparent than the armies of anonymous Indian collectors and illustrators who collaborated with William Roxburgh and the colonial botanists who followed him. Similarly, the prominent role played by Madhusudana in producing the English version of the *Bhagavata Purana* discussed in Chapter 3 was rarely replicated in later works of Orientalist scholarship, in which the works of numerous 'intellectual labourers' were claimed as the sole discoveries of pioneering Orientalists.[39] Such scholarly go-betweens, deprived of the earlier multiplicity of opportunities for patronage, possessed less power to assert their own agendas than, for example, Samuel Baron had at the beginning of his career.[40]

Despite their growing invisibility, the exchanges of information that went on outside the official purview of colonialism between and beyond the Company's settlements were some of the most important and lasting. I explored some of these in Chapter 6 by discussing the movements of slaves and the techniques they transferred during their transportation, such as the yams brought to St Helena from Madagascar or the rice crops introduced to Aceh from India. Early modern circulations of people provided the template for the movement of workers and convicts in the later British Empire: a movement that continues to impact the modern world, as post-colonial trajectories of migration follow old paths before branching off in new directions. Although the colonial archive is essential for writing the history of those who remain largely unnamed despite their ever-presence in its records, particularly slaves and women, it is also inadequate.[41] Perhaps for this reason, for the most part, only works of fiction have so far been able to provide insights into the lives of the lascars, slaves, coolies, and members of the Company, Americans and Africans as well as Europeans, Asians, and Eurasians.[42] However, studies of material culture and environmental change also have potential for uncovering hidden histories of circulation.

The task that I set out with in this book, to write a cultural history of the production of 'useful' or 'natural' knowledge in the early EIC settlements, is complicated by the fact that the experiences of the members of each of the settlements varied immensely based on their local environment and the particular knowledge networks that connected them to other places. It is through the concept of hybridity – encapsulating diverse information drawn together through ways of knowing that included openness, collaborative work, collecting or prospecting, classification, and circulation – that I have suggested that both the global scientific revolution and the making of knowledge in the early EIC settlements might be understood. 'Useful' knowledge – whether botanical, medical, cartographic, ethnographic, or linguistic – was by its nature not invented by metropolitan elites. Instead, it emerged from the multicultural and multi-directional interactions and conflicts that took place in and between the settlements of the early EIC and its rivals.

Notes

Note on Transcription and Transliteration

1. Michael Hunter, 'How to Edit a Seventeenth Century Manuscript', *Seventeenth Century*, 10 (1995), 277–310.

Introduction

1. John V. Pickstone, *Ways of Knowing: A New History of Science, Technology and Medicine* (Manchester: Manchester University Press, 2000).
2. Steven Shapin, *A Social History of Truth: Civility and Science in Seventeenth-Century England* (Chicago: University of Chicago Press, 1994), xxvi.
3. Nuala C. Johnson, *Nature Displaced, Nature Displayed: Order and Beauty in Botanical Gardens* (London: I. B. Tauris, 2011), 197, argues for botanical gardens as 'hybrid spaces' that challenge ideas of centre and periphery.
4. Paul Gilroy, *The Black Atlantic: Modernity and Double Consciousness* (London: Verso, 1993); Homi K. Bhabha, *The Location of Culture* (London and New York: Routledge, 1994); Robert Young, *Colonial Desire: Hybridity in Theory, Culture, and Race* (London: Routledge, 1995).
5. Daniela Bleichmar, *Visible Empire: Botanical Expeditions and Visual Culture in the Hispanic Enlightenment* (Chicago: The University of Chicago Press, 2012), 13.
6. For a succinct argument for the latter viewpoint, Gyan Prakash, *Another Reason: Science and the Imagination of Modern India* (Princeton: Princeton University Press, 1999), 12–14.
7. An important exception is Minakshi Menon, 'British naturalists in eighteenth century India' (Unpublished doctoral thesis, University of San Diego, CA, 2013). Philip J. Stern, *The Company-State: Corporate Sovereignty and the Early Modern Foundations of the British Empire in India* (New York: Oxford University Press, 2011), 10, notes that 'the governance of the "English" empire, at home and abroad was a layered and hybrid affair, resting on multiple constitutional foundations and constantly negotiated among a variety of royal agencies, local governments, councils, assemblies, courts, and corporate and legal communities'.
8. D. A. Washbrook, 'Orients and Occidents: Colonial Discourse Theory and the Historiography of the British Empire', in William R. Louis, Robin W. Winks, and Alaine M. Low (eds.) *The Oxford History of the British Empire*, 5 vols., *V: Historiography* (Oxford: Oxford University Press, 1999), 609; Sanjay Subrahmanyam, *Mughals and Franks: Explorations in Connected History* (New Delhi: Oxford University Press, 2005).
9. On networks as local at all points, Bruno Latour, *We Have Never Been Modern*, Catherine Porter trans. (Hemel Hempstead: Harvester Wheatsheaf, 1993), 117. For a networked approach to early English oversees expansion, Alison

Games, *The Web of Empire: English Cosmopolitans in an Age of Expansion, 1560–1660* (Oxford: Oxford University Press, 2008).

10. Geoffrey Parker and Lesley M. Smith (eds.) *The General Crisis of the Seventeenth Century* (London: Routledge, 1997); Geoffrey Parker, *Global Crisis: War, Climate Change and Catastrophe in the Seventeenth Century* (New Haven: Yale University Press, 2013); Anthony Reid, *Southeast Asia in the Age of Commerce*, 2 vols., *II: Expansion and Crisis* (New Haven: Yale University Press, 1993). Jan de Vries, 'The Economic Crisis of the Seventeenth Century after Fifty Years', *Journal of Interdisciplinary History* 40:2 (2009), 151–194, links the crisis to the overseas expansion of Northern Europe.
11. John Richards, 'The Seventeenth Century Crisis in South Asia', *Modern Asian Studies*, 24:4 (1990), 625–638.
12. Victor B. Lieberman, *Strange Parallels: Southeast Asia in Global Context, C 800–1830* (New York: Cambridge University Press, 2003).
13. Angela Schottenhammer (ed.) *Trade and Transfer across the East Asian Mediterranean* (Wiesbaden: Harrassowitz Verlag, 2005).
14. Jonathan R. Israel, *Radical Enlightenment: Philosophy and the Making of Modernity, 1650–1750* (Oxford: Oxford University Press, 2001).
15. Sheldon I. Pollock (ed.) 'Introduction', in *Forms of Knowledge in Early Modern Asia: Explorations in the Intellectual History of India and Tibet, 1500–1800* (Durham, NC: Duke University Press, 2011), 1–15.
16. Muzaffar Alam and Sanjay Subrahmanyam, *Indo-Persian Travels in the Age of Discoveries, 1400–1800* (Cambridge: Cambridge University Press, 2007).
17. Preetha Savithri Nair, 'Native Collecting and Natural Knowledge (1798–1832): Raja Serfoji Ii of Tanjore as a "Centre of Calculation" ', *Journal of the Royal Asiatic Society*, 15:3 (2005), 279–302; *Raja Serfoji II: Science, Medicine and Enlightenment in Tanjore* (London: Routledge, 2012).
18. Laura Hostetler, 'Qing Connections to the Early Modern World: Ethnography and Cartography in Eighteenth-Century China', *Modern Asian Studies*, 34:3 (2000), 623–662.
19. Liebermann, *Strange Parallels*; Alan Strathern, 'Sri Lanka in the Long Early Modern Period: Its Place in a Comparative Theory of Second Millennium Eurasian History', *Modern Asian Studies*, 43:4 (2009), 815–869.
20. Sujit Sivasundaram, 'Sciences and the Global: On Methods, Questions, and Theory', *Isis*, 101:1 (2010), 146–158.
21. Examples include Liebermann, *Strange Parallels*; Sheldon Pollock, 'Introduction', in *Forms of Knowledge*; John Richards, 'Early Modern India and World History', *Journal of World History*, 8 (1997), 197–209; Lynn A. Struve (ed.) *The Qing Formation in World-Historical Time* (Cambridge, MA: Harvard University Asia Center, 2004); Sanjay Subrahmanyam, *From the Tagus to the Ganges* (New Delhi: Oxford University Press, 2005).
22. Jack Goody, *Renaissances: The One or the Many?* (Cambridge: Cambridge University Press, 2010).
23. Subrahmanyam, *Mughals and Franks*; Subrahmanyam, *From the Tagus to the Ganges*.
24. Examples include Clare Anderson, *Subaltern Lives: Biographies of Colonialism in the Indian Ocean World, 1790–1920* (Cambridge: Cambridge University Press 2012); Natalie Zemon Davis, *Trickster Travels: A Sixteenth-Century Muslim between Worlds* (New York: Hill and Wang, 2006); Sanjay

Subrahmanyam, *Three Ways to Be Alien: Travails and Encounters in the Early Modern World* (Waltham: Brandeis University Press, 2011).

25. For example, Avner Ben-Zaken, *Reading Hayy Ibn-Yaqzan: A Cross-Cultural History of Autodidacticism* (Baltimore: Johns Hopkins University Press, 2011).
26. Paula Findlen, *Early Modern Things: Objects and Their Histories, 1500–1800* (Abingdon: Routledge, 2013).
27. Washbrook, 'Colonial Discourse Theory', 610.
28. Alan Lester, 'New Imperial and Environmental Histories of the Indian Ocean', in Vinita Damodaran, Anna Winterbottom, and Alan Lester (eds.) *The East India Company and the Natural World* (Basingstoke: Palgrave Macmillan, 2015), 1–15.
29. K. N. Chaudhuri, *The Trading World of Asia and the English East India Company, 1660–1760* (Cambridge: Cambridge University Press, 1978); N. Robins, *The Corporation That Changed the World: How the East India Company Shaped the Modern Multinational* (London and Ann Arbor: Pluto Press, 2006), 15.
30. Alicia Schrikker, *Dutch and British Colonial Intervention in Sri Lanka, 1780–1815: Expansion and Reform* (Leiden: Brill, 2007).
31. Markus Vink, ' "The World's Oldest Trade": Dutch Slavery and Slave Trade in the Indian Ocean in the Seventeenth Century', *Journal of World History*, 14 (2003), 131–177; Kerry Ward, *Networks of Empire: Forced Migration in the Dutch East India Company* (Cambridge: Cambridge University Press, 2009).
32. Harold J. Cook, *Matters of Exchange: Commerce, Medicine, and Science in the Dutch Golden Age* (New Haven: Yale University Press, 2007), 1.
33. David Armitage and Michael J. Braddick, 'Introduction', in Armitage and Braddick (eds.) *The British Atlantic World: 1500–1800* (Hampshire: Palgrave Macmillan, 2002), 1.
34. Gilroy, *The Black Atlantic*; Vincent Brown, *The Reaper's Garden: Death and Power in the World of Atlantic Slavery* (Cambridge, MA: Harvard University Press, 2008); Mary Louise Pratt, *Travel Writing and Transculturation* (Routledge: London, 1992).
35. Judith A. Carney and Richard N. Rosomoff, *In the Shadow of Slavery: Africa's Botanical Legacy in the Atlantic World* (Berkeley, CA: University of California Press, 2011); Judith A. Carney, *Black Rice: The African Origins of Rice Cultivation in the Americas* (Cambridge, MA: Harvard University Press, 2001); James H. Sweet, *Domingos Álvares, African Healing, and the Intellectual History of the Atlantic World* (Chapel Hill: University of North Carolina Press, 2011).
36. Londa Sciebinger, 'Prospecting for Drugs: European Naturalists in the West Indies', in Londa Schiebinger and Claudia Swan (eds.) *Colonial Botany: Science, Commerce and Politics in the Early Modern World* (Pennsylvania: University of Pennsylvania Press, 2005), 119–133; Jorge Cañizares-Esguerra, *Nature, Empire, and Nature: Explorations of the History of Science in the Iberian World* (California: Stanford University Press, 2006). Júnia Ferreira Furtado, 'Tropical Empiricism: Making Medical Knowledge in Colonial Brazil'; Daniela Bleichmar, 'Atlantic Competitions: Botany in the Eighteenth-Century Spanish Empire'; 'Antonio Barrera-Osorio', 'Empiricism in the Spanish Atlantic World', all in Delbourgo and Dew (eds.) *Science and Empire in the Atlantic World.*

37. Neil Safier, *Measuring the New World: Enlightenment Science and South America* (Chicago: University of Chicago Press, 2008).
38. Michael T. Bravo, 'Mission Garden, Natural History and Global Expansion, 1720–1820', in Schiebinger and Swan (eds.) *Colonial Botany*, 49–65.
39. Mordechai Feingold (ed.) *Jesuit Science and the Republic of Letters* (Cambridge MA: MIT Press, 2003); Ines G. Zupanov, *Missionary Tropics: The Catholic Frontier in India (16th–17th Centuries)* (Ann Arbor: University of Michigan Press, 2005); M. N. Pearson, 'The Thin End of the Wedge. Medical Relativities as a Paradigm of Early Modern Indian-European Relations', *Modern Asian Studies*, 29 (1995), 141–170.
40. Joan-Pau Rubiés, *Travel and Ethnography in the Renaissance. South India through European Eyes* (Cambridge: Cambridge University Press, 2005).
41. Steven Shapin, *The Scientific Revolution* (Chicago: University of Chicago Press, 1996).
42. Philip J. Stern, *Company-State*; Philip J. Stern, 'History and Historiography of the English East India Company: Past, Present, and Future!', *History Compass*, 7 (2009), 1146–1180. As Stern notes, this is part of the wider issue of the separation of the study of imperialism from British history, which still tends to be regarded mainly as springing from Anglo-Saxon, classical and European precedents rather than shaped by the colonial experience. See also David Armitage (ed.) *Ideological Origins of the British Empire* (Cambridge: Cambridge University Press, 2000).
43. Chaudhuri, *The Trading World of Asia and the EIC*, 26, fig. 2.
44. Richard Hakluyt, *The Principall Nauigations, Voiages and Discoueries of the English Nation* (London: by George Bishop and Ralph Newberie, deputies to Christopher Barker, printer to the Queenes most excellent Maiestie, 1589); William Noel Sainsbury (ed.) 'East Indies: January 1601', in *Calendar of State Papers Colonial, East Indies, China and Japan, Volume 2: 1513–1616* (1864), 118–121. http://www.british-history.ac.uk/cal-state-papers/colonial/east-indies-china-japan/vol2/pp118-121 [accessed 30 October 2015].
45. Josiah Child, *A New Discourse of Trade* (London: John Everingham, 1693), *Selected Works 1668–1697: A Collection of Seven Rare Works by, or Attributed to, Sir Josiah Child, Republished from Originals in the Goldsmiths' Library of Economic Literature* (Farnborough: Gregg, 1968).
46. Ashin Das Gupta and Michael Pearson (eds.) *India and the Indian Ocean, 1500–1800* (Calcutta: Oxford University Press, 1987); Michael Pearson, *The World of the Indian Ocean, 1500–1800: Studies in Economic, Social and Cultural History* (Burlington: Ashgate, 2005); Sinnappah Arasaratnam, *Maritime India in the Seventeenth Century* (Delhi: Oxford University Press, 1994) and *Maritime Trade, Society and European Influence in South Asia, 1600–1800* (Aldershot: Variorum, 1995); K. N. Chaudhuri, *The English East India Company: The Study of an Early Joint-Stock Company 1600–1640* (London: Frank Cass, 1965); K. N. Chaudhuri, *The Trading World of Asia and the English East India Company, 1660–1760* (Cambridge: Cambridge University Press, 1978); K. N. Chaudhuri, *Trade and Civilisation in the Indian Ocean: An Economic History from the Rise of Islam to 1750* (Cambridge: Cambridge University Press, 1985); K. N. Chaudhuri, *Asia before Europe: Economy and Civilisation of the Indian Ocean from the Rise of Islam to 1750* (Cambridge: Cambridge University Press, 1990). For a critique of Chaudhuri's 'slavish Braudelianism', see

Subrahmanyam's introduction to Holden Furber, Sinnappah Arasaratnam, and Kenneth McPherson, *Maritime India* (Delhi: Oxford University Press, 2004).

47. Kenneth McPherson, *The Indian Ocean: A History of People and the Sea* (Delhi: Oxford University Press, 1998), 137.
48. Miles Ogborn, *Indian Ink: Script and Print in the Making of the English East India Company* (Chicago: University of Chicago Press, 2007); Philip J. Stern, ' "A Politie of Civill & Military Power": Political Thought and the Late Seventeenth-Century Foundations of the East India Company-State', *Journal of British Studies*, 47 (2008), 253–283; Stern, 'Politics and Ideology in the Early East India Company-State: The Case of St. Helena, 1673–1696', *Journal of Imperial and Commonwealth History*, 35 (2007), 1–23.
49. Miles Ogborn, *Indian Ink: Script and Print in the Making of the English East India Company* (Chicago: University of Chicago Press, 2007).
50. Stern, *Company State.*
51. Miles Ogborn, 'Writing Travels: Power, Knowledge and Ritual on the English East India Company's Early Voyages', *Transactions of the Institute of British Geographers, New Series*, 27 (2002), 155–171.
52. Stern, ' "A Politie of Civill & Military Power" ', fn. 5.
53. Bhavani Raman, *Document Raj: Writing and Scribes in Early Colonial South India* (Chicago: University of Chicago Press, 2012).
54. Chaudhuri, *The Trading World of Asia and the EIC*, 30, admits that the Company's aims were not always implemented in Asia but blames this on problems of communication and discipline.
55. Stern, *Company State*, 6.
56. S. Mentz, *The English Gentleman Merchant at Work: Madras and the City of London 1660–1740* (Copenhagen: Museum Tusculanum Press, University of Copenhagen, 2005); H. V. Bowen, *Elites, Enterprise, and the Making of the British Oversees Empire 1688–1775* (Macmillan: London, 1993).
57. Julia Adams, *The Familial State: Ruling Families and Merchant Capitalism in Early Modern Europe* (Ithaca: Cornell University Press, 2005).
58. Shafaat Ahmad Khan, *The East India Trade in the 17th Century in Its Political and Economic Aspects* (London: Oxford University Press, 1923), 82, notes that the Governor of the Company informed the King in 1641 that the Company's debt meant that the trade 'is likely to come to a standstill'. See also Chaudhuri, *The English East India Company*, 73.
59. H. V. Bowen, ' "No Longer Mere Traders": Continuities and Changes in the Metropolitan Development of the East India Company, 1600–1834', in Bowen, Lincoln and Rigby (eds.) *The Worlds of the East India Company*, 19–32, 25; P. J. Marshall, 'The English in Asia', in Nicholas P. Canny and Alaine M. Low (eds.) *The Origins of Empire: British Overseas Enterprise to the Close of the Seventeenth Century* (Oxford: Oxford University Press, 2001), 264–285.
60. P. Linebaugh and M. Rediker, *The Many-Headed Hydra: Sailors, Slaves, Commoners, and the Hidden History of the Revolutionary Atlantic* (Boston: Beacon Press, 2000).
61. Shafaat Ahmad Khan, *The East India Trade*, 82, fn. 2.

62. British Library (BL) India Office Records (IOR) Original Correspondence (OC), no. 3074, fol. 44, 'Deposition by Reverend Simon Smith against Nathaniel Foxcroft', Madras, 13 September 1665.
63. IOR OC 3088, fol. 58, 'Proceedings of an assembly of EIC servants at Madras', 19 September 1665.
64. IOR E/3/29 3095, fol. 71, 'Summary of agreement between Mirza Malik Beg, Governor of Hugli and VOC director Arnoldus van Wachtendonck, 24 September 1665'. For Mirza Malik Beg and his relationship with the VOC and EIC, Om Prakash, 'The European Trading Companies and the Merchants of Bengal 1650–1725', *Indian Economic and Social History Review*, 1:3 (1964), 37–63.
65. I. B. Watson, 'Winter, Sir Edward (1621/2–1686)', in *Oxford Dictionary of National Biography* (Oxford University Press: 2004) http://0-www.oxforddnb.com.catalogue.ulrls.lon.ac.uk:80/view/article/29763 (accessed 30 May 2007).
66. Love, Vestiges, 255. Henry Yule (ed.) *The Diary of William Hedges, Esq. (afterwards Sir William Hedges), During His Agency in Bengal: As Well As on His Voyage Out and Return Overland (1681–1687)*, 3 vols. (London: Hakluyt Society, 1887), II, cclxxxi.
67. Colin Newbury, *Patrons, Clients and Empire: Chieftaincy and Over-Rule in Asia, Africa, and the Pacific* (Oxford: Oxford University Press, 2003), 12.
68. Stern, *Company State.*
69. Mir Jumla II or Muhammad Said Ardestani was a former member of the court of Golconda who defected to the Mughals; during 1658, he was occupying Bihar and cutting off the Company's salt supply as a result of the dispute with Winter (Love, *Vestiges*, 1, 165 and 184–185).
70. EIC in London to Masulipatam, 12 October 1659; cf. William Foster, *The English Factories in India, 1655–1660* (Oxford: Clarendon Press, 1921), 270.
71. See, for example, Khan, *East India Trade*, 204: 'The Dutch and the Interlopers harassed the Company's factors, preyed upon its commerce, and intrigued against its servants with the Mogul's Governors. There is reason to think some of them were utterly corrupt, intensely partial, and totally negligent.'
72. Yule, *Diary of William Hedges, III: Documentary Contributions towards a Biography of William Pitt*, xxiv–xxv.
73. Bod. MS Rawl. 302, fol. 186, Abstracts of letters relevant to negotiations between the two companies, July 1699–January 1700. See also Yule, *Diary of William Hedges*, III, li.
74. MS Rawl. A. 302, fols. 190–195.
75. National Archives of India (NAI) Foreign Department, vol. 4, fols. 53–5, 30 May 1701 Surat to Bombay ('old' Company). The 'new' Company outward letter book of the period (Maharastra State Archives (MSA), 'Surat Diaries', vol. 2, 1699–1707) complains that their presents to the Emperor had been ridiculed.
76. See H. Das, S. C. Sarkar, and L S. Amery, *The Norris Embassy to Aurangzib: (1699–1702)* (Calcutta: Firma K. L. Mukhopadhyay, 1959), 48, for the declaration of 'civil war' in a letter of the 'old' Company to their Bengal factory.
77. Das, Sarkar, and Amery, *Norris Embassy*, 52 and 58–59.

78. MSA, volume catalogued as Surat, 'Diaries', II, 1669–1707, but in fact an 'outward' letter book of the 'new' Company beginning in 1699, fols. 345–349, letter dated 26 and 31 May 1704.
79. IOR A/1/40, Letters patent of James II, 12 April 1686.
80. For example, Michael C. W. Hunter (ed.) *Archives of the Scientific Revolution: The Formation and Exchange of Ideas in Seventeenth-Century Europe* (Woodbridge: Boydell Press, 1998).
81. A transcription of the 1662 charter and an English translation can be downloaded from The Royal Society, 'Royal Charters' https://royalsociety.org/about-us/history/royal-charters/ (accessed 14 October 2014).
82. Minutes of Council, vol. 2, 6 December 1682.
83. Lorenzo Magalotti, *Travels of Cosmo the Third, Grand Duke of Tuscany, through the England during the Reign of King Charles the Second (1669)*... (London: J. Mawman, 1821). I am grateful to Dr Antonia Moon for this reference. Presumably, the main collection was housed in Craven House, where the Company moved their headquarters in 1666 (William Foster, *The East India House* (London: Lane, 1924)).
84. Nehemiah Grew, *Musæum Regalis Societatis, or a Catalogue & Description of the Rarities Belonging to the Royal Society & Preserved at Gresham College* (London: W. Rawlins, 1681); Michael Hunter (ed.) 'Between Cabinet of Curiosities and Research Collection: The History of the Royal Society's "Repository" ', in *Establishing the New Science: The Experience of the Royal Society* (Woodbridge: Boydell, 1989), 123–155; Jenni Thomas, 'A "philosophical storehouse": The life and afterlife of the Royal Society's repository' (Unpublished doctoral thesis, Queen Mary, University of London, 2009).
85. Royal Society, Minutes of Council, I, meeting of 24 June 1675, fol. 262. Also noted by Jenni Thomas, 'A philosophical storehouse', 19.
86. Robert Hooke, 'Hooke folio', transcribed and digitised by the Centre for Editing Lives and Letters and the Royal Society, 'Hooke Folio', f. 441 (CELL/RS/HF_441). The Hooke folio is a recently rediscovered manuscript containing Robert Hooke's copies from the records of meetings of the Royal Society. It can be viewed online at University College, University of London, Hooke Folio, http://www.livesandletters.ac.uk/projects/hooke-folio-online (accessed 16 September 2014).
87. Thomas Sprat, *The History of the Royal Society of London, for the Improving of Natural Knowledge* (London: J. Martyn and others, 1667).
88. Michael C. W. Hunter, *Establishing the New Science: The Experience of the Early Royal Society* (Suffolk: Boydell, 1989); Marie B. Hall, *Promoting Experimental Learning: Experiment and the Royal Society 1660–1727* (Cambridge: Cambridge University Press, 1991).
89. The word 'bezoar' comes from the Persian *pādzahr*. Originally signifying an antidote to poison, the word came to refer specifically to the hard concretions found in the bodies of animals, which were applied to snake wounds, perhaps giving rise to the false belief about them being found in the heads of snakes. Henry Yule, *Hobson-Jobson: A Glossary of Colloquial Anglo-Indian Words and Phrases, and of Kindred Terms, Etymological, Historical, Geographical and Discursive*, new edn., ed. William Crooke (London: J. Murray, 1903), online edition by Digital Dictionary of South Asia http://

dsal.uchicago.edu/dictionaries/hobsonjobson/ (accessed 28 January 2015) (hereafter *Hobson-Jobson*), 90.

90. CELL/RS/HF_447. No date, page headed 'Answer to the East India company concerning the stone sent by them to the Royal Society'. Not in Hooke's hand (possibly that of Mr Aston, who is mentioned as reading the account of the stone sent to the EIC in HF 449, 1 July 1680).
91. Martha Baldwin, 'The Snakestone Experiments: An Early Modern Medical Debate', *Isis*, 86 (1995), 394–418.
92. H. Oldenburg, 'An Account of Some Books; 1. The History of the Royal Society of London for the Advancement of Experimental Philosophy, by Tho. Spratt', *Philosophical Transactions*, 2 (1667), 502.
93. A. Cook, *Edmond Halley: Charting the Heavens and the Seas* (Oxford: Clarendon Press, 1998).
94. Geographical works which relied on collecting observations include Moses Pitt, *The English Atlas* (Oxford: 1680–1683); Nathaniel Cutler and Edmond Halley, *Atlas Maritimus & Commercialis* (London: James & John Knapton, etc., 1728); and John Flamsteed, *Atlas Coelestis*, ed. Margaret Flamsteed and James Hodgson (London: 1729).
95. Shapin, *Social History of Truth*; Jan Golinski, *Making Natural Knowledge: Constructivism and the History of Science*, 2nd edn. (Chicago: The University of Chicago Press, 2005).
96. Cook, *Matters of Exchange*.
97. Pamela H. Smith and Paula Findlen (eds.) *Merchants & Marvels: Commerce, Science and Art in Early Modern Europe* (New York: Routledge, 2002), 4.
98. For the spices, Royal Society, *Journal Book* (*JB*), VII, fol. 108–109, 6 December 1682; for Taiwanese writing *JB*, VII, fol. 297, 25 February 1684. *JB*, XI, fol. 1–2, 28 October 1702, contains discussions of cinnabar from China, fixatives for dyes from the East Indies, and balances from China.
99. [Anon], 'A Narrative Concerning the Success of Pendulum-Watches at Sea for the Longitudes', *Philosophical Transactions*, 1 (1665–1666), 13–15, published 1 January 1665.
100. CELL/RS/HF_433, 22 April 1680.
101. John Gascoigne, *Science in the Service of Empire: Joseph Banks, the British State and the Uses of Science in the Age of Revolution* (Cambridge: Cambridge University Press, 1998); J. E. McClellan, III, and F. Regourd, 'The Colonial Machine: French Science and Colonization in the Ancien Régime', *Osiris*, 15 (2000), 31–50; Cook, *Matters of Exchange*.
102. Jan Golinski, *Making Natural Knowledge: Constructivism and the History of Science*, 2nd edn. (Chicago: The University of Chicago Press, 2005).
103. J.-P. Rubiés, 'Instructions for Travellers: Teaching the Eye to See', *History and Anthropology*, 9:2–3, (1996), 139–190; D. Carey, 'Compiling Nature's History: Travellers and Travel Narratives in the Early Royal Society', *Annals of Science*, 54 (1997), 23–50; David S. Lux and Harold J. Cook. 'Closed Circles or Open Networks: Communicating at a Distance during the Scientific Revolution', *History of Science*, 36 (1998), 179–211.
104. Christopher Bayly, *Empire and Information: Intelligence Gathering and Social Communication in India, 1780–1870* (Cambridge: Cambridge University Press, 1996).

105. Examples include David Arnold, *Colonising the Body: State Medicine and Epidemic Disease in Nineteenth-Century India* (Berkeley: University of California Press, 1993); Bernard S. Cohn, *Colonialism and Its Forms of Knowledge: The British in India* (Princeton: Princeton University Press, 1996); Matthew H. Edney, *Mapping an Empire: The Geographical Construction of British India, 1765–1843* (Chicago: University of Chicago Press, 1997).
106. R. W. Home, 'The Royal Society and the Empire: The Colonial and Commonwealth Fellowship Part 1. 1731–1847.' *Notes and Records*, 56:3 (2002), 307–332.
107. George Basalla, 'The Spread of Western Science', *Science*, 156 (1967), 611–622, is often cited as a classic example of the earlier diffusionist model.
108. Christopher Bayly, *Imperial Meridian: The British Empire and the World, 1780–1830* (London: Longman, 1989); Kapil Raj, 'Colonial Encounters and the Forging of New Knowledge and National Identities: Great Britain and India', *Osiris*, 15 (2000), 119–134; Kapil Raj, *Relocating Modern Science: Circulation and the Construction of Knowledge in South Asia and Europe, 1650–1900* (Hampshire: Palgrave Macmillan, 2007).
109. Bruno Latour, *Science in Action: How to Follow Scientists and Engineers through Society* (Cambridge, MA: Harvard University Press, 1987), 221–233.
110. Michael T. Bravo, 'Ethnographic Navigation and the Geographical Gift', in David N. Livingstone and Charles W. J. Withers (eds.) *Geography and Enlightenment* (Chicago: University of Chicago Press, 1999), 199–235.
111. Raj, *Relocating Modern Science.*
112. Robins, *The Corporation That Changed the World*, 15.
113. For a recent networked approach to the history of the British Empire, Alison Games, *The Web of Empire: English Cosmopolitans in an Age of Expansion, 1560–1660* (Oxford: Oxford University Press, 2008).
114. Gilroy, *Black Atlantic*, 39.
115. St Helena Archives (SHA), *Letters Sent 1716–1717*, fol. 57–57a, 1 July 1717, Memorandums for Capt. Graves relating to the yams and potato plants.
116. Stefan Halikowski Smith, 'Languages of Sub-Alternity and Collaboration: Portuguese in English Settlements across the Bay of Bengal, 1620–1800', in Kingsley Bolton, Samuli Kaislaniemi, and Anna Winterbottom (eds.) *The East India Company and Language in Asia* (Cambridge: Cambridge University Press, forthcoming 2016).
117. John Keay, *The Honourable Company: A History of the English East India Company* (New York: Macmillan Publishing Co., 1994), 134.
118. IOR E/3/90 fol. 80–80r. General letter to Surat – 14 July 1686: the agents are told to follow the Dutch example of not allowing these married women to wear European dress unless they become wealthy and pay a 'fine' to the Company to allow them to do so.
119. IOR E/3/91 fol. 144r, Letter to Fort St George, 8 April 1687.
120. Charles Lockyer, *An Account of the Trade in India* (London: printed for the author, and sold by Samuel Crouch, 1711), 14–15; cf. Halikowski Smith, 'Languages of Sub-Alternity and Collaboration'.
121. Vahe Baladouni and Margaret Makepeace, *Armenian Merchants of the Seventeenth and Early Eighteenth Centuries: English East India Company Sources* (Philadelphia: American Philosophical Society, 1998).

122. For example, Bhaswati Bhattacharya, 'Armenian European Relationship in India, 1500–1800: No Armenian Foundation for European Empire?', *Journal of the Economic and Social History of the Orient*, 48:2 (2005), 277–322, notes that although the presence of the European factories boosted Armenian trade in India, 'Armenian trade, based to a great extent on various forms of community-based network and partnership, was not "exclusive" in nature'.
123. Most recently, S. Greenblatt (ed.) *Cultural Mobility: A Manifesto* (Cambridge: Cambridge University Press, 2010).
124. Holden Furber, *Private Fortunes and Company Profits in the India Trade in the 18th Century* (Brookfield: Variorum, 1997).
125. For a recent comparative approach, see Maxine Berg (ed.) *Goods from the East, 1600–1800: Trading Eurasia* (Basingstoke: Palgrave Macmillan, 2015).
126. R. Guha, 'Not at Home in Empire', *Critical Inquiry*, 23:3 (1997), 482–493.

1 Curious Collectors and Infamous Interlopers: Samuel Baron (fl. 1645–1695) and the EIC Settlements in Southeast and East Asia

1. Samuel Baron (1686), 'A Description of the Kingdom of Tonqueen', in Awnsham and J. Churchill (eds.) *A Collection of Voyages and Travels*, 6 vols. (London: printed for the editors, 1705–1732), VI, 1732, 1–40. Reprinted in Olga Dror and K. W. Taylor, *Views of Seventeenth Century Vietnam* (Ithaca, NY: Cornell Southeast Asia Program, no. 41, 2006). Page numbers given below refer to the Dror and Taylor edition unless otherwise indicated.
2. Sloane 1039, fol. 133r. Samuell Barron to Sir John Hoskins and Mr Robert Hooke, 2 February 1686. I am grateful to William Poole for his transcription of this letter.
3. BL IOR E/3/91, fol. 71 (143), London to Fort St George – 6 June 1686.
4. Jonathan D. Spence, 'The K'ang-hsi Reign', in *The Ch'ing Empire to 1800* ed. Willard J. Peterson, Series: The Cambridge History of China, 9(1) (Cambridge: Cambridge University Press, 2002).
5. For the profits made by the Portuguese as go-betweens in this trade in the period 1570–1620, see John E. Wills, 'Maritime China from Wang Chih to Shih Lang', in Jonathan D. Spence and John E. Wills (eds.) *From Ming to Ch'ing: Conquest, Region, and Continuity in Seventeenth Century China* (New Haven and London: Yale University Press, 1979), 213.
6. Derek Massarella, 'Chinese, Tartars and "Thea" or a Tale of Two Companies: The English East India Company and Taiwan in the Late Seventeenth Century', *Journal of the Royal Asiatic Society*, Third Series, 3 (1993), 393–426.
7. Wills, 'Maritime China'.
8. For the collections of scientific equipment including telescopes, timepieces, globes, and astronomical tables amassed by King Narai of Siam, see Ian Hodges, 'Western Science in Siam', *Osiris*, 13 (1998), 80–95. For a comparison with southern India in the later Enlightenment period, Nair, *Raja Serfoji*.
9. Royal Society, *Record Book*, VI, fols. 143–144, 23 January 1684, records Hooke experimenting with Japanese scales and weights. For the Chinese compass, see Chapter 2 and Timothy Brook, *Mr Selden's Map of China: The*

Spice Trade, a Lost Chart and the South China Seas (Profile Books: London, 2013), 100–105.

10. Timon Screech, ' "Pictures (the Most Part Bawdy)": The Anglo-Japanese Painting Trade in the Early 1600s', *Art Bulletin*, 87 (2005), 50–72; Ashok Kumar Srivastava, *Mughal Painting: An Interplay of Indigenous and Foreign Traditions* (Delhi: Munshiram Manoharlat, 2000). Ronald S. Love, 'Monarchs, Merchants, and Missionaries in Early Modern Asia: The Missions Étrangères in Siam, 1662–1684', *International History Review*, 21 (1999), 1–27; Ito Shiori, 'Western and Chinese Influences on Japanese Painting in the Eighteenth Century', in Masashi Haneda (ed.) *Asian Port Cities, 1600–1800: Local and Foreign Cultural Interactions* (Tokyo: University of Tokyo, 2009); Donald S. Lach and Edwin J. van Kley, *Asia in the Making of Europe,* 2 vols. in 5 books (Chicago: University of Chicago Press, 1965–1970), I.2: *A Century of Wonder* (1970), 73–74.
11. Love, 'Monarchs, Merchants, and Missionaries', 23. For demands made by the Brahmins at the court of the King of Golconda for animal products including a civet cat and alligator gall, IOR G/19/26, [n.f.], dated 2 September 1670.
12. IOR G/21/7, fols. 36–37 Siam to Bantam, 28 November 1678. See also IOR G/21/6(1) fol. 100r, transcribed in Anthony Farrington, Hsiu-Jung Chang, Huang Fu-San and Wu Mi-Tsa (eds.) *The English Factory in Taiwan, 1670–1685* (Taipai: National Taiwan University, 1995), 393–394.
13. James Delbourgo, 'Exceeding the Age in Every Thing: Placing Sloane's Objects', *Spontaneous Generations: A Journal for the History and Philosophy of Science*, 3 (2009), 41–54.
14. Cook, *Matters of Exchange*, 17.
15. Willis, *Maritime China*, 219.
16. Geraldine Barnes, 'Curiosity, Wonder, and William Dampier's Painted Prince', *Journal for Early Modern Cultural Studies*, 6:1 (2006), 31–50.
17. Sanjay Subrahmanyam, 'Connected Histories: Notes towards a Reconfiguration of Early Modern Eurasia', *Modern Asian Studies*, 31 (1997), 735–762; Alam and Subrahmanyam, *Indo-Persian Travels*; Hostetler, 'Qing Connections'.
18. Bayly, *Empire and Information.*
19. Louse Bénat Tachot and Serge Grunzinski (eds.) *Passeurs culturels: mécanismes de métissage* (Paris: Presses universitaires de Marne-la-Vallée & Éditions de la Maison des Sciences de l'Homme, 2001); Innes Zupanov, 'Goan Brahmans in the Land of Promise: Missionaries, Spies and Gentiles in Seventeenth- and Eighteenth-Century Sri Lanka', in Jorge Flores (ed.) *Re-exploring the Links: History and Constructed Histories between Portugal and Sri Lanka*, South China and Maritime Asia Series (Wiesbaden: Harrassowitz, cop. 2007), 171–210; Simon Schaffer, Lissa Roberts, Kapil Raj, and James Delbourgo (eds.) *The Brokered World: Go-betweens and Global Intelligence, 1770–1820* (Sagamore Beach, MA: Science History Publications, 2009).
20. Minhao Zeng, 'Subaltern Cosmopolitanism: Concept and Approaches', *Sociological Review*, 62:1 (2014), 137–148; Anderson, *Subaltern Lives.*
21. Raj, 'Mapping Knowledge Go-Betweens in Calcutta, 1770–1820', in Schaffer, Roberts, Raj, and Delbourgo (eds.) *The Brokered World,* 105–150, 111.

22. 'Introduction', in Schaffer, Roberts, Raj, and Delbourgo (eds.) *The Brokered World*, ix–xxxviii.
23. Gruzinski, 'Un honnête homme, c'est un homme mêle', in Bénat Tachot and Gruzinski (eds.) *Passeurs Culturels*, 1–22, for the potentials and limitations of concepts such as 'hybridism', 'melange', and cultural and biological mixing.
24. Stephen Greenblatt, *Renaissance Self-Fashioning from More to Shakespeare* (Chicago: University of Chicago Press, 1980), 3.
25. Stephen Greenblatt, *Marvelous Possessions: The Wonder of the New World* (Chicago: University of Chicago Press, 1991).
26. Anthony Pagden, *European Encounters with the New World: From Renaissance to Romanticism* (New Haven: Yale University Press, 1993).
27. Shapin, *Social History of Truth*, 126–192.
28. Zemon Davis, *Trickster Travels.*
29. Kumari Jayawardena, *Erasure of the Euro-Asian: Recovering Early Radicalism and Feminism in South Asia* (Colombo: Social Scientists' Association, 2007); Amrita Sen, 'Early Liaisons: East India Company, Native Wives and Inscription in the Seventeenth Century', *South Asian Review*, 33:2 (2012), 101–116; Ann L. Stoler, *Carnal Knowledge and Imperial Power: Race and the Intimate in Colonial Rule* (Berkeley: University of California Press, 2002).
30. Serge Gruzinski, *The Mestizo Mind: The Intellectual Dynamics of Colonization and Globalization* (New York: Routledge, 2002).
31. Anthony Reid, 'Merchant Princes and Magic Mediators', *Indonesia and the Malay World* 36:105 (2008): 253–267, 259, gives the example of Francis L'Etoile, a Eurasian born or naturalised in the Danish settlement of Tranquebar who entered the service of Sultan Jauhar al-Alam. Nair, *Raja Serfoji*, xxx, notes that Eurasians played an important role as experts in the Raja's court.
32. David Lambert and Alan Lester (eds.) *Colonial Lives across the British Empire: Imperial Careering in the Long Nineteenth Century* (Cambridge: Cambridge University Press, 2006); Anderson, *Subaltern Lives.*
33. Schaffer, Roberts, Raj, and Delbourgo, 'Introduction', in Schaffer, Roberts, Raj, and Delbourgo (eds.) *The Brokered World*, xxii.
34. Insun Yu, *Law and Society in Seventeenth Century Vietnam* (Asiatic Research Centre, Korea University, 1990).
35. Baron, *Description*, 210. For a Chinese description, Arnold J. A. Vissière, 'Ngan-Nan Ki Yeou: relation d'un voyage au Tonkin', *Bulletin de Geographie Historique et Descriptive* 4 (1889), 70–86.
36. W. J. M. Buch, 'La Compagnie des Indes néerlandaises et l'Indochina', *Bulletin de l'Ecole français d'Extrême-Orient*, 36 (1936), 97–196, 103–104.
37. Alexandre de Rhodes, *Histoire du Royuame du Tonkin* ed. Jean-Pierre Duteil (Paris: Éditions Kimé, 1999).
38. Buch, 'II. La Compagnie des Indes néerlandaises et l'Indochine', *Bulletin de l'Ecole française d'Extrême-Orient*, 37 (1937) 121–237, and Maybon, 'Un factorie Anglais', 169, fn. 1.
39. Buch, 'La Compagnie des Indes néerlandaises et l'Indochina'.
40. Sen, 'Early Liaisons'; Timothy P. Barnard, 'Metizos as Middlemen: Tomas Días and His Travels in Eastern Sumatra', in Peter Borschberg (ed.) *Iberians*

in the Singapore-Melaka Area (16th to 18th Century) (Weisbaden: Harrassowitz, 2004), 147–160.

41. William Dampier, *Voyages and Descriptions Vol. II, in Three Parts* (printed for James Knapton, London, 1700), 51, 60; Insun Yu, *Law and Society in Seventeenth Century Vietnam* (Asiatic Research Centre, Korea University, 1990).
42. John Kleinen, *Lion and Dragon: Four Centuries of Dutch-Vietnamese Relations* (Amsterdam: Boom, 2008), 58.
43. Roger Machin and Shimizu Hirokazu (eds.) *Experiment and Return. Documents Concerning the Japan Voyage of the English East India Company, 1671–1673* (Kyoto: Richard Cocks Society, 1978), 51.
44. Maybon, 'Un factorie anglais', 169, fn. 1.
45. Baron, *Description*, 205, 213, 223, 244, and 245.
46. Court Minutes, 3 March 1670, vol. 27, fol. 104, transcribed in Peter Pratt, *History of Japan, Compiled from the Records of the English East India Company at the Instance of the Court of Directors* (London: Curzon Press, 1972, first published in Kobe, 1931), 146–147.
47. Court Minutes, 10 March 1670–1671, vol. 27, fol. 105b, fol. 106, transcribed in Pratt, *History of Japan*, 147.
48. Derek Massarella, *A World Elsewhere: Europe's Encounter with Japan in the Sixteenth and Seventeenth Centuries* (New Haven: Yale University Press, 1990); Machin and Hirokazu, *Experiment and Return.*
49. IOR E/3/33 fol. 42 (OC 3646) S. Baron at Bantam to London, 4 June 1672 and Sloane 998, fol. 17r, Sunday 18 August.
50. Instructions from Henry Dacres and Council at Bantam to David Stephens, Samuel Baron, Simon Delboe and other factors for their settlement at Taiwan, 9 June 1672, repr. in Farrington et al. *English Factory in Taiwan*, 121–132, 131.
51. Anthony Farrington and Dhiravat na Pombejra, *The English Factory in Siam 1612–1685*, 2 vols. (London: The British Library, 2007), I, 383–407.
52. IOR G/12/17, fol. 9v.
53. Sloane 998, fol. 27r.
54. Sloane 998, fol. 42v.
55. Machin, *Experiment and Return.*
56. IOR E/3/36, OC 4105, Henry Dacres and Council to Bombay, Bantam, 11 August 1675
57. IOR E/3/33, fol. 42, OC 3646.
58. London to Bantam 'Sent on the Experiment Return & Zant Frigatt, London 21 September 1671'. Transcribed in Machin, *Experiment and Return*, 51. 'Portuguese' or 'Black Portuguese' was commonly used in the English records of the period to refer to people of mixed Portuguese and Asian descent.
59. IOR E/3/33, fol. 42, OC 3646.
60. Machin, *Experiment and Return*, 51.
61. Maybon, 'Un factorie anglais', 169, fn. 1.
62. Kleinen, *Lion and Dragon*, 57.
63. Michael Keevak, *The Pretended Asian: George Psalmanazar's Eighteenth-Century Formosan Hoax* (Detroit: Wayne State University Press, 2004).
64. Keevak, *Pretended Asian*, 12.

65. Keevak, *Pretended Asian*, 13.
66. On seventeenth-century identities as 'fluid constructions' in the Indian context, Gijs Kruijtzer, *Xenophobia in Seventeenth-Century India* (Leiden: Leiden University Press, 2009).
67. IOR E/3/29, OC 3144, fols. 152–162.
68. SHA, *Consultations*, II, fol. 175, 8 June 1685.
69. Henry W. Robinson and Walter Adams (eds.) *The Diary of Robert Hooke, M.A., M.D., F.R.S., 1672–1680* (London: Wykeham Publications, 1968), 293, Tues, 29 May 1677 – 'at Toothes with Mr. Barron of Funquin'; 314, Fri 21 Sept 1677 for the debate about petrification and Sat 22 Sept 1677 for the plan to visit Boyle. Boyle's work diaries mention both a conversation concerning petrification on the island of Hainan and a conversation about the height of the elephants of the King of Tonkin with one well travelled in the region. These entries conceivably refer to Baron's visit, although they are placed among materials dated to the late 1660s and early 1670s. CELL, *Boyle Workdiaries*, BP 27, 79 http://www.livesandletters.ac.uk/wd/view/text_ed/WD21_ed.html (accessed 23 November 2014).
70. Robinson and Adams (eds.) *Diary of Robert Hooke*, entries for Mon 20 August 1677 and Thurs 1 Oct 1677, which refer to meeting Gyfford at Sion College (a London-based organisation for priests) and procuring 'workmen' to assist him.
71. Benjamin Breen, 'No Man Is an Island: Early Modern Globalization, Knowledge Networks, and George Psalmanazar's Formosa', *Journal of Early Modern History*, 17.4 (2013), 391–417.
72. IOR E/3/33, fols. 11–12, OC 3637, 10 May 1672, Samuel Baron to the Agent and Council.
73. Summaries of letters from Bantam, G/21/7, fol. 22 '160 d[ollars] paid Mr Barron who is gone for Tunqueen'.
74. Baron, *Description*, 217. The 'prince', whose name is not given, was the son of Trịnh Cân.
75. G/12/17(4), Abel Payne and Council in Bantam to the factors in Tonkin, 29 May 1677; G/12/17(5) fols. 491r–491v – 16 December 1678, Thomas James and Council in Tonqueen to Bantam.
76. Kleinen, John. *Lion and Dragon*, 56.
77. BL RP 4713, item 3. I am grateful to Simon Schaffer for bringing this document to my attention. 'Gentile' implies that Baron's wife was not Christian.
78. IORG/12/16 fol. 200 – Peter Crouch and John Thomas at the mouth of the Siam river to 'Mr Samuel Barron, merchant in Syam'. (no date)
79. See Reid, *Southeast Asia in the Age of Commerce, II: Expansion and Crisis*, 49.
80. EW Hutchinson, *Adventurers in Siam in the Seventeenth Century* (London: The Royal Asiatic Society, 1940), 27.
81. John Anderson, *English Intercourse with Siam in the Seventeenth Century* (London: Kegan Paul; Trench, Trübner & co., 1890), 249, gives the name of the Ambassador to Persia as Ali Selim. He left Ispahan in early 1685 accompanied by the Persian Ambassador Ebraim Beague and arrived in Siam on English ships via Fort St George.
82. IOR G/12/17/1 fols. 40–41, transcribed in Farrington and Pombejra, *English Factory in Siam*, I, 418–419. 'A Syam junck arrived with two ambassadors from the King of Syam, but the commander an Englishman resident there.'

83. Anderson, *English Intercourse with Siam*, 224–252.
84. There is very little information about this embassy. Anderson, *English Intercourse with Siam*, 243, mentions, citing Alexander Hamilton, *New Account of the East Indies*, 2 vols. (Edinburgh, 1727), ii, 181, that one of these Ambassadors crossed the channel and signed a treaty with Charles II. Also, as Anderson notes, the treaty has never been found. However, there is a reference to a present made from the King of Siam to Charles II in *Out Letters* (*General*) VII, 424. 'Entry Book: September 1684, 23–30', *Calendar of Treasury Books, Volume 7: 1681–1685* (1916), 1335–1345.
85. Reid, *Southeast Asia in the Age of Commerce*, II, 235.
86. Hutchinson, *Adventurers in Siam*, 33, on the first treaty signed between Siam and the Dutch in 1617; Bhawan Ruangsilp, 'Dutch Interaction with Siamese Law and the City Rules of Ayutthaya in the Seventeenth and Eighteenth Centuries', in Masashi (ed.) *Asian Port Cities*, 139–161.
87. Alain Forest, *Falcon: L'imposteur De Siam: Commerce, Politique Et Religion Dans La Thaïlande Du Xviie Siècle* (Paris: Les Indes savantes, 2010), 22.
88. M. Ismail Marcinkowski, 'The Iranian-Siamese Connection: An Iranian Community in the Thai Kingdom of Ayutthaya', *Iranian Studies*, 35 (2002), 23–46.
89. Anderson, *English Intercourse with Siam*, 297. Hutchinson, *Adventurers in Siam*, 23; Love, 'Monarchs, Merchants, and Missionaries'.
90. Farrington and na Pombejra, *The English Factory in Siam 1612–1685;* IOR OC 107; OC 125; OC 1130; OC 3197.
91. The idea of 'friction' is borrowed from Anna Lowenhaupt Tsing, *Friction, an Ethnography of Global Contact* (Princeton: Princeton University Press, 2005).
92. Hodges, 'Western Science in Siam, 87.
93. David R. M. Irving, 'Lully in Siam: Music and Diplomacy in French – Siamese Cultural Exchanges, 1680–1690', *Early Music*, 40:3 (2012), 393–420.
94. IOR E/3/42, OC 4885 William Hodges in Tonkin to Nathaniel Cholmley at Madras, 27 December 1682, transcribed in Farrington and na Pombejra, *The English Factory in Siam*, II, 709.
95. Baron, *Description*, 217. Temporary 'insanity' was a known ruse for Vietnamese princes to employ in difficult situations.
96. Buch, 'II. La Compagnie des Indes néerlandaises et l'Indochine', 181.
97. Baron, *Description*, 254.
98. Samuel Baron at Siam, 15 November 1684, to William Gifford at Fort St George, Letters to Fort St George 1681–1795 (Madras: Superintendent Government Press, 1916–1946), hereafter, *Letters to FSG, 1684–5*, 42–44. Baron had set off with a Company ship, the *Smerniott*, which had been forced to turn back to Tonkin.
99. Journal of Peter Crouch, supercargo of the *Delight*, 25 September 1683–17 April 1684, entry for 3 March 1684, transcribed in Farrington and na Pombejra, *The English Factory in Siam*, II, 834–835.
100. Anderson, *English Intercourse with Siam*, 164–165, states that Phaulkon was originally the cabin boy of George White and arrived at Ayutthaya in 1675, eventually purchasing the ship *Mary*. Forest, *Falcon*, 25, 34, claims that Phaulkon arrived in 1678 in the company of George White and Burnby

and that he had been engaged in Bantam in 1674. Precisely how he entered Narai's service is not known.

101. Love, 'Monarchs, Missionaries and Merchants'.
102. For a concise account, John Guy, 'Siamese Embassies to the Court of the Sun King', in Anna Jackson and Amin Jaffer (eds.) *Encounters: The Meeting of Asia and Europe, 1500–1800* (London: V & A, 2004), 88–89. There are numerous French accounts of the encounter.
103. Forest, *Falcon*, 26.
104. IOR/G/12/16, fol. 172v. For an account of the reciprocal performance of theatre during the embassies between France and Siam, Irving, 'Lully in Siam', 410.
105. Rabī M. I. Muḥammad, *The Ship of Sulaimān*, John O'Kane trans. and ed. (New York: Columbia University Press, 1972), 58–59.
106. Samuel Baron at Siam, 15 November 1684 to William Gifford at Fort St George, *Letters to FSG 1684–1685*, 43.
107. E.G.E. Hall, 'From Mergui to Singapore, 1686–1819: A Neglected Chapter in the Naval History of the Indian Ocean', *Journal of the Siam Society*, 5 (1953), 1–18.
108. IORG/12/16 fol. 200, Peter Crouch and John Thomas to 'Mr Samuel Barron, merchant in Syam', 2 April 1684.
109. IORG/12/16 fols. 163–192, London to Fort St George, 2 April 1684.
110. Anderson, *English Intercourse with Siam*, 220–221.
111. Fort St George, Consultation of 2 August 1685, transcribed in Love, *Vestiges*, 488.
112. Letter of Yale, referenced as 'Oriental Repertory, II, 189', cf. Anderson, *English Intercourse with Siam*, 275.
113. Stuart Rankin, 'White, Samuel (*c.*1650–1689)', *Oxford Dictionary of National Biography*, Oxford University Press, 2004 http://0-www.oxforddnb.com.catalogue.ulrls.lon.ac.uk:80/view/article/53891, accessed 19 February 2008.
114. Anderson, *English Intercourse with Siam*, 333.
115. Michael Smithies, 'Tachard's Last Appearance in Ayutthaya, 1699', *Journal of the Royal Asiatic Society*, Third Series, 12:1 (2002), 67–78, attributes this to Tachard and not Phaulkon himself.
116. See Anon., *A Full and True Relation of the Great and Wonderful Revolution That Happened Lately in the Kingdom of Siam in the East-Indies* (London: Printed for Randal Taylor, 1690); Anderson, *English Intercourse with Siam*, Chapter X.
117. Hutchinson, 170.
118. For contrasting accounts of Ayutthaya in the 1690s, see Guy Tachard, 'Relation de Voyage aux Indies 1690–1699', BN MS Fr. 19030, fols. 137r–196r, 192r, translated in Smithies, 'Tachard's Last Appearance in Ayutthaya', 73, who describes it as 'a desert' emptied of traders of all nations, and Engelbert Kaemphfer, *A History of Japan ... Together with a Description of the Kingdom of Siam* (1727), who described the kingdom as flourishing and still host to a number of foreign traders.
119. For contemporary accounts, Michael Smithies, *Witness to a Revolution: Siam 1688* (Bankok, Siam Society, 2004).
120. Love, 'Merchants and Missionaries'; for wider parallels, Walter Lieberman, 'Transcending East-West Dichotomies: State and Culture Formation in Six Ostensibly Disparate Areas', *Modern Asian Studies*, 31:3 (1997), 463–546.

121. John Locke, *An Essay Concerning Human Understanding*, 2nd edn. (London: Printed by Eliz. Holt, for Thomas Basset, 1690), i.3.15 'Gross ideas of God'.
122. Jean-Baptiste Tavernier, *A Collection of Several Relations and Treatises Singular and Curious* (London: A Godbid & J. Playford for Moses Pitt, 1680).
123. Baron, *Description*, 194–195.
124. Robert Boyle, *General Heads for the Natural History of a Country Great or Small: Drawn Out for the Use of Travellers and Navigators* (London: John Taylor, 1692).
125. Jyotsna G. Singh, *Colonial Narratives/Cultural Dialogues: 'Discoveries' of India in the Language of Colonialism* (London: Routledge, 1996). See also Andrew Hadfield, *Literature, Travel, and Colonial Writing in the English Renaissance, 1545–1625* (Oxford: Clarendon Press, 1998); Rubieś, *Travel and Ethnology in the Renaissance*.
126. Fred Halliday, 'Orientalism and Its Critics', *British Journal of Middle Eastern Studies*, 20.2 (1993), 145–163; David A. Washbrook, 'Orients and Occidents: Colonial Discourse Theory and the Historiography of the British Empire', in Robin W. Winks (ed.) *The Oxford History of the British Empire* V: *Historiography*, (Oxford: Oxford University Press, 1999) 596–611, 606.
127. Alam and Subrahmanyam, *Indo-Persian Travels*.
128. Pratt, *Imperial Eyes*, 7.
129. BL Sloane 1039, fol. 133r.
130. Baron, *Description*, 193.
131. Baron, *Description*, 214.
132. Baron, *Description*, Chapter VI, 214. G17/17(7), fol. 552, Consultation of 15 October 1681 among other passages.
133. Hostetler, 'Qing Connections to the Early Modern World'.
134. I used Vissière's French translation of *P'an Ting-kouei* text's 'Ngan -Nan Ki Yeou'. References to Vissière below refer to the introduction while those to P'an refer to the translation.
135. Claudine Salmon and Trọng Hiêp Tạ, 'Les Récits de voyage chinois comme source pour l'étude du Vietnam (X[e]–XXe siècle)', *Bulletin de l'Ecole Française d'Extrême-Orient*, 83 (1996), 67–87.
136. Baron, *Description*, Ch XI, 236; Vissière 'Ngan -Nan Ki Yeou', 72. Doror and Taylor, *Views of Seventeenth Century Vietnam*, 16, note that this description also appears in Gio. Filippo de Marini, *Relation nouvelle et curiouse des royuames de Tunquin et de Lao* (Paris: Gervais Clouzier, 1666), 2. P'an Tong-Kouei describes the converging toes as a sign of closeness to 'other Indians' and racial difference from (presumably Han) Chinese.
137. Robert Knox, *An Historical Relation of Ceylon, Revised, Enlarged & Brought to the Verge of Publication as the Second Edition*, 2 vols. ed. J. H. O. Paulusz (Dehiwala, 1989), ii, 235 (British Museum MS M10836, fols. 495–496).
138. *Boyle Workdiaries*, XXVI, 79. As noted above, whether the gentleman in question was Baron is uncertain.
139. Baron, *Description*, 211.
140. Baron, *Description*, 229.
141. Muḥammad, *The Ship of Sulaimān*, 56–57. According to Reid, *Southeast Asia in the Age of Commerce*, II, 226–227, the Siamese rulers relied on the traditional levies supplemented with foreign mercenaries until a permanent army was established by Phaulkon.

142. Baron, *Description*, 214.
143. P'an, 'Ngan -Nan Ki Yeou', 84–85
144. Niu Junkai and Li Qingxin, 'Chinese "Political Pirates" in the Seventeenth-Century Gulf of Tongking', Nola Cooke, Li Tana, and James A. Anderson (eds.) *The Tongking Gulf through History* (Philadelphia: University of Pennsylvania, 2011), 133–142.
145. Vissière, 'Ngan -Nan Ki Yeou'.
146. Yu, *Law and Society in Seventeenth Century Vietnam.*
147. IOR G/12/17(8) fols. 17–18 3 September 1682, transcribed in Farrington and Pombejra, *The English Factory in Siam*, 672, notes Baron's arrival over the bar at Tonkin with a cargo of rice.
148. BL MS Add 34123 'Copybook of Henry Vansittart, Governor of Bombay, 1756', fols. 40–42: 'Samuel Baron's account of the trade of India, written from Fort St George in 1695'.
149. Muḥammad, 'The Ship of Sulaiman', 99.
150. Screech, ' "Pictures (the Most Part Bawdy)" ', 55.
151. Tavernier, *Six Voyages*, iv:8.
152. Tavernier, *A Collection of Several Relations and Treatises*, 2–3.
153. Bodleian MS Smith 57, fol. 27, Thomas Smith to Edward Bernard, London, 11 September 84 and fol. 29, 16 September 1684.
154. Kaempfer, *History of Japan.*
155. Michael North (ed.) *Artistic and Cultural Exchanges between Europe and Asia, 1400–1900: Rethinking Markets, Workshops and Collections* (Farnham: Ashgate, 2010).
156. William R. Sargent, 'Asia in Europe: Chinese Paintings for the West', in Jackson and Jaffer (eds.) *Encounters*, 272–281.
157. Graham Parlett, 'Company School Painting', in Jackson and Jaffer (eds.) *Encounters*, 282–283. Henry J. Noltie, *Robert Wight and the Botanical Drawings of Rungiah & Govindoo* (Edinburgh: Royal Botanic Garden Edinburgh, 2007).
158. The original drawings are in BL Add MS 5232 and Sloane MS 3060. The manuscript of Kaempfer's journal was acquired by Hans Sloane and published as Kaempfer, *The History of Japan*. Scheuchzer added additional maps and images, drawing on Japanese originals. See Jörg Schmeißer, 'Changing the Image: The Drawings and Prints in Kaempfer's *History of Japan*', in Beatrice M. Boddart-Bailey and Derek Massarella (eds.) *The Furthest Goal: Engelbert Kaempfer's Encounter with Tokugawa Japan* (Folkstone, Kent: Japan Library, 1995), 132–151.
159. Baron, *Description*, 198–199.
160. Royal Society Archives MS/828. The original illustrations were recently discovered in the library of the Royal Society, having been returned to the Royal Society from the Society of Antiquaries in 1998. Presumably they strayed into the Antiquaries' collections some time during the long association of the two organisations at Somerset House or Burlington House. They sat in a plan chest drawer, the envelope marked 'Chinese Watercolours', until identified by Keith More in 2007 (I am grateful to Keith Moore for this information).
161. Baron, *Advertisement*, 194, and letter to Hooke and Hoskins, 190.

162. Đặng Nam, *Tranh dân gian Việt Nam: Vietnamese Folk Pictures* (Hanoi: Nationalities Culture Publishing House, 1995).
163. Collections housed in the Vietnam National Fine Arts Museum (Bao Tang My Thuat Viet Nam), Hanoi.
164. Hostetler, 'Qing Connections'.
165. The illustrations were made in 1683, at which point the English were finally offered a settlement in the city.
166. The Seldon map is a rare example of a Chinese map containing a compass, but the compass in Baron's map is a Western compass with 16 points rather than the Chinese version with 24 that appears on the Seldon map. (Brook, *Mr Selden's Map of China*, 100–105.)
167. The collection of Dutch and English maps were previously kept in the library of Corpus Christi College, Oxford, but have now been transferred to the Maritime Museum, Rotterdam. See Sjoerd de Meet and Frits Loomeijer, *De schat van Corpus Christi: VOC-kaarten boven water* (Zutphen: De Walburg Pers, 2007), 'Jambee town' is item 20.
168. This technique was in use in China as early as the first century BCE in both military maps and landscape paintings. See Wilfred H. Wells, *Perspective in Early Chinese Painting* (London: Edward Goldston Ltd., 1935); Hsin-Mei Agnes Hsu, 'Structured Perceptions of Real and Imagined Landscape in Early China', in Kurt A. Raaflaub and Richard J. A. Talbert (eds.) *Geography and Ethnography: Perspectives of the World in Pre-Modern Societies* (Oxford: Blackwell, 2010). There is little literature dedicated to the history of painting in Northern Vietnam, for which problem see Nora Annesley Taylor, *Painters in Hanoi, an Ethnography of Vietnamese Art* (Honolulu: University of Hawaii Press, 2004), 'Introduction'.
169. Schmeißer, 'Changing the Image', 134–137.
170. For early borrowings of Asian techniques in Western painting, Lach and van Kley, *Asia in the Making of Europe*, II.1.73–74; Schmeißer, 'Changing the Image', 143.
171. de Meer and Loomeijer, *De schat van Corpus Christi: VOC-kaarten boven water*: the 1663 map is item no. 7 in this facsimile edition.
172. Baron, *Description*, 199.
173. Tonio Andrade, 'Pirates, Pelts, and Promises: the Sino-Dutch Colony of Seventeenth Century Taiwan and the Aboriginal Village of Favorolang', *Journal of Asian Studies* 64 (2005), 295–321; Massarella, 'Chinese, Tartars and "Thea" '.
174. On the problems of trade in Taiwan, see IOR G/12/4B fols. 150–151, Simon Delboe and Council at Bantam, 12 February 1673, transcribed in Farrington et al. *English Factory in Taiwan*, 170–173.
175. Massarella, 'China, Tartars, and "Thea" ', 408.
176. IOR E/3/16, fols. 134–135, Massarella, 'China, Tartars, and "Thea" ', 416. See also IOR G/12/16, fols. 133v–139, Thomas Angeir and Thomas Woolhouse at Taiwan to the Agent and Council in Siam, 20 December 1683, transcribed in Farrington et al., *English Factory in Taiwan*, 550–567.
177. Massarella, 'China, Tartars, and "Thea" ', 419.
178. IOR G/12/16, fols. 175r–192v, Journal of Peter Crouch, supercargo of the *Delight*, 24 April 1684–20 March 1685, transcribed in Farrington et al. *English Factory in Taiwan*, 580.

179. IOR G/19/4, fol. 145, Consultation, 8 April 1686.
180. Massarella, 'China, Tartars, and "Thea" ', 422; IOR G/19/4, fol. 89, 1 January 1687, Consultation, *Letters to FSG 1686–1687*, 6–7, Baron, Mose, and Dubois to William Gyfford, 31 May 1686.
181. Henry Mose and Charles Du Bois aboard the *Shrewsbury* to Fort St George, 30 December 1686, *Letters to FSG 1686–1687,* 21. IOR G/19/4, Samuel Baron in Amoy to Fort St George, 23 November 1686; *Letters to FSG 1686–1687,* 24.
182. Henry Burton to Fort St George, 22 February 1687, Cuddalore, *Letters to FSG 1686–1687,* 77–78.
183. IOR G/19/5 fol. 1v, 20 February 1688.
184. Henry Burton to Fort St George, 22 February 1687, Cuddalore, *Letters to FSG 1686–1687,* 77–78.
185. IOR E/3/91 fol. 104, London to Fort St George, 22 October 1686.
186. *FSG Diary and Consultations*, XXV, 9 February 1688.
187. *FSG Diary and Consultations*, XXVII, 11 February 1688 and IOR G/19/5, fol. 30v, Consultation, 16 April 1688.
188. IOR E/3/91 fol. 104, London to Fort St George, 22 October 1686.
189. *FSG Diary and Consultations*, XXII, 6 February 1688.
190. FSG, *Diary and Consultations*, Census, 214–215 (2 Feb 1687/8, from *Public Consultations*, XIII, 261–265).
191. IOR G/19/5, 2 August 1688 fols. 120–121.
192. Baron is said to have gone to a 'Portuguese parrier house in the Copang'. For the meaning of 'parrier house' and 'copang' respectively see, *Hobson-Jobson*, 678, and Yule, *Vestiges*, I, 445.
193. Anjali Chatterjee, *Bengal in the Reign of Aurangzib, 1658–1707* (Calcutta: Progressive Publishers, 1967), 24. '*Firingi*' is from 'Frank', here meaning Eurasian. See also Chapter 3.
194. MS Add 34123 – 'Copybook of Henry Vansittart, Governor of Bombay, 1756', fols. 40–42: 'Samuel Baron's account of the trade of India, written from Fort St George in 1695'.
195. Baron, *Description,* 267. For the common practice of feasting at the graves of dead ancestors in Southeast Asia, Reid, *Southeast Asia in the Age of Commerce*, II, 165–168.
196. Jayawardena, *Erasure of the Euro-Asian*, 3–5; Soren Mentz, 'Cultural Interaction between the British Diaspora in Madras and the Host Community', in Masashi, *Asian Port Cities*, 162–174.
197. C. H. Nightingale, 'Before Race Mattered: Geographies of the Color Line in Early Colonial Madras and New York', *American Historical Review*, 113:1 (2008), 48–71.
198. Nightingale, 'Before Race Mattered', 67; Love, *Vestiges*, II, 425–426.
199. Shapin, *A Social History of Truth.*
200. Tavernier, *A Collection of Voyages*, 9; Baron, *Description*, 207; Massarella, 'China, Tartars, and "Thea" ', 404.
201. See also Zemon Davis, *Trickster Travels.*
202. Massarella, 'China, Tartars, and "Thea" ', 396. The assumption that Japan was closed to all outside influence from this period onwards has been challenged by works including Donald Denoon, Gavan McCormack and

others, *Multicultural Japan: Palaeolithic to Postmodern* (New York: Cambridge University Press, 1996).

203. Willis, 'Maritime China from Wang Chih to Shih Lang', 233.
204. Raj, 'Mapping Knowledge', 122; Susan Neild-Basu, 'The Dubashes of Madras', *Modern Asian Studies*, 18 (1984), 1–31.
205. Massarella, 'China, Tartars, and "Thea" ', 426–427.

2 Linguistic Landscapes: Early English Studies of Malay and the EIC in Maritime Southeast Asia

1. R. Mee, 'An Old Malay Dictionary', *Journal of the Malayan Branch, Royal Asiatic Society*, 7 (1929), 316–326. A copy of the work remains among the 'Raffles Library', Singapore (*Catalogue of the Raffles Library* (Singapore, printed at The American Mission Press, 1905)).
2. Thomas Bowrey, *Dictionary English and Malayo, Malayo and English* (printed by Sam. Bridge for the Author, London, 1701). The *Dictionary* has no page numbers; references given below are therefore to sections.
3. John Considine, 'Wordlists of Exotic Languages in Seventeenth-Century England' in John Considine (ed.) *The Seventeenth Century*. Ashgate Critical Essays on Early English Lexicographers, IV (Farnham: Ashgate), 370.
4. Bernard Cohn [1985], 'The Command of Language and the Language of Command', in Bernard Cohn (ed.) *Colonialism and Its Forms of Knowledge: The British in India* (Princeton, NJ: Princeton University Press, 1996), 16–56.
5. M. Shapiro, 'A Political Approach to Language Purism', in B. Jernudd and M. Shapiro (eds.) *The Politics of Language Purism* (Berlin: Mouton de Gruyter, 1989), 21–30, 28.
6. Cohn, *Colonialism and Its Forms of Knowledge*, 16–51; Joseph Errington, 'Colonial Linguistics', *Annual Review of Anthropology*, 30 (2001), 19–39; Geoffrey C. Gunn, *First Globalization: The Eurasian Exchange, 1500–1800* (Lanham, MD: Rowman & Littlefield, 2003), Chapter 9, 'Language, Power and Hegemony in European Oriental Studies'; Joseph Errington, *Linguistics in a Colonial World: A Story of Language, Meaning, and Power* (Malden, MA: Blackwell, 2008).
7. Anthea Fraser Gupta, 'The Imagined Learner of Malay', in Jean-Marc Dewaele, Alex Housen and Li Wei (eds.) *Bilingualism: Beyond Basic Principles* (Clevedon: Multilingual Matters, 2003).
8. Anthony Reid, 'The Structure of Cities in Southeast Asia, Fifteenth to Seventeenth Centuries', *Journal of Southeast Asian Studies*, 11:2 (1980), 235–250.
9. Reid, *Southeast Asia in the Age of Commerce*, II, 132–136.
10. For example, Bodleian MS Douce Or e. 5 I is a copy of the end of an Arabic letter from 'Ala-uddin, Sultan of Aceh, to Queen Elizabeth, 1011 AH/1602 AD, accompanied by a copy of a trading permit issued by the Sultan in Malay (MS Douce Or e. 5 II). This and other early royal correspondence are described in W. G. Shellabear, 'Some Old Malay Manuscripts', *JSBRAS*, 31 (1901), 110–111. See also Anna Winterbottom and Samuli Kaislaniemi, 'Perverted through Ambition: Early English Scholarship on Malay', in

Kingsley Bolton, Samuli Kaislaniemi and Anna Winterbottom (eds.) *The East India Company and Language* (Cambridge: Cambridge University Press, forthcoming) for more details.

11. Annabel T. Gallop and Bernard Arps, *Golden Letters: Writing Traditions of Indonesia = Surat Emas: Budaya Tulis Di Indonesia* (London: British Library, 1991).
12. W. G. Shellabear, '*Hikayat seri Rama*: Introduction to the text of the MS, in the Bodleian Library at Oxford', *Journal of the Straits Branch, Royal Asiatic Society*, 70 (1917).
13. Richard Greentree and Edward Williams Byron Nicholson, *Catalogue of Malay Manuscripts and Manuscripts Relating to the Malay Language in the Bodleian Library* (Oxford: Clarendon Press, 1910), 3–4.
14. Femme Gastra, 'War, Competition and Collaboration: Relations between the English and Dutch East India Companies in the Seventeenth and Eighteenth Centuries', in Bowen, Lincoln and Rigby (eds.) *The Worlds of the East India Company*, 49–58.
15. Ogburn, *Indian Ink*, 120–129.
16. William Foster, 'An English Settlement in Madagascar', *English Historical Review*, 27 (1912), 239–250.
17. A list of the company personnel required in 1662 includes four persons 'to be employed when occasion shall require in voyaging to Atcheen etc...' (London to Surat, 18 August 1662, copy in MSA, 'Collections of Papers Received from the India Office', fols. 9–16, from original Letter Book, III, fol. 113). The Courteen Association had had a factory at Aceh but it was short-lived and impoverished (William Foster, 'An English Settlement in Madagascar', *English Historical Review*, 27 (1912), 239–250).
18. Thomas Bowrey, *A Geographical Account of the Countries round the Bay of Bengal 1669 to 1679*, Sir Richard Carnac Temple (ed.) (Hakluyt Society: Cambridge, 1935), p. 306.
19. Annabel Teh Gallop, 'Gold, Silver, and Lapis Lazuli: Royal Letters from Aceh in the Seventeenth Century', 105–139, in R. M. Feener, Patrick T. Daly and Anthony Reid (eds.) *Mapping the Acehnese Past* (Leiden: KITLV Press, 2011).
20. Reid, 'Structure of Cities in Southeast Asia', p. 237.
21. Gaastra, 'War, Competition and Collaboration'.
22. William Foster, *John Company* (London: Bodley Head, 1926).
23. Foster, *John Company*, pp. 84–85, notes that the court minutes record the men's return on 8 February 1676. The elephant is described in an eight-page pamphlet; Anon., *A True and Perfect Description of the Strange and Wonderful Elephant Sent from the East Indies and Brought to London on Tuesday the Third of August* ([London]: printed for J. Conniers, 1675).
24. Russell Jones, 'The First Indonesian Mission to London', *Indonesia Circle*, 28 (1982), 9–19, 10, gives selections from the Dutch account of the visit. See also Foster (ed.) *John Company*, 97. Stern, *Company State*, 69–70, and Reid, *Southeast Asia in the Age of Commerce*, II, 241–243, both give accounts of the embassy and reproduce the paintings of the two Ambassadors.
25. Foster, *John Company*, 102–103.
26. The original Arabic letter and translation are in the National Archives and in Photostat in BL IOR photo 149, fols. 45–47.

27. IOR MS Eur photo 149/13, fol. 52, Thomas Hyde to Lord Hide, 25 May 1682, Oxford.
28. IOR MS Eur photo 149/13, fol. 32 – 'Abstract of Letters from Bantam Dated the 28th of February 1680/1 23th of July & 19th of August 1681'.
29. IOR MS Eur photo 149/13, fol. 52, Josiah Child to Lord Hide, 2 June 1682, London.
30. Reid, *Southeast Asia in the Age of Commerce*, II, 72, for estimates of the size of the Bantam army.
31. Foster, *John Company*, 118–120.
32. Reid, *Southeast Asia in the Age of Commerce*, II, 280–281.
33. For details, David Veevers, ' "The Company as Their Lords and the Deputy as a Great Rajah": Imperial Expansion and the English East India Company on the West Coast of Sumatra, 1685–1730', *Journal of Imperial and Commonwealth History*, 41:5 (2013), 687–709.
34. John Bastin, *The British in West Sumatra* (University of Malaya Press, Kuala Lumpur, 1965).
35. Bastin, *British in West Sumatra.*
36. William Dampier, *A New Voyage round the World. Describing Particularly, the Isthmus of America, Several Coasts and Islands in the West Indies, the Isles of Cape Verd , the Passage by Terra del Fuego …* First ed. (London: printed for James Knapton, at the Crown in St Pauls Church-Yard, 1697). Also issued with vol. 2, *Voyages and Descriptions*, 1699, and vol. 3, *A Voyage to New Holland &c. in the Year 1699.*
37. Geraldine Barnes, 'Curiosity, Wonder, and William Dampier's Painted Prince', *Journal for Early Modern Cultural Studies*, 6:1 (2006), 31–50.
38. [Thomas Hyde] *An Account of the Famous Prince Giolo, Son of the King of Gilolo, Now in England: An Account of His Life, Parentage, and His Strange and Wonderful Adventures: The Manner of His Being Brought for England: With a Description of the Island of Gilolo, and the Adjacent Isle of Celebes: Their Religion and Manners* (London: printed and sold by R. Taylor by Amen-Corner, 1692).
39. Ogborn, *Indian Ink.*
40. MSS Eur D1076 fols. 1–12 petitions of the 'new' Company and lists of stockholders.
41. William Marsden, *A Dictionary of the Malayan Language: To Which Is Prefixed a Grammar, with an Introduction and Praxis* (London: printed by Cox and Baylis, 1812), iv, lists 'the vocabulary collected by the Dutch navigators at *Ternatī*'.
42. Russell Jones, 'Malay Studies and the British: 1: An Outline History to the Early Twentieth Century', *Archipel*, 28 (1984), 117–148.
43. Frederick de Houtman van Gouda. *Spraeck ende woord-boeck, Inde Maleysche ende Madagaskarsche Talen, met vele Arabische ende Turcsche woorden* (Amsterdam, 1603).
44. For details of these Dutch works and their relation to English scholarship, Winterbottom and Kaislaniemi (forthcoming).
45. According to Kees Groneneboer, 'The Dutch Language in Maluku under the VOC', *Cakalele*, 5 (1994), 1–10, this work was written by the VOC merchant Albert Corneliszn Ruyll and published in 1612.

46. C. Wiltens and S. Danckaerts, *Vocabularium ofte Woort-boek naer order vanden Alphabet in 't Duysch-Maleysch-Duytch....* ('s-Gravenhage: H. Jz. van Wouw, 1623). See also John M. Echols, 'Dictionaries and Dictionary Making: Malay and Indonesian', *Journal of Asian Studies*, 38 (1978), 11–24, 15. Echols also notes that a Latin version was issued in 1631. W. Lineham, 'The Earliest Word-Lists and Dictionaries of the Malay Language', *Journal of the Royal Asiatic Society Malayan Branch*, 22 (1949), 183–187, identifies this as David Haex, *Dictionarium Malaico-Latinum et Latinum-Malaicum* (Rome, Congr. de Propag. Fide, 1631).
47. F. Gueynier, *Vocabulaer, ofte Woorden-boeck, naer ordre van den Alphabet, in't Duytsch ende Maleys* (Batavia, 1677); cf. Echols, 'Dictionaries and Dictionary Making', 15, n. 19. For the production of Malay-language materials in Batavia during the eighteenth century, Gunn, *First Globalisation*, 233–234.
48. James T. Collins, 'Language Death in Maluku: The Impact of the VOC', *Bijdragen Tot De Taal-, Land En Volkenkunde*, 159 (2003), 247–289.
49. TNSA, *Sundry Book*, VII, fol. 22.
50. London to Surat, 'Encouragement to Those Learning the Banian Language', London to Surat, 15 March 1677, copied in TNSA, *Public Despatches from Court*, I, fols. 186, 194.
51. MSA, *Secretariat Outwards*, V, 3 January 1693, fols. 144–156.
52. Madras Record Office, *Diary and Consultation Book 1672–1679* (Madras: Government Press, 1910), 82. Papers delivered to the Agent and Council by William Puckle, 16 February 1676.
53. IOR G/35/2, fol. 125, London to Bencoulen, 30 December 1687.
54. William Dampier, *Voyages and Descriptions Vol. II, in Three Parts* (London: printed for James Knapton, 1699), 58. Moody's proficiency in Malay as well as his experience in the pepper trade are also noted in the letter from Fort St George appointing him second in command (IOR G/35/2, fol. 216, FSG to Bencoulen, sent 15 May 1690, received 1 November 1690). Moody was swiftly sent to Manjuta to negotiate concerning opening the pepper trade to Indrapora. Dampier's presence is noted in the Bencoulen records for 1690 (G/35/2).
55. Augustus Spalding, *Dialogves in the English and Malaiane langvages* (London: imprinted by Felix Kyngston for William Welby, 1614).
56. For details, see Winterbottom and Kaislaniemi, 'Perverted through Ambition'.
57. Yule, *Diary of William Hedges*, II, cccliii. Yule notes this is from the *Court Book* c. 1660 but that the original has been lost. The Arabic translation of Grotius' 1640 work was made in 1660 by Edward Pococke, and a copy is retained in the Bodleian library (Bod MS Arab e. 3). The translation is also noted by Stephen Neill, *A History of Christianity in India* (Cambridge University Press: Cambridge, 1984), I, 364, who claims that the work was also translated into Malay, Chinese, and Persian.
58. Stern, *Company State*, 107.
59. MSA, *Surat 'Outwards'*, I.2, Surat to London 28 January 1664 fols. 4–37.
60. BL MS Add 4293, fols. 45–47, Robert Boyle to Robert Thomson, 5 March 1677.

61. Bodleian MS Tanner 36*, fol. 57, John Fell to William Sancroft, 19 June 1681, and fol. 61, John Fell to William Sancroft, 21 June [1681].
62. Bodleian MS Tanner 36*, fol. 69, [2] May 1682. [The first part of the date is obscured by damage to the edge of the page.]
63. Stern, *Company State,* 115–118.
64. BL IOR G/35/3, fols. 33–37 Consultation of 15 July 1695; the 'Malay padre' for Bencoulen complains that he has not been paid. The word 'padre' was being applied to Protestant as well as Catholic priests by the early eighteenth century (*Hobson-Jobson*, 651).
65. For a report on converts to Christianity sent from the Bantam factory to Tonquin, Sloane 998, fol. 17r, Sunday 18 August.
66. Jan Hasel, Justus Heurnius, and Albert Ruyl (trans.), Thomas Hyde, and Thomas Marshall (eds.) *Jang Ampat Evangelia … Daan Berboatan Derri Jang Apostoli Bersacti, Bersalin Dallam Bassa Malayo* (H. Hall: Oxford, 1677).
67. K. Dekker, 'Marshall, Thomas (1621–1685)', *Oxford Dictionary of National Biography* (Oxford University Press, 2004), http://0-www.oxforddnb.com.catalogue.ulrls.lon.ac.uk/view/article/18149, accessed 2 May 2010.
68. Horace Hart, *Notes on a Century of Typography at the Oxford University Press, 1693–1794.* Repr. with notes and intro by Harry Carter (Oxford: Clarendon Press, 1970).
69. Stern, *Company State*, 115. According to Anthony Farrington, *Trading Places: The East India Company and Asia 1600–1834* (London: British Library, 2002), the work appears among a list of 185 books in the Bantam factory. The distribution of the book is noted in IOR G/21/7, fol. 23. The print run was 500 copies (Russell Jones, 'Malay Studies and the British. I. An Outline History to the Early Twentieth Century', *Archipel*, 28 (1984a): 117–148).
70. Bodleian MS Marshall OR 70 and 77.
71. Bod Marshall MS OR 70, f. 41–42.
72. Ashmole 1808, fol. 16 and Bod MS Marshall OR 77 (throughout) both give 'ga' (ڬ) in its archaic form with three dots below.
73. Bod MS Marshall OR 70, f. 47.
74. Greentree and Nicholson, *Catalogue of Malay Manuscripts*, 17–18, speculate that the annotator was one of the factors expelled from Bantam, but given that the factors only left around the time the Ambassadors returned to Bantam, this would push the date of this second draft (OR 77) back to after March 1683. However, it seems just as likely that the annotator was one of the former factors who attended the Bantam embassy of 1682. For Burnett, Martin Greig, 'Burnet, Gilbert (1643–1715)', *Oxford Dictionary of National Biography*, Oxford University Press, 2004; online edn., September 2013 http://www.oxforddnb.com/view/article/4061, accessed 19 March 2015.
75. Ashmole 1808.
76. Considine, 'Wordlists of Exotic Languages in Seventeenth-Century England', p. 370.
77. Michael C. W. Hunter, Antonio Clericuzio, and Lawrence Principe, *The Correspondence of Robert Boyle, 1636–1691* (London: Pickering & Chatto, 2001). Digital edition by *Electronic Enlightenment* (Oxford, last update 2013). Sir William Mainston to Robert Boyle, 19 December 1682 (Gregorian:

29 December 1682), a request for the return of the draft copy, http://dx.doi.org/10.13051/ee:doc/boylroPC0050369a1c and 15 May 1683 (25 May 1683), enclosing the final copy, http://dx.doi.org/10.13051/ee:doc/boylroPC0050411a1c.

78. Hunter et al., *Electronic Enlightenment*, Thomas Hyde to Robert Boyle, 23 September 1684 (Gregorian: 3 October 1684), ed. Robert McNamee et al. Ver. 2.5. University of Oxford, 2014. Web. 19 March 2015. http://dx.doi.org/10.13051/ee:doc/boylroPC0060042a1c. Hyde's opinion here was based on that of 'the Bishop' (perhaps Bishop John Fell), but it is not clearly justified. Hyde promises to discuss the matter further with Thomas Marshall on his return to Oxford.
79. Thomas Hyde to Robert Boyle, Oxford, 23 February 1688/9 (5 March 1689), *Electronic Enlightenment*, ed. Robert McNamee et al. Ver. 2.5. University of Oxford. 2014. Web. 19 March 2015. http://dx.doi.org/10.13051/ee:doc/boylroPC0060284a1c.
80. Thomas Hyde and Edward H. Clarendon. *Ğadāwil-i Mawāḍī'-I Ṯawābit Dar ṭūl Wa-'araḍ kih Bi-Raṣ ad Yāfta Ast Uluġ Bīk Ibn-Šāhruḫ Ibn-Tīmūr Sive Tabulae Long. Ac. Lat. Stellarum Fixarum* (Oxonii: Hall, Henry, Hyde Thomas, Davis, Richard, 1665).
81. Hart, *Notes on a Century of Typography*, 36, 37.
82. University of Oxford, *Pietas Universitatis Oxoniensis in obitum augustissimae & desideratissimae Reginae Mariae* (Oxford: Oxford University Press, 1695).
83. Annabel Teh Gallop, 'A Seventeenth-Century Miscellany: Two New Discoveries in the British Library', 71–86, in *Tradisi Penulisan Manuskrip Melayu* (Perpustakaan Negara Malayasia, Kuala Lumpur, 1997), 73–76.
84. Teh Gallop, 'A Seventeenth-Century Miscellany'.
85. A specimen of the several sorts of letter given to the university by Dr John fell late Lord Bishop of Oxford: to which is added the letter given by Mr. F. Junius. Oxford: The Theater, 1695.
86. Annabel Teh Gallop, personal communication.
87. Hunter et al., *Electronic Enlightenment*, Thomas Hyde to Robert Boyle, Oxford, 23 February 1688/9 (Saturday, 5 March 1689), http://dx.doi.org/10.13051/ee:doc/boylroPC0060284a1c.
88. Thomas Bowrey, *A Geographical Account of the Countries round the Bay of Bengal 1669 to 1679*, Sir Richard Carnac Temple (ed.) (Hakluyt Society: Cambridge, 1935).
89. BL MS Sloane 5222, fols. 6–17.
90. LMA MS 03041(iii).
91. National Archives PROB 11/532/176. The drafts of the African journal are with Bowrey's collections in the London Metropolitan Archives (MS 3041/8; Bowrey's papers were transferred to the LMA from the Guildhall Library). For the African journals, Arne Bialuschewski, 'Thomas Bowrey's Madagascar Manuscript of 1708', *History in Africa* 34:1 (2007): 31–42.
92. Bowrey, *A Geographical Account of the Countries round the Bay of Bengal*.
93. H. Christof Wolfart and David H. Pentland, 'The "Bowrey" Dictionary of Henry Kelsey', in William Cowan (ed.) *Papers of the Tenth Algonquian Conference* (Ottawa: Carleton University Press, 1979), 37–42.

94. Bowrey, *Preface.*
95. BL India Office Private Papers (IOPP), MSS Eur A. 33.
96. Ursula Simms-Williams, 'Zoroastrian Manuscripts in the British Library, London', in A. Cantera (ed.) *The Transmission of the Avesta* (Wiesbaden: Harrassowitz, 2012), 173–194.
97. IOPP MS Eur. E. 192. 2 (a–d).
98. Guildhall MS 3041/8. Receipts from engraver John Sturt dated 10 September 1702 and 9 February 1703. A letter from George Grierson to Richard Temple enclosed with IOPP MSS Eur E. One hundred and ninety-two states that the Sanskrit letters are copied from Athanasius Kircher's *China Illustrata.*
99. Manuscript copies remain in British Library MS Or.70.bb.9 (OMS 16 B. XVI) and IOPP MSS Eur E. 192. The printed copies are in Thomas Hyde, *Syntagma dissertationum*, 2 vols., Thomas Sharpe (ed.) (Oxford, 1760), II, 533–559.
100. BL IOPP MS Eur E. 192 [c] item 17.
101. SOAS special collections, EB760.10/11608. This manuscript was transferred along with other parts of Marsden's collection from King's College to SOAS.
102. Richard Carnac-Temple (ed.) *The Papers of Thomas Bowrey 1669–1713*, Second Series, No. LVIII (London: Hakulyt Society, 1927), xxx, fn. 1; and Richard Carnac-Temple (ed.) *New Light on the Mysterious Tragedy of the 'Worchester' 1704–5* (London: Ernst Benn Ltd., 1930), 382–383.
103. He may be the same Henry Smith whom Cunningham mentions as a collector (Chapter 4).
104. Annotations in SOAS EB760.10/11608 following the 'English-Malayo' section. Marsden's comments on the annotated version, which he says has come 'accidentally' into his possession, are in Marsden, *A Dictionary of the Malayan Language: To Which Is Prefixed a Grammar*, xli. He adds that 'nothing further respecting this Henry Smith has ever come to my knowledge'.
105. London Metropolitan Archives (LMA), MS 24176/1-2 [item numbers, pages are not foliated], Henry Smith to Thomas Bowrey dated 9 May 1700, 3 June 1700, and 29 June 1700. Smith's letters are unclear about the exact nature of his crime but contain several complaints about Captain Gullock. According to a letter from Lord Bellomont to the Board of Trade, dated Boston the 29 November 1699 transcribed in John Franklin Jameson, *Privateering and Piracy in the Colonial Period: Illustrative Documents* (New York: Macmillan, 1923), Gullock's ship, the *America*, was taken by pirates with the aid of mutinous crew members. It seems likely that Smith was among the latter.
106. For Smith's negotiations in Scotland, Temple, *New Light on the Mysterious Tragedy of the 'Worcester'*, 382–406.
107. Marsden, *A Dictionary of the Malayan Language: To Which Is Prefixed a Grammar*, p. xl.
108. Bowrey, *Dictionary*, 'Preface'. For the importance of the trade from India to Manila and the Pacific, Holden Furber, 'The India Trade in the Pacific through Two Centuries 1600–1800', item XII, 1–17, in Holden Furber, *Private Fortunes and Company Profits in the India Trade in the 18th Century* (Aldershot: Variorum, 1997).
109. Temple, 'Introduction' in Thomas Bowrey, *A Geographical Account.*

110. LMA MS 3041/2, not foliated. Beginning 'By all European nations who have arrived to be most considerable in the Trade to East India, it has always bin justly esteemed as absolutely necessary to have some considerable settlement in the South seas of India . . . '.
111. Add MS 5222.10. For the early modern interest in Singapore as a target of settlement by several European nations, Peter Borschberg, *The Singapore and Melaka Straits: Violence, Security, and Diplomacy in the 17th Century* (Singapore: NUS Press, 2010).
112. LMA MS 3041/2, not foliated, headed 'Articles of agreement made and concluded between . . . of London Merchants of the one part & Thomas Bowrey of London Merchant of the other part'.
113. Dampier, *New Voyage around the World* (1699 ed.), fold-out map before title page. Brook, *Mr Seldon's Map of China*, p. 106.
114. BL Add MS 5222.15.
115. Add MS 5222.6.
116. Add MSS 5222.9 and 5222.7.
117. Add MS 5222.10.
118. Add MS 5222.17.
119. Add MS 5222.14.
120. Add MS 5222.8.
121. Add MS 5222.12.
122. Thomas Bowrey to the Honourable Directors of the East India Company, IOPP MSS Eur. E. 192 (a), Item 31, fol. 1. No date but c. 1701–1704.
123. Reid, *Southeast Asia in the Age of Commerce*, II, 44–47.
124. IOPP MSS Eur. E. 192 (a), Item 31, fol. 1.
125. IOPP MSS Eur. E. 192 (a), item 27, fol. 1, 2 August 1702, and item 28, fol. 1, 7 August 1702. For the sago tree, item 8 fol. 1, Hyde to Bowrey, 7 February 1700.
126. Marshall, *Jang Ampat Evangelia*, 5–6.
127. IOPP MSS Eur. E. 192, item 7, fol. 1 Hyde to Bowrey 29 November 1700.
128. Raffaele Simone, 'The Early Modern Period', in Giulio Lepschy (eds.) *History of Linguistics* [5 vols.], *Vol. III: Renaissance and Early Modern Linguistics* (London: Longman, 1998), 149–188.
129. Bowrey, *Dictionary*, grammar section.
130. On the grammatical differences between 'low' and 'high' Malay, K. Alexander Adelaar, 'Malay: a Short History', *Oriente Moderno*, new series 19(80).2, Alam Melayu Il Mondomalese: Lingua, Storia, Cultura (2000), 225–242: 232–233.
131. Vivian Salmon, *The Study of Language in Seventeenth Century England*, Studies in the History of the Language Series, vol. 17 (John Benjamins Publishing Company: Amsterdam/Philadelphia, 1988), 67–68.
132. Salmon, *The Study of Language*.
133. Pieter A. M. Seuren, *Western Linguistics: An Historical Introduction* (Oxford, UK: Blackwell Publishers, 1998), 44–50.
134. Noam Chomsky, *Cartesian Linguistics* (New York: Harper & Row, 1966); Salmon, 'Pre-Cartesian Linguistics', repr. in *The Study of Language in Seventeenth Century England*.
135. Seuren, *Western Linguistics*, 56.

136. MS Ashmole 1808, Dedication.
137. MS Marshall OR 77, fol. 33.
138. Errington, *Colonial Linguistics*; Richard Steadman-Jones, *Colonialism and Grammatical Representation* (Oxford: Blackwell, 2007); Gunn, *First Globalisation*, 246–247.
139. Errington, *Colonial Linguistics*.
140. Gunn, *First Globalisation*, 224. John Considine, *Academy Dictionaries 1600–1800* (New York: Cambridge University Press, 2014).
141. See also Kader, 'Some Aspects of Seventeenth-Century Malay'.
142. William Marsden, *History of Sumatra* (London: printed for the Author, and sold by Thomas Payne and Son and others), 161.
143. Marsden, *A Dictionary of the Malayan Language: To Which Is Prefixed a Grammar*, 28–29; Ashmole 1808, fol. 26.
144. For further discussion, see Winterbottom and Kaislaniemi 'Perverted through Ambition'.
145. Bowrey, *Preface*.
146. Gupta, 'The Imagined Learner of Malay', 145.
147. Ashmole 1808, 'To the Candid Reader, the Preface'.
148. Marsden, *A Dictionary of the Malayan Language: To Which Is Prefixed a Grammar*, v.
149. K. A. Adelaar, 'The Malay Language (Historical)', map 67, in S. A. Wurm, Peter Mühlhäusler, and D. T. Tryon (eds.) *Atlas of Languages of Intercultural Communication in the Pacific, Asia, and the Americas* (Berlin: Mouton de Gruyter, 1996).
150. Ogilby, *Asia*, 134.
151. Ashmole 1808, 'To the Candid Reader, the Preface'. Marsden, *A Dictionary of the Malayan Language: To Which Is Prefixed a Grammar*, viii, similarly considers the Malay style of Malacca and Johor the most cultivated.
152. Dampier, *New Voyage*, 515.
153. This probably derives from the Malay name for Phuket: 'Ujung Salang'. *Hobson-Jobson*, p. 473.
154. Bowrey, *A Geographical Account*, 235–237.
155. Alice H. Nah, '(Re)Mapping Indigenous 'Race'/Place in Postcolonial Peninsular Malaysia', *Geografiska Annaler. Series B, Human Geography*, 88 (2006), 285–297. As Nah notes, 'Orang Asli' was in fact a name bestowed to denote their indigenous status.
156. Russell Jones, 'Introduction', to William Marsden *A Dictionary and Grammar of the Malayan Language* (Singapore: Oxford University Press, 1984).
157. Gupta, 'The Imagined Learner of Malay'.
158. George Brian Souza, 'Opium and the Company: Maritime Trade and Imperial Finances on Java, 1684–1796', *Modern Asian Studies*, 43 (2009), 113–133.
159. Carl A. Trocki, 'Opium and the Beginnings of Chinese Capitalism in Southeast Asia', *Journal of Southeast Asian Studies*, 33:2 (2002), 297–314.
160. David Grigg, 'The Worlds of Tea and Coffee: Patterns of Consumption', *GeoJournal*, 57:4 (2002), 283–294.
161. Royal Society, *Register Book*, vol. 9. fol. 71 (no author or date, placed between documents dated 1701 and 1703), 'A Memoriall to live under the Equinoctial and between the Tropicks; the Common Distemper Incident to

the Inhabitants of Those Climates with Ye Manner of Cure by the Natives and Likewise by Other European Practitioners in Physick.'

162. Reports of the revenue received from the tobacco, betel, *'ganjee'* (cannabis), and liquor farmers occur frequently throughout the Fort St George and Fort St David records. See, for example, Fort St George, *Public Consultations*, vol. XXIX (TNSA), fols. 215–216. *Public Despatches from Court* (TNSA), vol. XVII, fols. 1–6.
163. Anthony Reid, 'From Betel-Chewing to Tobacco-Smoking in Indonesia', *Journal of Asian Studies*, 44:3 (1985), 529–547.
164. On Southeast Asian spirits and their persistence among converts to world religions, Anthony Reid, *Southeast Asia in the Age of Commerce*, II, 136–143.
165. See also OR. 70, fol. 17, where *bakti* باكتي is translated as 'munificent', from which is derived *kabaktien*, 'benefactions'.
166. Collet's diary is in BL IOPP MSS Eur D1153.
167. Reid, *Southeast Asia in the Age of Commerce*, II, 157.
168. For example, IOR G/35/3, consultations of the Bencoulen factory, 1695, fols. 92, 199, 15–17. See also Veevers, 'The Company as Their Lords and the Deputy as a Great Rajah', 696.
169. Reid, *Southeast Asia in the Age of Commerce*, II, 316.
170. Reid, *Southeast Asia in the Age of Commerce*, II, 298–303, 319.
171. BL IOR MS Eur photo 149, fols. 45–47 (Hyde's translation).
172. Cf. Gunn, *First Globalisation: The Eurasian Exchange, 1500–1800*, 41.
173. John Ogilby, *Atlas Japannensis (from the Dutch of Arnoldus Montanus)* (London, 1670).
174. Paullette Dellios, 'A Lexical Odyssey from the Malay World', *Studia Universitatis Petru Maior – Philologia*, 4, 141–144: 144. Marsden gives 'Orang-utan' as 'man of the woods' under the entry for 'Orang' ('man').
175. Marsden, *A Dictionary of the Malayan Language: To Which Is Prefixed a Grammar.*
176. Kader, 'Some Aspects of Seventeenth-Century Malay', 91.
177. Marsden, *A Dictionary of the Malayan Language: To Which Is Prefixed a Grammar*, 48
178. Ulug Beg, Hyde, *Ǧadāwil-i Mawāī'*. For the involvement of the Royal Society, *Minutes of Council*, I, 1663–1681, May 1664.
179. BL APAC Printed Book 15298 a.30; Thomas Hyde, *Historia Religionis Veterum Persarum* (Oxonii: e Theatro Sheldoniano, 1700), cap. XVIII.
180. Golvers, *Chinese Heaven*, p. 18.
181. Hyde, *Historia Religionis Veterum Persarum* (Cap. XVII for the connection between the ancient Persian and Chinese calendars and Cap. I, 23, for the transmission of the ancient Persian religion to the island of Gilolo).
182. Reid, *Southeast Asia in the Age of Commerce*, II, 180.
183. The original prints are in Hyde's manuscripts in BL Or.70.bb.9. Some were also printed in Hyde, *Syntagma dissertationum*, ii, 533–559.
184. Brook, *Mr Seldon's Map of China.*
185. Brook, *Mr Seldon's Map of China*, 101–102.
186. Bl Sloane Or. 853a fol. 23.
187. Brook, *Mr Seldon's Map of China*, plates 10 and 12.
188. Sloane Or. 853a fol. 23.

189. IOPP MSS E. 192, item 8 fol. 1 – Thomas Hyde to Thomas Bowrey, 7 February 1700.
190. Sloane Or. 853a fol. 2.
191. Sloane Or. 853a fol. 11.
192. Brook, *Mr Seldon's Map of China*, 90.
193. Reid, *Southeast Asia in the Age of Commerce*, II, 43.
194. Reid, *Southeast Asia in the Age of Commerce*, II, 10–16.
195. Reid, *Southeast Asia in the Age of Commerce*, II, 308, 311.
196. K. A. Adelaar, 'Chinese: First and Contact Languages (Indonesia, Malaysia)', map 73, in Wurm, Mühlhäusler, and Tryon, *Atlas of Languages of Intercultural Communication*.
197. Reid, *Southeast Asia in the Age of Commerce*, II, 312.
198. Anthony Reid, 'Chinese on the Mining Frontier in Southeast Asia', in Eric Tagliacozzo and Wen-Chin Chang (eds.) *Chinese Circulations: Capital, Commodities, and Networks in Southeast Asia* (Duke University Press: Durham and London, 2011), 21–36.
199. Blussé, *Strange Company*.
200. IOR G/21/7, fols. 100–101 – Bencoulen, 8 October 1686 and G/35/1(3) fol. 72. See also Anthony Farrington, 'Bengkulu: An Anglo-Chinese Partnership', in Bowen, Lincoln and Rigby (eds.) *Worlds of the East India Company*, 111–117. For the collaboration between the English and Chinese in Bantam: Bassett, 'The Factory of the English East India Company at Bantam', 226–227.
201. Farrington, 'Bengkulu: An Anglo-Chinese Partnership', 115.
202. London Metropolitan Archives, Bowrey MS 3041/2.
203. Arne Bialuschewski, 'A True Account of the Design, and Advantages of the South-Sea Trade: Profits, Propaganda, and the Peace Preliminaries of 1711', *Huntington Library Quarterly*, 73:2 (2010), 273–285.
204. For more detail on the later uses of the *Dictionary,* see Winterbottom and Kaislaniemi, 'Perverted through Ambition'.
205. Keevak, *The Pretended Asian*; Breen, 'No Man Is an Island'.
206. IOPP MSS Eur. E. 192 (a), Item 3, fol. 1r. G. P. Salmanaazaar to Thomas Bowrey at the Garter Coffee House.
207. Terrien de Lacouperie and E. Colborne Baber, 'Formosa Notes on MSS. Races and Languages', *Journal of the Royal Asiatic Society of Great Britain and Ireland*, new series, 19 (1887), 413–494.
208. For example, Kader, 'Some Aspects of Seventeenth-Century Malay', and Gupta, 'The Imagined Learner of Malay'.
209. See also Kate Teltscher, 'Hobson-Jobson: The East India Company Lexicon', forthcoming in Bolton, Kaislaniemi, and Winterbottom (eds.) *The East India Company and Language*.

3 Toleration and Translation: English Versions of Two Hindu Texts from Bengal

1. 'A familiar & free Dialogue betwixt John Marshall and Muddoosoodun Rauree Brahmin at Cassimbazaar in Bengall in East India, began 18 March 1674/5', BL MS Harley 4253, fol. 2.

2. Ravi M. Gupta and Kenneth R. Valpey eds. *The Bhāgavata Purāṇa: Sacred Text and Living Tradition* (New York: Columbia University Press, 2013).
3. BL MS Harley 4256, fols. 1–230v.
4. Stern, *Company State.*
5. Carolien Stolte, *Philip Angel's Deex-Autaers: Vaiṣṇava Mythology from Manuscript to Book Market in the Context of the Dutch East India Company, c. 1600–1672.* Dutch Sources on South Asia c. 1600–1825, V (New Delhi: Manohar, 2012).
6. The original was composed in Marathi in 1616 and published in three early editions, no copies of which survive. Thomas Stephens, *Kristapūraṇa*, Fr. Nelson Falcao trans. (Bangalore: Kristu Jyoti College, 2012), was transcribed from several Marathi manuscripts with a parallel English translation.
7. Ludo Rocher, *Ezourvedam, A French Veda of the Eighteenth Century* (Amsterdam: John Benjamins Publishing Company, 1984).
8. See especially Rosalind O'Hanlon, 'Performance and the World of Paper: Puranic Histories and Social Communication in Early Modern India', *Past and Present*, 219 (2013). Sumit Guha, 'Serving the Barbarian to Preserve the Dharma: The Ideology and Clerical Elite in Peninsular India, c. 1200–1800', *Indian Economic and Social History Review*, 47:4 (2010), 497–525; Raman, *Document Raj.*
9. Chaudhuri, *Trading World of Asia and the EIC*, 52.
10. Anjali Chatterjee, *Bengal in the Reign of Aurangzib, 1658–1707* (Calcutta: Progressive Publishers, 1967), 92–94.
11. Chatterjee, *Bengal in the Reign of Aurangzib*, 204.
12. Richard M. Eaton, *The Rise of Islam and the Bengal Frontier, 1204–1760* (Berkeley: University of California Press, 1993); S. K. Abdul Latif, *The Muslim Mystic Movement in Bengal 1301–1550* (Calcutta: KP Bacgchi & Co., 1993).
13. Shafaat Ahmad Khan, *John Marshall in India: Notes and Observations in Bengal (1668–1672)*, (Oxford: Oxford University Press, 1927), 7–8. Eaton, *Rise of Islam*, 270.
14. Syed Jamil Ahmed, 'Mimicry and Counter-Discourse on the Palimpsest of Nagori: The Play of Saint Anthony and His Double, Sādhu Āntoni', *Asian Theatre Journal*, 27:1 (2010).
15. Chatterjee, *Bengal in the Reign of Aurangzib*, 226.
16. Muzaffar Alam and Sanjay Subrahmanyam, *Writing the Mughal World: Studies on Culture and Politics* (New York: Columbia University Press, 2012), 286.
17. See, for example, Katherine Butler Brown, 'Did Aurangzeb Ban Music? Questions for the Historiography of His Reign', *Modern Asian Studies*, 41:1 (2007).
18. Satish Chandra, '*Jizyah* and the State in India during the 17th Century', *Journal of the Economic and Social History of the Orient*, 12:3 (1969), 322–340.
19. Chandra, '*Jizyah* and the State'.
20. Jadunath Sarkar, *Shivaji and His Times* (London: Longmans, Green and Co, 1920); Bhat Viśveśvara and Vasudeo S. Bendrey, *Coronation of Shivaji the Great* (Bombay: P. P. H. Bookstall, 1960).
21. Rosalind O'Hanlon and Christopher Minkowski, 'What Makes People Who They Are? Pandit Networks and the Problem of Livelihoods in Early Modern Western India', *Indian Economic and Social History Review*, 45:3 (2008), 381–416; O'Hanlon, 'Performance in a World of Paper'.

22. Stephens, *Kristapurāṇa.*
23. Rev. Bishop Thomas Dabre, 'Foreword' in Stephens, *Kristapurāṇa*, li–liii.
24. Stephens, *Kristapurana*, 'Preface', x–xi.
25. Ines G. Županov, 'Conversion Historiography in South Asia: Alternative Indian Christian Counter-Histories in Eighteenth Century Goa', *Medieval History Journal*, 12:2 (2009), 303–325; Stolte, *Deex-Autaers*, 24–29.
26. His Tamil translations of scripture remain in the Panaji Central Library in Goa. Roberto Nobili, *Cathecismo Mandurense*, pts. 1 and 2 (Tranvancorensi, 3 October 1674): some of vol. 1 is in print and some handwritten. The introductions to both parts are in Latin. For the criticism of Nobili by Gonçales Fernandes Trancosa, see Stolte, *Deex-Autaers*, 27.
27. Ines G. Županov ' "The Wheel of Torments": Mobility and Redemption in Portuguese Colonial India (Sixteenth Century)', in Greenblatt (ed.) *Cultural Mobility*, 24–74.
28. Ahmed, 'Mimicry and Counter-Discourse on the Palimpsest of Nagori'.
29. There is less literature on the 'Malabar' rites controversy than on the Chinese. Ines G. Županov, *Disputed Mission: Jesuit Experiments and Brahmanical Knowledge in Seventeenth-Century India* (New Delhi: Oxford University Press, 1999), 199, fn 13.
30. Israel, *Radical Enlightenment.*
31. Robert Iliffe, *Newton: The Making of a Politician* (2013). Newton Project, University of Sussex http://www.newtonproject.sussex.ac.uk/view/contexts/CNTX00002 (accessed 4 March 2014).
32. B. A. Gerrish, 'Natural and Revealed Religion', in Knud Haakonssen (ed.) *The Cambridge History of Eighteenth-Century Philosophy* (Cambridge: Cambridge University Press, 2006), 641–665.
33. Stern, *Company State*, 103.
34. Kapil Raj, 'The Historical Anatomy of a Contact Zone', *Indian Economic and Social History Review*, 48:1 (2011), 55–82, 60, makes this point for the later period.
35. Stern, *Company State*, 103–104.
36. Records of a court held at the Town Hall, 22 January 1717, Tamil Nadu State Archives (TNSA), Mayoral Court Records, II, 1716–17, fols. 29–32.
37. Geo Augier, Charles James, Cesar Chamberlain in Surat to Bombay, c. 1677, MSA, Surat Outwards Correspondence Book, III, fols. 22–23.
38. Henry Lord, *A Display of Two Forraigne Sects in the East Indies Vizt: The Sect of the Banians the Ancient Natiues of India and the Sect of the Persees the Ancient Inhabitants of Persia* (London: Francis Constable, 1630).
39. Kate Teltscher, ' "Maidenly and Well Nigh Effeminate": Constructions of Hindu Masculinity and Religion in Seventeenth-Century English Texts', *Postcolonial Studies*. 3.2 (2000): 159–170. (But note the confusion of *banians* and *dubashes* on 160 based on a misreading of Bayly.)
40. See Amrita Sen and Jyotsna G. Singh, 'Classifying the Natives in Early Modern Ethnographies: Henry Lord's A Display of Two Foreign Sects in the East Indies (1630)', *Journeys*, 14:2 (2013), 69–84.
41. Lord, *Two Forraigne Sects*, II, 2–4.
42. Lord, *Two Forraigne Sects*, I, 41–45.
43. Lord, *Two Forraigne Sects*, I, 94.

44. Stolte, *Deex-Autaers.*
45. Stolte, *Deex Autaers,* 78.
46. Peter Burke, 'The Philosopher as Traveller: Bernier's Orient', in Jaś Elsner and Joan-Pau Rubiés (eds.) *Voyages and Visions: Towards a Cultural History of Travel* (London: Reaktion Books, 1999), 124–137.
47. Stotle, *Deex Autaers.*
48. Joan-Pau Rubiés, 'Reassessing "the Discovery of Hinduism": Jesuit Discourse on Gentile Idolatry and the European Republic of Letters', in Anand Amaladass and Ines G. Ẑupanov (eds.) *Intercultural Encounters and the Jesuit Mission in South Asia (16th–18th centuries)* (Bangalore: Asian Trading Corporation, 2014), 11.
49. Athanasius Kircher, *China Illustrata* (1677), trans. by Charles D. Van Tuyl (Muskogee, OK: Indian University Press, Bacone College, 1987).
50. Kircher, *China Illustrata;* Stotle, *Philip Angel's Deex-Autaers,* 32.
51. Paula Findlen, 'The Last Man Who Knew Everything…Or Did He?', in Paula Findlen ed. *Athanasius Kircher: The Last Man Who Knew Everything* (New York and London: Routledge, 2004).
52. Subrahmanyam, *Mughals and Franks.*
53. Isaac Pyke, 'The Stringers Journall at & from Bombay Continued', British Library APAC, MS Eur D5. Pyke was at different times Governor of St Helena and Bencoulen, as discussed in Chapter 6.
54. Chatterjee, *Bengal in the Reign of Aurangzib,* 109–110. The grant (*nishan*) is translated in Temple (ed.) *Diary of Streynsham Master,* II, 21–22.
55. Jagadish Narayan Sarkar, 'Mīr Jumla's Overseas Commercial Activities', *JOBRS* (1945), 262–264; *The Life of Mir Jumla, the General of Aurangzab* (Calcutta: Thacker, Spink, 1951).
56. Chatterjee, *Bengal in the Reign of Aurangzib,* 124–151.
57. Job Charnock, Chuttanuttee, 16 February 1688 to Madras. Printed in National Archives of India, *Bengal and Madras Papers, 1670–1785* (Government Press: Calcutta, 1928).
58. Rila Mukherjee, 'The Story of Kasimbazar: Silk Merchants and Commerce in Eighteenth-Century India', *Review* (Fernand Braudel Center) 17:4 (1994), 499–554.
59. Chatterjee, *Bengal in the Reign of Aurangzib,* 109–121.
60. Khan, *John Marshall,* 67–68.
61. British Library cataloguing notes to Kasimbazar factory records G/23/1.
62. Khan, *John Marshall,* 67–8, 70–71, 78. Khan's *parwāna* is translated in Richard C. Temple, *The Diaries of Streynsham Master, 1675–1680: And Other Contemporary Papers Relating Thereto,* 2 vols. (London: J. Murray, 1911), II, 22–24.
63. For the continuation of this practice into the colonial period, see Bayly, *Empire and Information.*
64. Temple, *Diary of Streynsham Master,* I, 339 and II, 318.
65. Khan, *John Marshall in India,* 21. *Diary of Streynsham Master,* I, 370–377.
66. MS Harley 4253, fol. 1.
67. Khan, *John Marshall in India,* 29; Ayesha Irani, personal communication. 'Rāḍha' may also be used as a variation of 'Rāṛh' in reference to the region.
68. MS Harley 4253, fol. 1.

69. Ferdinando Sardella and Abhishek Ghosh, 'Text History: Modern Reception and Text Migration of the Bhāgavata Purāṇa', in Gupta and Valpey (eds.) *The Bhāgavata Purāṇa*, 221–247.
70. Kunal Chakrabarti, *Religious Process: The Pūraṇas and the Making of a Regional Tradition* (New York: Oxford University Press, 2001).
71. C. Mackenzie Brown, *The Triumph of the Goddess* (Delhi: Sri Satguru Publications, 1990).
72. Henry More, *A Collection of Several Philosophical Writings of Dr Henry More: As Namely, His Antidote against Atheism, Appendix to the Said Antidote, Enthusiasmus Triumphatus, Letters to Des-Cartes, &c., Immortality of the Soul, Conjectura Cabbalistica* (London: printed by James Flesher for William Morden, 1662).
73. Sarah Hutton, 'More, Henry (1614–1687)', *Oxford Dictionary of National Biography*, Oxford University Press, 2004; online edn, Jan 2008 http://www.oxforddnb.com/view/article/19181 (accessed 13 June 2013).
74. John Covel's diary of his travels is published in J. T. Bent, *Early Voyages and Travels in the Levant* (Farnham, Surrey: Ashgate, 2010).
75. E. S. de Beer ed., *The Correspondence of John Locke: Letters nos. 462–848*. 8 vols. (Oxford: Oxford University Press, 1976–1989); vol. 2, p. 22–23 (Letter 471). Digital edition by Electronic Enlightenment. http://www.e-enlightenment.com/item/lockjoOU0070177_1key001cor/eds/999/ (accessed 14 October 2014).
76. Khan, *John Marshall*, 2.
77. Ethel Bruce Sainsbury and William Foster, *Court Minutes of the East India Company 1668–1670* (Oxford: Clarendon Press, 1929) 4 (Court of Committees, 6 January 1668 (Court Book, XXVI, 175)); 7 (Court of Committees, 13 January 1668 (Court Book XXVI, 181)) and 10 (Court of Committees, 20 January 1668 (Court Book XXVI, 189)). Craven is probably William, Earl of Craven (1608–1697). Marshall's younger brother Robert, a Company factor in Bantam, accompanied the Malay embassy of 1682 to London (Chapter 2).
78. Khan, *John Marshall*, 5–7. The journal of the *Unicorn*'s voyage kept by Captain Thomas Harman was apparently retained by Marshall and sent with some of his other papers to Covel, eventually becoming MS Harley 4252.
79. Khan, *John Marshall*, 7–9.
80. Khan, *John Marshall*, 20.
81. Khan, *John Marshall*, 11–12.
82. John Marshall, Last will and testament, National Archives, Kew, Surrey, UK (Probate 11/360).
83. BL Add MS 22910, fol. 109, John Marshall to John Covel, Kazimbazaar, 29 November 1685.
84. John Marshall, 'A Letter from the East Indies, of Mr John Marshal to Dr Coga, Giving an Account of the Religion, Rites, Notions, Customs, Manners of the Heathen Priests Commonly Called Bramines. Communicated by the Reverend Mr Abraham de La Pryme', *Philosophical Transactions (1683–1775)*, 22 (1700–1701), 729–738.
85. Gupta and Valpey, 'Churning the Ocean of *Līlā*'.
86. O'Hanlon, 'Performance and the world of paper', 117.

87. Washbrook and O'Hanlon, 'Introduction', in David Washbrook and Rosalind O'Hanlon (eds.) *Religious Cultures in Early Modern India: New Perspectives* (New Delhi: Routledge, 2011).
88. Khan, *John Marshall*, 22.
89. Khan, *John Marshall*, 36–37.
90. Stern, *Company State.*
91. Marshall, Last will.
92. Marshall, Last will. According to Streynsham Master (*Diary of Streynsham Master*, II, 237), the executor was in fact another brother, William Marshall.
93. MS Harley 7199 seems to be a later copy of part of MS Harley 4256.
94. MS Harley 4253, fols. 1r–8v 'A familiar & free Dialogue betwixt John Marshall and Muddoosoodun Raure [Madhusūdana Rāṛhi] Brahmin at Cassumbuzar in Bengall in East India began the 18th March 1674/5'; MS 4254–5 'Notes and Observations' and MS Harley 4256 'the Hindu book called Sreebaugabat poram' (which mirrors the text in MS Harley 4253 up until fol. 29 and continues the story up until the final folio 230v) are in Marshall's hand.
95. 'An accot. of the hindoos Book called Seebaugabatporam [Śri Bhāgavata Pūraṇa]', MS 4253 (2), fols. 9r–40r and BL IOR MS Eur. C 461 are not in Marshall's hand.
96. MS Eur C. 461, fol. 7.
97. MS Harley 4256, 190.
98. Marshall, Last will.
99. MS Harley 4253, fol. 32; Khan, *John Marshall*, 422–423.
100. BL Oriental Manuscripts (OMS), Add 7038.
101. For comparison, see O'Hanlon, 'Performance and the World of Paper', 118–9, on the verses recited by Raghoba Mahadevrao and written down by Arthur Crawford. See also Christian Lee Novetzke, 'The Brahmin Double: The Brahminical Construction of Anti-Brahminism and Anti-Caste Sentiment in the Religious Cultures of Precolonial Maharashtra', in Washbrook and O'Hanlon eds. *Religious Cultures.*
102. From their neater script, it seems more likely that the amanuensis made the copies from Marshall's originals.
103. See Edward C. Dimock, *The Place of the Hidden Moon: Erotic Mysticism in the Vaisnava-Sahajiyā Cult of Bengal* (Chicago: University of Chicago Press, 1966), 27, for discussion of an ancient Bengali saying that whoever listens to the 18 *mahapuranas* recited in the vernacular would be cast into hell and the shift in this attitude after the fifteenth century.
104. Sindhu S. Dange, *The Bhāgavata Purāṇa: Mytho-Social Study* (Delhi: Ajanta Publications, 1984). D. Dennis Hudson, *Krishna's Mandala: Bhagavata Religion and Beyond* (New Delhi: Oxford University Press, 2010), suggests that books 1–6 were composed as early as the seventh century BCE.
105. Wendy Doniger, *The Hindus: An Alternative History* (New York: Penguin Press, 2009). For a table of rival dates, Ganesh Vasudeo Tagare, 'Introduction', in *The Bhāgavata Pūraṇa*, translated and annotated by Ganesh Vasudeo Tagare (Delhi: Motilal Banarsidass, 1976), xxxxiv.
106. Personal communication, Ayesha Irani. For these forms of vernacular Bengali devotional literature focused on Krishna, see Eaton, *The Rise of Islam and the Bengal Frontier, 1204–1760*, 110.
107. MS Harley 4256, fols. 22v–23r.

108. MS Harley 4256, fols. 24r–25v.
109. MS Harley 4256, fols. 52–53v.
110. MS Harley 4256, fol. 93.
111. MS Harley 4256, fols. 108v–109r and fols. 121 v–123r.
112. MS Harley 4256, fols. 109v–110r.
113. Tagare, 'Introduction', l–li, notes that Rādha being the *raison d'être* of some Vaiṣṇava sects; they interpret the verses in Book 10 of the BhP as indicating her presence.
114. Ayesha Irani, personal communication.
115. MS Harley 4256, fol. 29 v.
116. O'Hanlon, 'Performance and the World of Paper', 120.
117. MS Harley 4256, fols. 52–53v.
118. MS Harley 4256, fols. 53–53v.
119. MS Harley 4256, fol. 108.
120. IO MS Eur. C 461, fol. 15.
121. MS Harley 4256, fols. 102, 104, 109, 111.
122. MS Harley 4256, fol. 112v.
123. IOR MS Eur. C 461, fol. 7.
124. Devi Chand trans., *The Sāmaveda* (Delhi: Munashiram Manoharlal, 1981).
125. MS Harley 4256, fol. 141.
126. IOR MS Eur. C 461, fols. 7, 20.
127. IOR MS Eur C. 461, fol. 7.
128. Slightly different versions of the story of the cosmic man appear on IOR MS Eur. C 461, fol. 18 and fol. 26.
129. IOR MS Eur. C 461, fols. 17 and 25.
130. IOR MS Eur. C 461, fol. 17.
131. IOR MS Eur. C 461, fol. 25.
132. Christian Lee Novetzke, 'The Brahmin Double: The Brahminical Construction of Anti-Brahminism and Anti-Caste Sentiment in the Religious Cultures of Precolonial Maharashtra', *South Asian History and Culture*, 2:2 (2011), 232–252.
133. MS Harley 4256, fol. 216v.
134. On the identities of Śuka ('the parrot') and Parīksit ('the examiner'), see Wendy Doniger, 'Echoes of the Mahābhārata: Why Is a Parrot the Narrator of the Bhāgavata Pūraṇa and the Devībhāgavata Pūraṇa?', in Wendy Doniger ed., *Purāna Perennis: Reciprocity and Transformation in Hindu and Jaina Texts* (Albany, NY: State University of New York Press, 1993), 31–57.
135. Henry More, D. D., *Divine Dialogues, Containing Disquisitions Concerning the Attributes and Providence of God*, 4 edn., 3 vols. (Glasgow: Robert Foulis and John Paton, 1743), I, 16–17.
136. Marshall, 'A Letter from the East Indies', p. 729.
137. Gerrish, 'Natural and Revealed Religion'.
138. Compare Daniel Stolzenburg, 'Four Trees, Some Amulets, and the Seventy-Two Names of God: Kircher Reveals the Kabbalah', in Paula Findlen ed. *Athanasius Kircher*, 149–169, 161, where he notes that most Christian interpretations of the Kabbalah traced it to Moses, whereas Kircher traced both the Kabbalah and the Bible to Adam. As he notes, the attribution of a

common wisdom of mankind to Adam left the importance of revelation questionable.

139. MS Harley 4253, fol. 3.
140. Jonathan B. Edelmann, 'Cosmology – Dialogues on Natural Theology: The Bhāgavata Purāṇa's Cosmology as Religious Practise', in Gupta and Kenneth R. Valpey (eds.) *The Bhāgavata Purāṇa*, 48–62; Tagore, 'Introduction', xviii.
141. I am grateful to Dr Ayesha Irani for identifying 'Beedy' as *bīja*.
142. Marshall, 'A Letter from the East Indies', 730.
143. MS Harley 4253, fol. 6.
144. E. H. 'Rick' Jarow, 'A *Maṇḍala* of Remembrance: The Bhāgavata Purāṇa's Time-Transcending Narrative of Līlā', 36–47, in Gupta and Valpey ed. *The Bhāgavata Purāṇa*.
145. I am grateful to Ayesha Irani for identifying 'teas' as *tejas*. In Tagare ed., *Bhāgavata Pūraṇa*, God's power is identified instead with *māyā* (illusion) but *tejas* is assigned a creative role in II:5, 27–28.
146. MS Harley 4253, fol. 3.
147. Sardella and Ghosh, 'Text History', pp. 222–3.
148. MS Harley 4256, fol. 104.
149. MS Harley 4256, fol. 225.
150. Plato, *Phaedo*, R.S. Bluck trans. (Cambridge: Cambridge University Press, 1955).
151. René Descartes, *Meditations on First Philosophy in Focus*, Stanley Tweyman (ed.) (London: Routledge 1993).
152. Margaret J. Osler (1998). 'Gassendi, Pierre', in E. Craig (ed.) *Routledge Encyclopedia of Philosophy*. London: Routledge. http://0-www.rep.routledge.com.catalogue.ulrls.lon.ac.uk:80/article/DA036SECT4 (accessed 27 March 2007).
153. John Henry, 'A Cambridge Platonist's Materialism: Henry More and the Concept of Soul', *Journal of the Warburg and Courtauld Institutes*, vol. 49 (1986), 172–195.
154. No Sanskrit equivalents are given for the English terms used in these passages.
155. MS Harley 4253, fol. 11.
156. Tagare, 'Introduction', xxxiii and lviii. Tagare makes the argument that the BhP attempts a harmonisation of the Vedantic concept of *antahkarana* (the levels of mind) with *vyūtha*, but that it mainly fails to achieve this.
157. *Paradise Lost*, VII, 163–71. Cf. Anna Baldwin, 'Platonic Ascents and Descents in Milton', in Anna P. Baldwin and Sarah Hutton (eds.) *Platonism and the English Imagination* (Cambridge: Cambridge University Press, 1994), 151–162.
158. More, *Antidote to Atheism* (Chapter IV).
159. More, *Divine Dialogues*, I, 84.
160. Ralph Cudworth, *True Intellectual System of the Universe: The First Part; Wherein, All the Reason and Philosophy of Atheism Is Confuted; and Its Impossibility Demonstrated* (London: printed for R. Royston, 1678).
161. More, 'Introduction', *Divine Dialogues*.
162. Henry, 'A Cambridge Platonist's Materialism'.
163. Gupta and Valpey, 'Churning the Ocean of *Līlā*'; Selva J. Raj and Corinne G. Dempsey, *Sacred Play: Ritual Levity and Humor in South Asian Religions* (Albany: State University of New York Press, 2010).

164. MS Harley 4253, fol. 7.
165. Jarow, 'A *Maṇḍala* of Remembrance'.
166. More, *Divine Dialogues*, 51–52. The implication here that nature is changeable is radical for the early modern period, when most European philosophers regarded natural history as stable, in contrast to human history.
167. Mark Goldie, 'Cambridge Platonists (Act. 1630s–1680s)', *Oxford Dictionary of National Biography*, Oxford University Press. http://www.oxforddnb.com/view/theme/94274 (accessed 17 February 2014).
168. More, *Antidote to Atheism* (Chapter V).
169. More, *Divine Dialogues*, III, 334.
170. More, 'Preface', *Antidote to Atheism.*
171. Graham M. Schweig, *Dance of Divine Love: The* Rāsa Līla *of the* Bhāgavata Purāna (Princeton and Oxford: Princeton University Press, 2005).
172. MS Harley 4256, fol. 125.
173. MS Harley 4256, fol. 118v.
174. Damaris Masham, *A Discourse Concerning the Love of God* (London: printed for Awnsham and John Churchill, 1696).
175. Masham, *Discourse*, 22.
176. Dimmock, *The Place of the Hidden Moon.*
177. MS Harley 4256, fol. 12.
178. For a summary of the European debates, Rob Iliffe, 'The Masculine Birth of Time: Temporal Frameworks of Early Modern Natural Philosophy', *British Journal for the History of Science*, 33 (2000).
179. MS Harley 4256, fols. 22v–23r, on Vishnu's incarnation as Varaha. Marshall writes in the margin: 'since which time till the 2nd of July 1674 is . . . years', apparently with the intention of filling in the number of years later.
180. Khan, *John Marshall,* 190 and 265. MS Harley 4256, fols. 223v–224.
181. Stolte, *Deex-Autaers*, 110 and 212.
182. The age of the Earth was considered to be 4,004 years according to Archbishop James Ussher (1581–1656). Jonathan B. Edelmann, *Hindu Theology and Biology: The Bhagavata Purana and Contemporary Theory* (Oxford: Oxford University Press, 2012), 45.
183. Khan, *John Marshall*, 181.
184. MS Harley 4256 fols. 223v–224r. Compare Ebenezer Burgess, 'Translation of the Sûrya-Siddhânta, A Text-Book of Hindu Astronomy; with Notes, and an Appendix', *Journal of the American Oriental Society*, 6 (1858–1860), 141–498.
185. IOR MS Eur. C. 461, fol. 16r.
186. MS Harley 4254, fol. 11r.
187. MS Harley 4254, fol. 224.
188. On the themes of death and their relationship to the BhP's philosophy of time, see Jarow, *Tales for the Dying.*
189. IOR B/ 35, f. 139, Court Minutes, 12 September 1679.
190. W. Salmon to J. Covel, Streatham, 19 April [no year given]. Letter enclosed with IOR MS Eur. C 461.
191. Cyril Ernest Wright, *Fontes Harleiani: A Study of the Sources of the Harleian Collection of Manuscripts Preserved in the Department of Manuscripts in the British Museum* (London: The British Museum, 1972).
192. MS Harley 4256 fol. 1; Hyde, *Historia Religionis Veterum* Persarum.
193. MS Harley 4256 fol. 1; Hyde, *Historia*, pp. 169–171, 177, 181.

194. Iliffe, 'Newton: The Making of a Politician'.
195. H. W. Turnbull (ed.) *The Correspondence of Isaac Newton,* II, 1676–1687 (Cambridge: Cambridge University Press for the Royal Society, 1960).
196. Thomas Hyde, *Historia,* 30: *'Nam Indoram Idololatricorum (in suâ vix intelligibili Religione mysteriosâ, quae tam specioso nomini non respondet) Antistes olim primarius perhibetur* برهمة *Brahma, seu Brâhama, qui ab Abrahami nomine Hebraico sic vocatus creditor, ipsum hoc nomine indigitando.'* See also the entry on Hyde in Bayle, Pierre and Pierre Desmaizeaux, *The Dictionary Historical and Critical of Mr Peter Bayle* (London: printed for J. J. and P. Knapton, 1743), 343, fn. D.
197. Isaac Newton, *Chronology of Ancient Kingdoms Amended*... (London: Tonson, Osborn and Longman, 1728), 348–349. See also Bruce Lincoln, 'Isaac Newton and Oriental Jones on Myth, Ancient History, and the Relative Prestige of Peoples', *History of Religions,* 42:1 (2002), 1–18.
198. John Locke to John Covel, Oates 25 Oct 1700 (5 November 1700). Beer ed. *Correspondence of John Locke,* Digital edition by Electronic Enlightenment. McNamee et al. eds. http://dx.doi.org/10.13051/ee:doc/lockjoOU00 70172a1c (accessed 21 February 2014).
199. John Covel to John Locke: Wednesday, Oct. 30. 1700 (10 November 1700), Beer ed. Digital edition by Electronic Enlightenment. McNamee et al. eds. http://dx.doi.org/10.13051/ee:doc/lockjoOU0070 177a1c (accessed 21 February 2014).
200. John Locke to John Covel, Oates 3 Nov. 1700 (14 November 1700). De Beer ed. Digital edition by Electronic Enlightenment. McNamee et al. eds. http: //dx.doi.org/10.13051/ee:doc/lockjoOU0070182b1c (accessed 21 February 2014).
201. P. L. Heyworth. *Letters of Humfrey Wanley: Palaeographer, Anglo-Saxonist, Librarian, 1672–1726: With an Appendix of Documents* (Oxford: Clarendon Press, 1989), item 169, Humphrey Wanley to Thomas Tudway, 23 February 1716.
202. Heyworth. *Letters of Humfrey Wanley,* item 162, 324, Humphrey Wanley to John Covel, 24 January 1716.
203. BL Add MSS. 7037–7040.
204. Baldwin and Hutton (ed.) *Platonism,* 74.
205. Pierre Martin, ' "Lettre du Pere Pierre Martin, Missionaire de la Compagnie de Jesus, au Pere de Villette de la même Compagnie", Balassor, Bengale, le 30 Janvier 1699', *Lettres Edifiantes,* 10 (1730–1732), 22–32.
206. Pierre Martin, ' "Lettre du Père Pierre Martin...au Père le Gobien de la même Compagnie", Camien-naixen-patty, Maduré, le 1er de Juin 1700', *Lettres Edifiantes,* 10 (1730–1732), 33–81, 54. The original of the passage quoted reads: *'lire & transcrire les livres que le vénérable Pere Robert de Nobilibus & nos autres Peres ont composés'.*
207. Rocher, *Ezourvedam,* introduction.
208. J. B. P. More (ed.) *Bagavadam OU BHAGAVATA PURANA Ouvrage religieux et philosophique indien traduit par Maridas Poullé de Pondichéry en 1769* (Paris: Éditions IRISH, 2004).
209. For Gentil's maps, many containing illustrations of Hindu deities, Susan Gole (ed.) *Maps of Mughal India, Drawn by Colonel Jean-Baptiste Joseph Gentil,*

Agent for the French Government to the Court of Shuja-ud-Daula at Faizabad in 1770 (Kegan Paul International: London and New York, 1988);

210. More (ed.) *Bagavadam*, 8.
211. Rocher, *Ezourvedam*, 110–111 – *'tu as inventé différentes incarnations que tu attribue toutes a ce* Vichenou, *tu as entretenu le monde de ces reveries… tu as fait oublier aux hommes jusqu'au même nom de Dieu, tu as plongé dans l'idolatrie et ils y ont même pris gouter'.*
212. Rubiés, 'Reassessing "the Discovery of Hinduism" ', 15, notes the closeness of many of the methods employed by the Tranquebar missionaries to their Jesuit predecessors.
213. For discussion of the relevance of the BhP for modern scientific concepts, Edelmann, *Hindu Theology and Biology*.
214. Paper (now lost) read by Prof. E. B. Cowell to the Cambridge Philological Society, 17 April 1872, cf. John Peile, *Biographical Register of Christ's College, 1505–1905, and of the Earlier Foundation, God's House, 1448–1505*, 2 vols. (Cambridge: Cambridge University Press, 1910), I, 592.

4 Botanical and Medical Networks: Madras through the Collections of Two EIC Surgeons

1. D. E. Allen, 'Petiver, James (c. 1665–1718)', in *Oxford Dictionary of National Biography*, Oxford University http://www.oxforddnb.com/view/article/22041 (accessed 26 March 2015).
2. D. G. Crawford, *History of the Indian Medical Service, 1600–1913*, 2 vols. (London: W. Thacker & Co., 1914), I, 91. Browne was previously the surgeon of a ship called the *Dragon*.
3. The first letter from Petiver to Browne is dated 61 March 1689, BL MS Sloane 3332, fols. 6–7. Petiver sends a book on English botany and requests a correspondence. Browne's burial on 22 December 1698 is recorded in the 'Book of Marriages, Burials and Christenings' kept in the museum at the site of Fort St George in Chennai.
4. Crawford, *History of the IMS*, I, 245.
5. The specimens sent by Samuel Baron are numbered consecutively in the NHM volumes, and in his descriptions of the plants, James Petiver designated them by these numbers and the initial 'SB' below, which system is followed below.
6. Nair, 'Native Collecting and Natural Knowledge', *Raja Serfoji*.
7. Kapil Raj, 'Surgeons, Fakirs, Merchants and Craftspeople: Making L'Empereur's Jardin in Early Modern South Asia', in Swan and Schiebinger (eds.) *Colonial Botany*.
8. John Gascoigne, 'The Royal Society, Natural History and the Peoples of the "New World(s)", 1660–1800', *British Journal for the History of Science*, 42 (2009), 539–562; Carey, *Asian Travel in the Renaissance*; Lux and Cook, 'Closed Circles or Open Networks?'; Thomas, 'A "Philosophical Storehouse" '.
9. Richard Grove, *Green Imperialism: Colonial Expansion, Tropical Island Edens, and the Origins of Environmentalism, 1600–1860* (Cambridge: Cambridge University Press, 1995), 79. Mark Harrison, *Climates and Constitutions:*

Health, Race, Environment and British Imperialism in India (New Delhi: Cambridge University Press, 2002), 8–9, argues that any clear division between 'Eastern' and 'Western' medicine is illusory.

10. Chakrabarti, *Materials and Medicine*, 9.
11. Chakrabarti, *Materials and Medicine*, 9.
12. Chakrabarti, *Materials and Medicine*, 12.
13. See Natural History Museum, 'Reconstructing Sloane', Centre for Arts and Humanities Research, Projects and Fellowships, http://www.nhm.ac.uk/research-curation/science-facilities/cahr/projects-partnerships/sloane/index.html (accessed 26 March 2015).
14. David Arnold, *Warm Climates and Western Medicine: The Emergence of Tropical Medicine, 1500–1900* (Amsterdam: Rodopi, 1996), 11.
15. Richard H. Drayton, *Nature's Government: Science, Imperial Britain, and the 'Improvement' of the World* (New Haven: Yale University Press, 2000).
16. Cook, *Matters of Exchange* (chapter 1).
17. BL MS Add 34123 – 'Copybook of Henry Vansittart, Governor of Bombay, 1756', fols. 40–42: 'Samuel Baron's account of the trade of India, written from Fort St George in 1695'.
18. Chaudhuri, *The Trading World of Asia and the East India Company*, 51.
19. J. F. Richards, *Mughal Administration in Golconda* (Oxford: Clarendon Press, 1975), 72.
20. Chakrabarti, *Matters of Exchange*.
21. Richards, *Mughal Administration in Golconda*.
22. Crawford, *History of the IMS*, 127–8. S. Alavi, *Islam and Healing: Loss and Recovery of an Indo-Muslim Tradition, 1600–1900* (Basingstoke: Palgrave Macmillan), 3.
23. Cook, *Matters of Exchange*, 180.
24. BL IOR E/3/29, OC 3144, fols. 152–162, Sir George Oxenden and Council to EIC in London, 1 January 1666.
25. For this grant, Sudipta Sen, *Empire of Free Trade: The East India Company and Making of the Colonial Marketplace* (Philadelphia: University of Pennsylvania Press, 1998), and Stern, *Company State*, 202.
26. Crawford, *History of the IMS*, I, 91. R. K. Gupta and S. R. Bakshi, *Rajasthan through the Ages* (New Delhi: Sarup & Sons, 2008), 79, mention a famous Mughal general of this same name who was defeated by the Maratha armies in 1695.
27. *Hobson-Jobson*, 412–413, for the use of the word in the military and as the holder of a *ḥawāla*, a tenure beween zamindar and ryot. See also Francis Joseph Steingass, *A Comprehensive Persian-English Dictionary* (London: Routledge & K. Paul, 1892). Digital editions of both dictionaries are available from Digital Dictionaries of South Asia, University of Chicago. Last updated 10 June 2013. URL: http://dsal.uchicago.edu/dictionaries/ (last accessed 21 May 2014). I am grateful to Minakshi Menon and Prashant Keshavmurthy for their advice about this term.
28. Letter from Sir John Goldsborough at Chuttanutti dated 14 October 1693, *Letters to FSG*, V, 164–168.
29. J. Bruce, *Annals of the Honourable East India Company* (London: Black, Parry and Kingsbury, 1810), III, 154, reproduced in Crawford, *History of the IMS*, I, 94–95, who notes that Blackwall later returned to EIC employment and

was later sent as a surgeon to Bencoulen (a fate which was considered a punishment close to death given its 'unhealthful' reputation, for which see Chapter 6). Love, *Vestiges*, and Crawford, *History of the IMS*, also note a previous episode in which Browne drunkenly challenged Blackwall to a duel. On the Mughal camp, see Jos J. L. Gommans, *Mughal Warfare: Indian Frontiers and Highroads to Empire, 1500–1700* (London: Routledge, 2003), 100–111.

30. Tamil Nadu State Archives (hereafter TNSA), *Letters from Fort St George (FSG)*. Volume and page numbers here indicate the original volumes. Most entries are also transcribed in Madras Record Office, *Letters from Fort St George 1679–1765* (Madras: Government Press 1915–1941), IV, 102–108; a letter of 1 December 1693 identifies Johannes Potuliet as 'Physician to Prince Azem Tarra' (probably Azim Shah).
31. TNSA, *Letters from FSG*, VIII, 99–100, fol. 882, 23 April 1698, and 111–114, 13 May 1698. Higginson sends a letter in Persian appealing on the doctor's behalf and 2,000 fanams – and four gold *mohurs* – to pay the debt.
32. TNSA, *Public Consultations*, XXVI, 43–44, 11 October 1697. The incident is also noted in Crawford, *History of the IMS*, I, 91–92.
33. TNSA, *Public Consultations*, XX, 43–44, 1 August 1693; Browne returns from the Mughal camp with a French soldier who had defected from the EIC's army and procures a pardon for him.
34. J. F. Richards, 'The Hyderabad Karnatik, 1687–1707', *Modern Asian Studies*, vol. 9, No. 2 (1975), 241–260, 248.
35. See Rao Bahadur K. V. Rangaswami Iyangar, 'Manucci in Madras', in *The Madras Centenary Commemoration Volume* (Delhi: Asian Educational Services, 1994, first published Madras, 1939), 150. Letter, undated but c. 1699 from John Pitt (of the 'new' Company) to 'Manuchi', BL IOR OC 6685 and 6790.
36. Despatch from England dated 16 April 1697 and quoted in Love, *Vestiges of Old Madras*, II, 88.
37. MS Sloane 3321 fol. 213, Bulkley FSG, 12 February 1707. The letter implies that this was his first visit.
38. For example, SB228 on the different types of myrobalans and 229 on sandal. Browne also identifies various species of *nux-vomica* according to local classifications.
39. BL MS Sloane 4062, fol. 290, Browne to Petiver, FSG, 30 September 1698. It is clear that Bulkley had previously assisted Browne in making some of collections the latter sent to Petiver; see, for example, Petiver and Browne, 'Sam Browne His Second Book of Plants', 708.
40. BL MS Sloane 3321, fol. 28, Bulkley to Petiver, 23 February 1699–1700.
41. BL MS Sloane 3321, fol. 67, Bulkley to Petiver, FSG, 30 January 1700/1 (Recd. 30 January 1701/1702).
42. BL MS Sloane 3321, fols. 110–111, Bulkley to Petiver, FSG, 12 February 1702/3 (Recd. 20 November 1703).
43. BL MS Sloane 3321, fols. 84–5, Bulkley, FSG, 9 November 1701 (Recd. 4 June: 1702, marked 'no XI').
44. Sloane Herbarium, vol. 32, fols. 135–150 contain plants gathered in Bengal. Bulkley reports sending them in MS Sloane 3321, fol. 169 – an unsigned

letter to Petiver from 'your humble servant' at Fort St George 7 February 1704/1705 (Recd. 13 November 1705).

45. BL MS Sloane 3321 fol. 171 – 1 March 1704/1705 (Recd. 13 March 1705/1706). It is in this letter that the reference to tigers occurs. Collections from Pegu sent to Petiver by Bulkley are in HS 32, fols. 119–153. None of these specimens are named.
46. V. N. Rao, 'Coconut and Honey: Sanskrit and Telugu in Medieval Andhra', *Social Scientist*, 23:10/12 (1995), 24–40.
47. V. N. Rao, David Shulman and Sanjay Subrahmanyam, *Symbols of substance: court and state in Nāyaka period Tamilnadu* (Delhi: Oxford University Press, 1992).
48. P. Hymavathi, *History of Āyurvēda in Āndhardēśa AD 14thc–17thc.* (Warangal: Bhargava Publishers, 1993). R. Rao, 'Medical Allusions in the Inscriptions of AP', *Bulletin of the Indian Institute for the History of Medicine*, 5 (1975), 198–206 and 6 (1976), 28–36, notes that the school of medicine is referred to as Parahita in inscriptions from Andhra. See also R. Farooqui, 'Medicine in the Deccan', *Bulletin of the Indian Institute for the History of Medicine*, 17 (1986), 31–52. Rao, Shulman and Subrahmanyam, *Symbols of substance*, 113, also note the wealth of medical texts produced around Thanjavur.
49. As R. S. Weiss, 'Divorcing Ayurveda: *Siddha* Medicine and the Quest for Uniqueness', in D. Wujastyk and F. M. Smith (eds.) *Modern and Global Ayurveda: Pluralism and Paradigms* (Albany: State University of New York Press, 2008), 77–99, argues, the modern distinction between *siddha* and *ayurveda* should not be allowed to conceal historical overlaps and exchanges between the two traditions.
50. W. Roxburgh, *Plants of the Coast of Coromandel* (London: printed by W. Bulmer and Co. for George Nicol, 1795–1819), distinguishes between the practises of 'Telinga' and 'Malabar' physicians. See, for example, page 8 on the uses of *Oldenlandia umbellata*.
51. Hymavathi, *History of Āyurvēda*, 25, mentions the Telugu compositions of Srinatha, including details of the drug trade to other countries.
52. Love, *Vestiges*, II, 558–559.
53. N. Dirks, 'Colonial Histories and Native Informants: The Biography of an Archive', in Carol A. Breckenridge and Peter van de Veer (eds.) *Orientalism and the Postcolonial Predicament: Perspectives on South Asia* (Philadelphia: University of Pennsylvania Press, 1993); Bayly, *Empire and Information* (Cambridge: Cambridge University Press, 1996); Schaffer, Roberts, Raj, and Delbourgo (eds.) *The Brokered World*.
54. Raj, *Relocating Modern Science*, and Sujit Sivasundaram, 'Tales of the Land: British Geography and Kandyan Resistance in Sri Lanka, c. 1803–1850', *Modern Asian Studies*, 41 (2007), 925–961, 927.
55. Browne occasionally refers to local practitioners borrowing from Portuguese customs, for example in their treatment of smallpox. Petiver and Browne, vol. 6, specimen 234 (1054), 'Tumba maraum' or 'Carpa maraum Malab'.
56. Crawford, *History of the IMS*, I, 414.
57. BL IOR, G/19/5, 22 March 1688, fols. 20v–21v.
58. Lockyer, c.f. Love, *Vestiges of Old Madras*, I, 83–84
59. Yule, *Vestiges*, I, 564. A *dubash* (from *do bhāṣī*, lit. 'two languages') also has the meaning of secretary or representative (*Hobson-Jobson*, 328). See

S. Neild-Basu, 'The *Dubashes* of Madras', *Modern Asian Studies*, 18 (1984), 1–31. 'Conicoply' is probably from Tamil *kanakkapillari*. Public hospitals were established by the Emperor Jahangir and were apparently common even in smaller towns during the reign of Aurangzeb. I. A. Khan, 'The Middle Classes in the Mughal Empire', *Social Scientist*, 5 (1979), 28–49.

60. Cf. 'John Fryer – A Traveller of XVII Century', *Bulletin of the Department of the History of Medicine* (later *Bulletin of the Indian Institute for the History of Medicine*), 2, (1964) (no author given, D. V. Subba Reddy ed.). 'Peruvian bark' refers to cinchona, the source of quinine.
61. TNSA, *Mayoral Court Records*, II, fol. 197, 3 December 1718.
62. SB228.
63. Chakrabarti, 'Neither Meate nor Drinke', 7.
64. Cook, *Matters of Exchange*, 306.
65. SB101.
66. MS Sloane 4062, fol. 290, Browne to Petiver, FSG, 30 September 1698.
67. Payments for drugs purchased by Browne in the bazaar for use in the Company hospital are recorded in 1695–1698 (TNSA, *Public Consultations*, XXII, fols. 265–272, 2 August 1695; XXII, fols. 265–272, 31 August 1696; and XXXIII, fols. 267–268, 16 February 1698). Pratik Chakrabarti, 'Medical Marketplaces beyond the West', in M. S. R. Jenner and P. Wallis (eds.) *Medicine and the Market in England and Its Colonies, c. 1450–c. 1850* (Basingstoke: Palgrave Macmillan, 2007), 199–200, notes that the bazaar was not only a static marketplace within a town, but could refer also to a group of traders who followed the EIC's armies.
68. MS Sloane 3333 fol. 201, Samuel Browne, FSG 17 October 1696. The definition of tutia is from William Lewis, *An Experimental History of the Materia Medica* (London: J. Johnson, 1791). Further discussions of the identity of tutia occur in BL MS Sloane 3332, fol. 67 – 'An abstract of Bontius his Animadversions upon Garcia da Orta with some additional remarks by Mr Alexander Brown', 12 June 1695.
69. SB709. Cape Comorin or Kanyakumari is on the southernmost point of the Indian peninsula. Agallochum is a fragrant wood used as incense since ancient times (H. H. Wilson, 'Notes on the Sabhá Parva of the Mahábhárata, Illustrative of Some Ancient Usages and Articles of Traffic of the Hindus', *Journal of the Royal Asiatic Society of Great Britain and Ireland*, 7:1 (1843), 137–144).
70. MS Sloane 3333, fol. 201, Samuel Browne to James Petiver, FSG, 17 October 1696. *Asafoetida* is much used in *ayurveda* in the treatment of simple digestive problems such as wind, bloating, indigestion, and constipation, and also for respiratory problems such as bronchitis, bronchial asthma, and whooping cough. It is also used as a circulatory stimulant, lowering blood pressure and thinning the blood. (Information from '*Ferula* Assa-Foetida', Plants for a Future: Edible, Medicinal and Useful Plants for a Healthier World, URL: http://www.pfaf.org/database/plants.php?Ferula+assa-foetida, accessed 14 October 2009.) *Asafoetida* had been known in the West for some time by the seventeenth century: it appears in Garcia da Orta's seventh colloquy, who in turn refers to Avicenna's descriptions (Garcia da Orta, *Colloquies on the Simples & Drugs of India*, Clements R. Markham trans. (London: Henry Sotheran, 1913).

71. Parkinson *Theatrum Botanicum,* 1650. van Rheede, *Hortus Malabaricus,* X, tab. 11, describes the use of the *Moringa* in the preparation of vegetables and to treat various types of inflammation. Marshall mentions the use of the seeds of *nirmalī* (*Strychnos potatorum*) to clean water (Khan, *John Marshall,* 337). Khan notes that this use appears in Sushruta.
72. Clopton Havers (1657–1702).
73. MS Sloane 3321, fol. 28, Bulkley to Petiver, 23 February 1699/1700.
74. Madras Record Office, *Records of Fort St George, Diary and Consultation Book of 1693* (Madras: Government Press, 1918), 65–66; 129.
75. Ibn Sina, *al-Qanun fi'l-tibb,* translated by L. Bakhtiar, *The Cannon on Medicine* (Chicago: Kazi Publications, 1999); J. M. Riddle, 'Ancient and Medieval Chemotherapy for Cancer', *Isis,* 76 (1985), 319–330.
76. Sloane MS 3321, fol. 19, Bulkley to Petiver, Fort St George (FSG), 13 October 1699.
77. The Royal Society's minutes and the *Philosophical Transactions* frequently record testing substances on dogs. Dissection was performed in medical schools like the famous one at Leiden University. Although dissection is mentioned in Suśruta (see K. G. Zysk, 'The Evolution of Anatomical Knowledge in Ancient India, with Special Reference to Cross-Cultural Influences', *Journal of the American Oriental Society,* 106 (1986), 687–705), it had fallen out of favour long before the seventeenth century.
78. Golinski, *Making Natural Knowledge,* xii–xiii.
79. For an overview of recent challenges to this model, E. Waltraud and P. B. Mukharji, 'From History of Colonial Medicine to Plural Medicine in a Global Perspective', *NTM International Journal of History and Ethics of Natural Sciences, Technology and Medicine,* 17 (2009), 447–458.
80. For example, MS Sloane 3321, f. 19 – Bulkley to Petiver, FSG, 13 October 1699; Bulkley requests instructions for collecting, asks for European seeds and for Petiver to popularise some roots and ground pearl, and sends a coffee plant from Mocha and collections from Vizigapatnam.
81. For the full map, commissioned by Thomas Pitt but completed shortly after his death, Stern, *The Company State,* 32–33.
82. P. Bowe, 'The Indian Gardening Tradition and the Sajjan Niwas Bagh, Udaipur', *Garden History,* 27:2 (1999): 189–205; J. Brookes, *Gardens of Paradise: The History and Design of the Great Islamic Gardens* (London: Weidenfeld and Nicolson, 1987); Grove, *Green Imperialism,* 73–94; J. L, Wescoat and J. Wolschke-Bulmahn, *Mughal Gardens: Sources, Places, Representations, and Prospects* (Washington, DC: Dumbarton Oaks Research Library and Collection, 1996).
83. Irfan Habib, 'Economic and Social Aspects of Mughal Gardens', in Westcoat and Wolschke-Bulmahn (eds.) *Mughal Gardens,* 127–138.
84. J. L. Wescoat and J. Wolschke-Bulmahn (eds.) *Mughal Gardens,* 'Introduction', 25. See also Chandra Mukerji 'Dominion, Demonstration and Domination: Religious Doctrine, Territorial Politics and French Plant Collection', in Schiebinger and Swan, *Colonial Botany,* 19.
85. Alam and Subrahmanyam, *Writing the Mughal State,* 381.
86. Madras 'Public Consultations', vol. 30, 12–15 July 1701; cf. Yule, *Vestiges,* II, 16. The Council were alarmed by this request since, at the time, the gardens left exposed the weakest defences of the city. Fortunately for them, Khan

was dissuaded with the aid of a lavish dinner, 'dancing wenches', and much alcohol.

87. Cf. Yule, *Vestiges*, II, 84.
88. Stern, *Company State*, 199.
89. See Chapter 6.
90. BL Sloane MS 3333 fol. 201, Browne to Petiver, FSG, 17 October 1696: 'The Company's gardener I know and have seen many of his collections, which he sold at very great rates.' HS, vol. 32 fol. 131, is marked 'Sent from Fort St George by my sister Mary du Bois... au Randalia, vol. 6, nodif. fol.', implying that Randal also made extensive collections. HS 188 contains a specimen from Randal on fol. 178 collected in 1700. See also J. E. Dandy, *The Sloane Herbarium: An Annotated List of the Horti Sicci Composing It* (London: British Museum, 1958), 52 and 189.
91. MS Sloane 4062, fol. 290, Browne to Petiver, FSG, 30 September 1698.
92. MS Sloane 4062, fol. 290, Browne to Petiver, FSG, 30 September 1698. See Anna Winterbottom, 'Of the China Root: A Case Study of the Early Modern Circulation of *Materia Medica*', *Social History of Medicine*, 28:1 (2015), 22–44.
93. 'Wild agallo' is agallochum (SB709), benjamin is incense (*Hobson-Jobson*, 86) and 'camphire' may refer to the fragrant henna plant *Lawsonia alba*.
94. MS Sloane 3321, fol. 185 Bulkley to Petiver, FSG, 24 January 1706. As he noted in an earlier letter (fol. 172, 15 March 1705), the gourds were gathered by sailors on the ship *Abingdon*.
95. MS Sloane 3321, fols. 84–85, Bulkley to Petiver, FSG, 9 November 1701.
96. MS Sloane 3321, fol. 19, Bulkley to Petiver, FSG, 13 October 1699: 'I desire that you will send me as many medicinall seeds as you thinke may be like to growe here, I should like to raise some Europe plants here.'
97. Tobacco cultivation was established in the Deccan by 1605. K. Srinath Reddy and Prakash C. Gupta (eds.) *Report on Tobacco Control in India* (New Delhi: Shree Om Enterprises, 2004) http://www.who.int/fctc/reporting/Annex6_Report_on_Tobacco_Control_in_India_2004.pdf (accessed, 19 June 2010), 19–32. Petiver notes in a commonplace book, BL MS Sloane 4020 fol. 216: 'Tobacco certainly is a proper native of America, though now it is gott in to the Easte Indies also where (in the province of Guzarat [Gujarat] especially) they sow it in abundance.' J. Crawfurd, 'On the History and Migration of Cultivated Narcotic Plants in Reference to Ethnology', *Transactions of the Ethnological Society of London*, 7 (1869), 78–91, notes that both the Emperor Jahangir and King James issued edicts or tracts attempting to curb the consumption of tobacco in their realms.
98. Cannabis is described in James Petiver, 'The Eighth Book of East India Plants, Sent from Fort St George ... ', *Philosophical Transactions*, 23 (1702/3), 1454, specimen nos. 14 and 15.
99. BL MS Sloane 3333, fol. 195, Petiver, 'A return to Mr Samuel Browne's remarks on the 4th book of Bontius his animadversions on Garcia da Orta' (undated).
100. See, for example, BL Add MS 34123, 'Copybook of Henry Vansittart, Governor of Bombay, 1756', a compilation of pieces of advice from the seventeenth and eighteenth centuries. Fol. 34 'Advice from Mr Pitt' includes advice on drugs that will sell in Persia including China root and pearl, and

drugs that may be purchased in Aceh, including marijuana. Fol. 36 is a similar list of goods proper for the Coromandel Coast, including China root, bezoar stones, lignum aloes, and opium. The advice of Samuel Baron in the same volume, fols. 40–42 (see Chapter 1), discusses the regional trades in China root, opium, and arsenic.

101. For example, OC 7880, Bulkley to Du Bois, 28 October 1698, transcribed in Yule, *Diary of William Hedges*, II, cccxx.
102. J. Parkinson, *Theatrum botanicum* (London: Thomas Coates, 1640). SB26, a febrifuge which Parkinson, *Theatrum Botanicum*, 1651, refers to as 'Nimbo'.
103. SB113 and 121 cross-reference to G. da Orta and C. Acosta, *Tractado de las drogas y medicinas de las Indias Orientales* (Burgos: Martin de Victoria, 1578).
104. H. A. van Rheede tot Draakestein et al., *Hortus Indicus malabaricus, continens regioni malabarici apud Indos celeberrimi omnis generis plantas rariones* (12 vols.) (Amsterdam, 1678–1703). For an English translation, Manilal, *Van Rheede's Hortus Malabaricus*. See also H. Y. Mohan Ram, 'On the English Edition of Van Rheede's Hortus Malabaricus by K. S. Manilal (2003)', *Current Science*, 89 (10), 2005, 1672–1680. A full comparison between Browne's plants, their medical uses, and those described in the *Hortus Malabaricus* to which they are cross-referenced is given in Anna Winterbottom, 'Company Culture: Information, Scholarship, and the East India Company Settlements, 1660–1720' (Doctoral thesis, University of London, 2011), Appendix 2.
105. See, for example, James Petiver, *Opera*, III (London: Smith and Bateman, 1695), 43–48: 'An abstract of what Collections I have received the last Twelve Months, and the Persons whom I am Obliged to for them'.
106. Ray, *Historia Plantarum*, 3 vols. Samuel Smith and Benjamin Walford, 1693–1704), i, 132, refers to Petiver's description of one of Browne's plants in *Philosophical Transactions*. The 'Icones Avium Madraspatnam' (Fig. 4.2) were printed in John Ray, *Synopsis methodica avium et piscium* (Londini: Prostant W. Innys ..., 1713), appendix.
107. MS Sloane 3321, fol. 211, Bulkley to Petiver, no date (Recd. 9 December 1706 from Ireland). The original paintings are in Add MS 5266, fols. 91–2. They were copied again by the engraver of four pages towards the back of the BL's copy of Petiver's *Opera* (Shelfmark 443.i.1.(8)K. Birds.) and in John Ray, Synopsis methodica avium et piscium: opus posthumous (London: William Innys, 1713).
108. The specimens are kept in the Sloane Herbarium in the NHM and described in Dandy, *Sloane Herbarium*. Manuscript notes in the front of the first Browne volume in the Sloane Herbarium note that the seven books were borrowed by James Petiver and Hans Sloane during 1699 from the Royal Society's repository, to which they had been transferred after a request to the EIC. A note by Francis Hawskbee reads 'The 1 2 3 & 4 books are in the Society's house, the rest are missing'. Presumably the rest were returned to the repository at a later stage.
109. The seven articles are all published in *Philosophical Transactions*, 22 (1700/1), 579–594, 699–721, 843–858, 933–946, 1007–1022, and 1023 (1702/1703), 1055–1056, 1251–1265. As Dandy, *Sloane Herbarium*, 101, notes, an eighth book is also described in *Philosophical Transactions*, *23*,

1450–1460, but this was sent after Browne's death and also contains specimens from Bulkley. For other volumes containing specimens collected by Browne, see Dandy, *The Sloane Herbarium*, 102.

110. James Petiver, *Gazophylacii Naturae & Artis* (Londini: ex officinâ Christ. Bateman, 1702), issues an appeal for any information concerning this still-mysterious substance, offering a guinea for any information. By 1704, he had acquired a description of it (J. Petiver, 'Advertisement', dated 15 March 1704, Aldersgate Street, in *Gazophylacii naturæ & artis decas tertia* (London: printed for Sam Smith and C. Bateman, 1704)), and it is depicted in his 'Hortus Peruviansus or South Sea Herbal' (1715), repr. in *Opera*, tab. 6, no. 7 'Kinquina or Jesuits Bark'. W. Oliver, 'A Letter from Dr William Oliver, Physician and Fellow of the Royal Society, to Mr James Petiver, F. R. S. concerning the Jesuits Bark', *Philosophical Transactions* 24 (1704–1705): 1596.
111. MS Sloane 3321, fol. 28 – Bulkley to Petiver, 23 February 1699/1700.
112. SB2, SB15, SB16, SB75, SB86, SB119, SB161, SB189, SB211, SB234.
113. SB64 and SB129.
114. Gonorrhoea: SB66, SB112, SB120, SB126, SB211. Marshall also gives remedies for 'the French pox' (Khan, *John Marshall in India*, 334).
115. SB167, SB224.
116. SB238. Used as a general counter-poison. Browne reports trying it against the bite of the 'Cobree de Capello' (*Naja tripudians*) without success.
117. SB70 and SB74 cure the whooping cough, and SB189 is a remedy for scabs and itches in young children.
118. For example, SB80 is listed as a cure for period pains, as is SB126.
119. SB212 'cures hypocondriack melancoly and hysterick passion'.
120. This is a bibliographic rather than scientific identification based on Petiver's references. See Winterbottom, 'Company Culture', Appendix 2.
121. SB125.
122. J. Ray, *Historia plantarum* (London: Mariæ Clark, 1686–1704).
123. SB199.
124. SB108.
125. R. Rao, 'Medico Historical Information from Non-Medical Sources', *Bulletin of the Indian Institute for the History of Medicine*, 16 (1986), 1–9, 5, notes that several commonly used items of *materia medica* such as the *nimba* (*Azadirachta indica*) are not mentioned in ayurvedic texts. Professor K. S. Manilal believes that some of the *materia medica* listed in the *HM* may be derived from earlier Buddhist pharmacopoeia (I am grateful to him for this personal communication).
126. Weiss, 'Divorcing Ayurveda', 77–99.
127. BL MS Sloane 3321, fol. 19, Bulkley to Petiver, FSG, 13 October 1699: 'pray send store of instructions for collecting &c.'.
128. BL MS Sloane 3321, fol. 104, Bulkley, FSG, 17 August 1702: Bulkley reports sending instructions to Alexander Read at Bencoulen.
129. BL MS Sloane 3321, fol. 19, Bulkley to Petiver, FSG, 13 October 1699.
130. BL MS Sloane 3321, fol. 112, Cunningham to Petiver dated Chusan 12 February 1703. He mentions sending collections to Sloane.
131. MS Sloane 3321, fol. 117, [Cunningham – a copy, not in his hand] 'Recd the 4th Aug of a China man att Emoy that came from Cochin his name

Watteo Aneos nephew the following letter directed to The Supracargoes & Capts in the service of the Honourable Company trading to the East Indies' describes his capture. Fol. 132, Cunningham to Petiver dated Pulo Condore, 8 January 1704, sends collections from Chusan and Pulo Condore to Petiver and Sloane. Collections from Cunningham in Pulo Condore include HS 278, 279, and 280.

132. HS 59 (Dandy, *Sloane Herbarium*, 32) contains Cunningham's collections from China made in 1698 and at the Cape of Good Hope in 1699 and HS 252 from Amoy, Chusan and the Crocodile Isles (Dandy, *Sloane Herbarium*, 62). Plants from Batavia and Pulo Condore are in HS 253 (Dandy, *Sloane Herbarium*, 63) and from the Cape in HS 257 (Dandy, *Sloane Herbarium*, 64); more plants from Batavia are in HS 289 (Dandy, *Sloane Herbarium*, 70). The herbarium of Charles du Bois is in the Department of Plant Sciences at the University of Oxford. It contained at least four large volumes of specimens from Bulkley and 194 seeds collected in 1702–1703 (MS Sherard 191).
133. BL MS Sloane 3321, fol. 112, Cunningham to Petiver dated Chusan 12 February 1703. This may be the same Henry Smith who was involved with the production of Malay dictionaries (Chapter 2).
134. The British Library's collection of Petiver's works with MS notes [443.i.1.] contains a picture of a plant 'Ageratum' that Petiver acquired from Oldenhand, 'a learned physician and very curious botanist at the Cape'. The journals of Alexander Brown and Isaac Pyke, both of whom report meeting him, are BL MS Sloane 1689 and IOR MS Mss Eur D 5, respectively.
135. BL Sloane MS 3321, fol. 103, Bulkley, FSG, 9 October 1702 (marked 'no 13', no date of receipt). See Cook, *Matters of Exchange*, 315–317.
136. Sloane MS 3321, fol. 133, Bulkley to Petiver 10 February 1703/1704, where he reports a discussion of the 'Punsaloy or fruit of the Panitsjaka mer'. In van Rheede, *Hortus Malabaricus*, III, tab. 41, 45, it is given as 'Panitsjika-Maram'. Bulkley follows the *HM* in noting the use of the fruit as glue, and adds Kamel's report of successfully using it to treat fever.
137. G. Camelli, 'A Description and Figure of the True Amomum, or Tugus. Sent from the Reverend Father George Camelli, at the Phillipine Isles, to Mr. John Ray and Mr. James Petiver, Fellows of the Royal Society', *Philosophical Transactions (1683–1775)*, 21, (1699), 2–4. See also R. A. Reyes, 'Botany and Zoology in the Late Seventeenth Century Philippines: The Work of Georg Josef Camel SJ (1661–1706)', *Archives of Natural History*, 36 (2009), 262–276.
138. Petiver notes in the discussion of SB57 that Kamel sent a drawing of indigo to him and Ray. There are several of Kamel's drawings among Petiver's published collections – for example *Opera*, tab. XXVIII (6). Petiver mentions Cunningham's sending him a specimen from 'the island of Mischowahi' in the discussion of SB106.
139. BL MS Sloane 3321, fols. 84–85, Bulkley, FSG, 9 November 1701, and MS Sloane 3333 fol. 201, FSG, 17 October 1696. I. Habib, *An Atlas of the Mughal Empire* (Delhi: Oxford University Press, 1982), map 0B 'The Mughal Empire, Economic', which marks these as 'aloe-wood forests' and 'lac-forests'.
140. BL MS Sloane 1689, fol. 11v.
141. Cook, *Matters of Exchange*.

142. BL MS Sloane 3321, fol. 133, Bulkley to Petiver, 10 February 1704.
143. BL MS Sloane 3321, fol. 18, Edward Bulkley to James Petiver, FSG 12 October 1699.
144. Petiver uses much of his commentary on Browne's collections to level criticism at his rival Leonard Plukenet.
145. Royal Society, *Journal Book*, X, meeting of 26 October 1698.
146. It seems that the initial request was triggered by Browne's volumes. Also several of these do contain small envelopes with seeds; the reference to the barber's shop implies that there were more.
147. Royal Society, *Journal Book*, X, 23 November 1698: 'Mr Pepys, Mr Pettiver, Dr Havers & Dr Sloane were appointed a Committee to looke after putting in order the East India Collection now received from the E. India Company & the same persons were ordered to return the thanks of the Society [to the EIC].'
148. Mary Somerset, the Duchess of Beaufort, was an avid gardener, and, assisted for a time by William Sherard, she amassed a collection of thousands of rare species from around the world at her estate in Badminton. P. E. Kell, 'Somerset, Mary, duchess of Beaufort (*bap.* 1630, *d.* 1715)', in *Oxford Dictionary of National Biography* http://0-www.oxforddnb.com.catalogue.ulrls.lon.ac.uk:80/view/article/40544 (accessed 26 November 2007).
149. du Bois' garden at Mitcham, Surrey, was filled with exotic plants, and his 13,000 specimens of dried plants in 74 volumes were given to the Oxford Botanical Garden. His *hortus sicci* remain in the herbarium at Oxford University. See B. D. Jackson, 'Dubois, Charles (*bap.* 1658, *d.* 1740)', rev. P. E. Kell, in *Oxford Dictionary of National Biography*, http://0-www.oxforddnb.com.catalogue.ulrls.lon.ac.uk:80/view/article/8113 (accessed 14 December 2007).
150. Royal Society, *Journal Book*, X, 1 March 1698/1699.
151. Royal Society, *Journal Book*, X, 20 December 1699.
152. J. Petiver, 'Some Farther Account of Divers Rare Plants, Lately Observed in Several Curious Gardens about London, and Particularly in the Company of Apothecaries Physick-Garden at Chelsey', *Philosophical Transactions*, 27 (1710–1712), 416–426. For example, Petiver comments on SB45 that Plukenet grew this plant in Chelsea.
153. BL IOR O.C. 7880, Bulkley to du Bois, 28 October 1698, transcribed in Yule, *Diary of William Hedges*, ii, cccxx.
154. BL MS Sloane 3321, fol. 18, Bulkley to Petiver, FSG 12 October 1699.
155. BL MS Sloane 3321, fol. 66 Bulkley to Petiver, FSG, 6 February 1700.
156. Chakrabarti, *Materials and Medicines,* 114–115.
157. Sloane 3322, fol. 41, Bulkley to Petiver,14 September 1713.
158. A. Cooper, *Inventing the Indigenous: Local Knowledge and Natural History in Early Modern Europe* (Cambridge: Cambridge University Press, 2007), 27.
159. BL Sloane 3322, fol. 41, Bulkley to Petiver, 14 September 1713.
160. BL Sloane MS 3321, fol. 19, 12 October 1699.
161. SB227.
162. Sloane Herbarium, vegetable substances drawers, marked Sloane 278. The note accompanying the specimen reads: 'The bark of the tree which the

Malabars call Ettee and which I think to be the Jesuit's bark but not collected in a good time, nor from a flourishing tree.' I believe this is from Browne on the basis of the handwriting and the use of the word 'Ettee' from the Tamil.

163. SB28.
164. J. Petiver, 'Some Attempts Made to Prove That Herbs of the Same Make or Class for the Generallity, Have the Like Vertue and Tendency to Work the Same Effects. In a Discourse Made before the Royal Society, by Mr. James Petiver Apothecary, and Fellow of the Said Society', *Philosophical Transactions (1683–1775)*, 21, (1699), 289–294.
165. Roxburgh, *Plants of the Coast of Coromandel*. In the preface, Russell acknowledges that much of Roxburgh's knowledge of the plants was 'collected from the natives' but gives no further detail.
166. Chakrabarti, *Materials and Medicines*.
167. BL IOR OC 7880, Bulkley to du Bois, 28 October 1698, transcribed in Yule, *Diary of William Hedges*, II, cccxx.
168. J. Richards, *The Mughal Empire, New Cambridge History of India* (Cambridge: Cambridge University Press, 1993), I.5, chapter 10.
169. For an overview of recent debates, R. Travers, 'The Eighteenth Century in Indian History: A Review Essay,' *Eighteenth-Century Studies*, 40:3 (2007): 492–508.
170. Irfan Habib, 'Economic and Social Aspects of Mughal Gardens', 135. Love, *Vestiges*, vol. II, 90, for Bulkley's garden tomb. See also Julian James Cotton, *List of Inscriptions of Tombs or Monuments in Madras Possessing Historical or Archaeological Interest* (Madras: Government Press, 1945). Entry 68 is for Edward Bulkley 'from his monument on the South Esplanade, near the hospital'; he died on 8 August 1714.
171. Browne was offered but declined the position of surgeon in Chutanuti (Calcutta) (Crawford, *History of the IMS*, I, 246).
172. K. Raj, 'Mapping Knowledge Go-Betweens', in Schaffer, Roberts, Raj, and Delbourgo (eds.) *The Brokered World*, 105–150.
173. Bayly, *Empire and Information* (Chapter 7) 'Colonial Controversies: Astronomers and Physicians'.
174. W. Jones, *The Works of Sir William Jones* (London: Stockdale, 1807), 13 vols., V, 1–164.
175. Chakrabarti, 113.
176. M. L. Pratt, *Imperial Eyes: Travel Writing and Transculturation* (London: Routledge, 1992). L. Schiebinger, 'Prospecting for Drugs: European Naturalists in the West Indies', in Schiebinger and Swan (eds.) *Colonial Botany*, 119–133, 128.
177. Breen, 'No Man Is an Island'.

5 Bio-Prospecting and Experimentation: Producing and Using *An Historical Relation of Ceylon*

1. Robert Knox, *An Historical Relation of the Island Ceylon* (London: Richard Chiswell, 1681), iv:12, 170–174 describes the voyage home. The account Knox and his fellow captive Steven Rutland gave to the Dutch in Batavia

on his arrival with Ryckloff van Goens Junior is translated in F. H. de Vos, 'Extracts Relating to Ceylon from the Dag-Register, Batavia, AD. 1678–1680', *Journal of the Ceylon Branch of the Royal Asiatic Society*, 16 (1918), 181–198.

2. Bodleian MS Rawl Q. c. 15, fol. 1 (Paulusz II, 527). The autobiography was first printed in *An Historical Relation of Ceylon Together with Somewhat Concerning Severall Remarkable Passages of My Life That Hath Hapned since My Deliverance out of My Captivity* ed. by J. Ryan (Glasgow: James MacLehose and Sons, 1911) and reprinted in Paulusz ed. *An Historical Relation of Ceylon*. Hereafter referred to as *Autobiography* with folio numbers for the manuscript and page numbers of Paulusz vol. II in brackets. Knox says that the drugster recognised him as someone who had been abroad for a long time from his old-fashioned beard and whiskers.
3. On 15 September 1680, Knox and Rutland were directed to attend the Committee for Shipping 'to communicate what they had observed of affaires in those parts' (Yule, *Diary of William Hedges*, II, ccclii).
4. *Autobiography*, 3 (528).
5. Hooke was known to Knox' brother James, from whom he learnt of the escape of Robert Knox. Hooke's diary entry on 12 September 1680 reads 'Knox his brother escaped out of Ceylon after 22 years detainder'. Robinson and Adams (eds.) *The Diary of Robert Hooke*, 454.
6. Since Said, *Orientalism*, travel accounts have been discussed in a number of contexts. Some have followed his interpretation that European travel accounts tend to perpetuate imperialistic images of the 'other', for example Ronald Inden, 'Orientalist Constructions of India', *Modern Asian Studies* (1986), 20, 401–446, and Kate Teltscher, *India Inscribed: European and British Writing* (Delhi: Oxford University Press, 1995). Others including Gerald MacLean, *The Rise of Oriental Travel: English Visitors to the Ottoman Empire, 1580–1720* (Basingstoke: Palgrave, 2004) and Rubiés, *Travel and Ethnography in the Renaissance*; J. Carrillo, 'From Mt Ventoux to Mt Masaya: The Rise and Fall of Subjectivity in Early Modern Travel Narratives' in Elsner and Rubiés, *Voyages and Visions*, 57–73, have sought evidence for more diversity in European writings about Asia than claimed by Said. Finally, some authors including Alam and Subrahmanyam, *Indo-Persian Travels*, have extended the frame of reference outside Europe. Works that discuss the practical use of travel narratives in the collection of botanical information and species include Raj, *Relocating Modern Science*; Cook, *Matters of Exchange*; Chandra Mukerji, 'Dominion, Demonstration and Domination: Religious Doctrine, Territorial Politics and French Plant Collection' in Schiebinger and Swan, *Colonial Botany*, 19–33.
7. Cañizares-Esguerra, *Nature, Empire, and Nation*; Joan-Pau Rubiés, 'Instructions for Travellers: Teaching the Eye to See', *History and Anthropology*, 9, 2–3, (1996), 139–190. See also J. Stagl, *A History of Curiosity: The Theory of Travel 1500–1800* (Chur: Harwood Academic Publishers, 1995), 78–81, for the sixteenth-century methodising of travel narratives within Europe; Peter C. Mancall (ed.) *Bringing the World to Early Modern Europe: Travel Accounts and their Audiences* (Leiden: Brill, 2007).
8. Boyle, *General Heads*. Michael Hunter, 'Robert Boyle and the Early Royal Society: A Reciprocal Exchange in the Making of Baconian Science', *British*

Journal for the History of Science, 40 (2007), 1–24; Carey, 'Compiling Nature's History'; Cook and Lux. 'Closed circles or open networks'.

9. Bayly, *Empire and Information*; Ogborn, *Indian Ink*.
10. F. Bacon 'On Travel', in *The Essays or Counsel, Civill and Moral (1625)*, ed. by M. Kiernan (Oxford: Clarendon Press, 1985), 41–42.
11. Kandy emerged as an independent kingdom in the late fifteenth century and remained under indigenous rule until finally falling to the British in 1815. See Kingsley M. de Silva, *History of Sri Lanka*, revised ed. (London: Vijitha Yapa Publications, 2005), and Chandra Wickremesekera, *Kandy at War: Indigenous Military Resistance to European Expansion in Sri Lanka 1594–1818* (Colombo: Vijitha Yapa Publications, 2004).
12. Knox' time in Kandy and the political background of the period are discussed in more detail by Paulusz, *Relation*, I.
13. The French versions were published by Paul Marret in Paris (1684) and Amsterdam (1693) and the German text appeared in Leipzig in 1689. The Dutch version, Simon de Vries trans, *t'Eyland Ceylon in syn binnenste, of 't koningrijck Candy: geopent en nauwkeuriger dan oyt te vooren ontdeckt* (Utrecht: by Wilhelm Broedelet, 1692), was based on the German version. For a contemporary review of the Dutch edition, *Bibliotheque Universalle et Historique*, 173:23 (1692), 219–244. Sumana D. Saparamadu, ' "Introduction" to *An Historical Relation of Ceylon*', *Ceylon Historical Journal* 5 (1958), vii–lv.
14. Locke refers to the 'late relation of Ceylon' in his discussion of the effects of absolute monarchy on civil society in the *Second Treatise of Government*, 7:92. He purchased a copy of Knox '*Relation*, soon after its publication in August 1681. (P. Laslett, The English Revolution and Locke's 'Two Treatises of Government', *Cambridge Historical Journal*, 12 (1956), 48–49.)
15. Philippus Baldaeus, *Beschrijving der Oost-Indische Kusten Malabar en Coromandel . . . [&] Ceylon* (Amsterdam: Johannes Janssonius van Waesberge, 1672).
16. João Ribeiro, *Fatalidade Historica da Ilha de Ceilão*, was not published in Portuguese until 1836, but according to Županov, 'Goan Brahmins in the Land of Promise', n. 4, it was composed some time after 1680, the preface being dated in Lisbon 1685. The French edition of 1701 is discussed below. I used João de Ribeiro, *The Historical Tragedy of the Island of Ceilão* trans. by P. E. Pieris (Columbo: 1930).
17. Saparamadu, 'Introduction'.
18. H. White, 'Knox in Its Literary Aspect', *Ceylon Branch of the Royal Asiatic Society*, 13 (1893), 23–35; Brian Yothers, 'Global Captivities: Robert Knox's An Historical Relation of Ceylon and the New England Captivity Narrative Tradition', *Journeys* 8 (2007), 1–2.
19. C. R. Boxer, 'Ceylon through Puritan Eyes: Robert Knox in the Kingdom of Kandy 1660–1679', *History Today*, 4 (1954), 660–667.
20. Donald W. Ferguson, 'Knox's Sinhalese Vocabulary', *Ceylon Branch of the Royal Asiatic Society*, 14 (1896), 155–200. Ferguson also includes the Sinhala words listed by Hooke in BL MS Sloane 1039. Richard Boyle, *Knox's Words. A Study of the Words of Sri Lankan Origin or Association First Used in English Literature by Robert Knox and Recorded in the Oxford English Dictionary* (Colombo: Visidunu Prakashakayo, 2004).

21. M. C. M. Iqbal, I. Raheem and K. Tennakone, 'The Royal Society, Robert Hooke and a captive from Ceylon: Fascinating Facets from the 17th Century', *Notes and Records of the Royal Society of London* (forthcoming).
22. Sarojini Jayawickrama, *Writing That Conquers: Re-Reading Knox's An Historical Relation of the Island Ceylon* (Colombo: Social Scientists' Association, 2004).
23. A recent biography is Katherine Frank, *Crusoe: Daniel Defoe, Robert Knox, and the Creation of a Myth* (New York: Pegasus Books, 2012). See also Donald W. Ferguson, *Captain Robert Knox... Contributions towards a Biography* (Columbo and Croyden: Printed for Private Circulation, 1896–1897), written before the discovery of Knox' autobiography or the interleaved copy of the *Relation*; H. A. I. Goonetileke, 'Robert Knox in the Kandyan Kingdom, 1660–1679. A Bio-Bibliographical Commentary', *Sri Lankan Journal of the Humanities*, 1 (1975), 81–151; I. B. Watson, 'Knox, Robert (1641–1720)', *Oxford Dictionary of National Biography*.
24. According to a letter to his cousin John Strype, the interleaved book was given to Knox by Richard Chiswell before the first voyage he made after returning from Ceylon and later returned to him for a final check before its intended publication as a second edition (Ferguson, *Captain Robert Knox*, 53). The interleaved version with its marginal notes is kept in the British Museum's Centre for Anthropology (M10836) and was published in Paulusz ed., *An Historical Relation*. Hereafter referred to as *Relation* (interleaved) with folio numbers and the page numbers of Paulusz vol. II in brackets.
25. Cambridge University Library MSS Add. 1 and Add. 9.
26. BL MS Sloane. 4067, fol. 6.
27. BL MS Lansdowne 1197, fols. 12–13v.
28. The will is dated 30 November 1711 and was published in Ferguson, *Captain Robert Knox*, 55.
29. Felicity Henderson, 'Unpublished Material from the Memorandum Book of Robert Hooke, Guildhall Library, MS 1758', *Notes and Records of the Royal Society of London* 61 (2007), 129–175, and in BL MS Sloane 4024.
30. Henderson, 'Memorandum Book of Robert Hooke', 144.
31. 'Talipot' or 'tallipot', as Knox has it, is from the Sinhala word *talapata*. The botanical name for the tree is *Corypha umbraculifera*. The tree is described and depicted in Knox, *Relation*, i.4.15. See also T. Birch, *The History of the Royal Society*, 4 vols., London, 1756–1757, iv, 64 (meeting of 12 January 1681). Hooke mentions the leaf again in his diary of 1682: Henderson, 'Memorandum Book of Robert Hooke', 153, 18 May 1682, and *Relation* (interleaved), part II, ch. 5 (Paulusz, II, 166). The talipot leaf was again produced in a Royal Society meeting in 1694: Royal Society, *Journal Book* (hereafter JB), IX, fol. 162, 6 June 1694.
32. Henderson, 'Memorandum Book of Robert Hooke', 150, Tuesday 16 August 1681.
33. Paulusz, *Relation*, II, xv.
34. BL MS Lansdowne 1197.
35. BL RP 4713, item 3.
36. *Autobiography*, 4–6 (530–533).
37. Royal Society, *Minutes of Council*, II, 24 November 1683. The objects were presented on 3 November 1683 and included a plough, pictures and medicinal herbs (Birch, *The History of the Royal Society*, iv, 226).

38. Ferguson, *Captain Robert Knox*. The journal containing these observations has not yet been identified. Hooke refers to it being in the hands of the Earl of Clarendon.
39. BL MS Sloane, 1039 fol. 156.
40. Frank, *Crusoe*, 5–7.
41. *Autobiography*, 8–24 (535–552) BL IOR, E/3/90, fols. 170–182-3. St Helena Archives (hereafter SHA), *Consultation Books* (hereafter CB), II, fols. 164–165, Knox recorded at a Council Meeting of 7 April 1685.
42. *Autobiography*, 31–32 (559–561), and BL IOR E/3/91, fols. 47r–48 (or 96–97), 20 March 1685/6.
43. *Autobiography*, 32 (561). There is little information about either journey to Barbados. Presumably Knox was trading in slaves. He claims in fol. 48 that the second voyage was 'by order'. Paulusz i, 425–426, transcribes a section from the Dutch records which note his arrival on 3 October 1688, an account he had given them of chasing an English pirate ship to Mascarenhas (Bourbon: Reunion Island) two years previously, and his departure on 13 October. SHA CB vol. 3 fol. 55 (69) notes that Knox left St Helena on 31 October 1688. A later passage mentions that Knox left behind five Portuguese men, whom according to a letter to Strype dated 1 July 1689 (cf. Ferguson, *Captain Robert Knox*, 42) he had picked up from Mascarenhas where they had been shipwrecked. A letter from Josiah Child to Governor Blackmore of St Helena dated 20 March 1689 says that he believes Knox is lost (cf. H. F. Janisch, *Extracts from the St Helena Records* (Jamestown St Helena: B. Grant, 1885)).
44. BL MS Sloane 4024.
45. Lisa Jardine, *The Curious Life of Robert Hooke* (London: Harper Perennial, 2003), 283.
46. *Autobiography*, 34–43 (563–572); Ferguson, *Captain Robert Knox*, 43–45. Knox appears in the Bencoulen records in 1691, IOR G/35/2, fol. 284r, 19 February 1691.
47. *Autobiography*, 43–44 (572–573). On 59 (590), Knox describes two of the sailors as slaves. See also Ferguson, *Captain Robert Knox*, 46.
48. Yule, *Diary of William Hedges*, ii, cccliii, notes that Knox appeared at the EIC's Court of Directors in January 1694 and informed the Directors of the death of Job Charnock.
49. Royal Society, JB, X, fol. 16, 24 February 1697, fol. 55, 15 December 1697, and fol. 205, 4 December, 1700.
50. *Autobiography*, 60c–62 (592–595).
51. *Autobiography*, 70 (603).
52. JB, XI, fol. 24, 9 June 1703 (Knox presents a Malayan tobacco pipe, a purple shell and a lump of petrified small pebbles, shells, etc., from Gombroone), and fol. 124, 12 November 1707 (presents part of a sail made from tree bark and used in China).
53. Knox, *Relation*, 'Dedication'.
54. Jayawickrama, *Writing That Conquers*, 255–287; Yothers, 'Global Captivities', for a comparison with captivity narratives from New England.
55. *Autobiography*, 5 (516).
56. Ferguson, *Captain Robert Knox*, 27, speculates that Hooke corrected the spelling and 'dressed' the language throughout.

57. Rubiés, 'Instructions for Travellers: Teaching the Eye to See', discusses Hooke's preface with reference to drawing up heads of inquiry for travellers and instructions for the authors of natural histories. Alam and Subrahmanyam, *Indo-Persian Travels*, note that the authorship of several European travel texts of this period is problematic, pointing to the co-authorship of the travel text of François Pyrard de Laval by Pierre Bergeron and the probable organisation of the notes of Jean-Baptiste Tavernier by the Protestant writer Samuel Chappuzeau.
58. Knox, *Relation* (interleaved), 194 (271). Paulusz (*Relation*, I, 383–392) speculates that the illustrator was Knox' brother who died shortly before the publication, perhaps explaining the unfinished appearance of some illustrations. Hooke's diary refers to purchasing Knox' prints from Edward Lascelles, Knox' brother-in-law, but noted that he was 'offended' with them (Henderson, 'memorandum book of Robert Hooke, 148, entry for 17 May 1681).
59. Frank, *Crusoe*, 185
60. These include J. J. Saar, *Ost Indianisge* (Nuremburg, 1662), which Paulusz, *Relation*, I, 'Introduction', thinks served as a model for the wrongly proportioned elephant depicted in one illustration. The talipot palm depicted in Knox, *Relation*, 15, is relatively accurate (M. C. M. Iqbal, personal communication), suggesting it was drawn with reference to the leaf first shown to the society in January 1681.
61. As noted by Saparamandu, 'Introduction'.
62. Knox, *Relation*, 'A new map of the Kingdom of Candy Uda in the Island of Ceylon'. The label reads: 'Eynde van de Caneel landen'.
63. Waesberge had in turn acquired plates from the map maker Jansson: E. G. R. Taylor, 'Robert Hooke and the Cartographical Projects of the Late Seventeenth Century (1660–1696)', *Geographical Journal*, 90 (1937), 529–540, 539. Hooke's involvement is evident from his notes on the project in BL MS Sloane 1039 and from his correspondence with Gottfried Wilhelm von Leibniz on the subject (Royal Society, *Early Letters*, L5, Mo Leibnitz to RH 18/20 Jan 78). Hooke also documents the project in his diary from 1675, in which he eventually referred to Pitt as a 'rascal' for his failure to deliver the promised payment for his involvement. See L. Rostenburg, *The Library of Robert Hooke and the Scientific Book Trade of Restoration England* (Santa Monica, CA: Modoc Press, 1989).
64. Knox, *Relation* (interleaved copy), 93, (481).
65. Autobiography, 97–102 (497–502).
66. Knox, *Relation* (interleaved), 1 (5).
67. *Knox, Relation* (interleaved), 235, (495–496). The passage is given in full in Iqbal, Raheem, and Tennakone, 'Fascinating Facets', 14.
68. John Fryer, *A New Account of East India and Persia 1672–1681* (London, 1698). Chiswell also presented the interleaved copy of the *Relation* to Knox.
69. Pitt, *English Atlas*. The fifth volume appeared as text only, Moses Pitt being imprisoned for debt shortly afterwards.
70. Moses Pitt, 'Advertisement about the English Atlas Now Printing at the Theatre in Oxford by Moses Pitt Bookseller in London' in J. B. Tavernier, *The Six Voyages of John Bapista Tavernier*, trans. by J. Philips, published by Daniel Cox (London: printed for William Gobbid and Moses Pitt, 1677).

71. Charles W. J. Withers, 'Encyclopaedism, Modernism and the Classification of Geographical Knowledge', *Transactions of the Institute of British Geographers*, New Series, 21:1 (1996), 275–298.
72. Knox, revised dedication to the EIC, published in Paulusz, *Relation*, II, xxxviii.
73. For example, Tavernier's dedication to Louis XVI reads: '*j'espère SIRE, que ces Relations exactes & fidèles que j'ai écrites depuis mon retour sur les Mémoires que j'avais recueillis, ne seront pas moins utiles à ma Nation que les riches marchandises que j'ai rapportées de mes voyages.*' J. B. Tavernier, *Les six voyages de Jean Bapiste Tavernier* (Paris, 1676), 4.
74. IOR E/3/29, OC 3144, fol. 152–162.
75. An example of a Dutch collection of maps acquired and copied by the English despite a VOC ban on releasing cartographic information is the collection of maps that were until recently in the library of Corpus Christi College, Oxford (de Meer and Loomeijer, *De schat van Corpus Christi: VOC-kaarten boven water*).
76. Knox, *t'Eyland Ceylon*, Simon de Vries trans. frontispiece. For a description of the process of making ola books and presenting them to great men, as is shown in the picture, Knox, *Relation*, iii:10,109–110.
77. Županov, 'Goan Brahmans', 181.
78. Knox, *Relation*, iv.8.150, describes being interviewed by the King to determine whether he could read and write English. Knox apparently pretended to have forgotten how to read and write in order to avoid the King's service but mentions another Englishman, Richard Varnham, who had entered the King's service and acted as a translator of English.
79. Paulusz, *Relation*, I, 442–445, identifies the map Knox was shown as a twin of No. 326 in the Ryksarchief and notes that Bandara-Kosvatta and Eladetta, unknown until Knox' work, appeared on Dutch maps soon after the publication of the *Relation*, and the text was also drawn on for the 'Map of Ceylon Surveyed along Its Length and Breadth by Order of the Late Rt Hon'ble and Austere Ryklof van Goens', David Mill, Professor of Theology at Utrecht, published by Johannes van Keulen, c. 1710–1712.
80. Robert Knox, *t'Eyland Ceylon*, de Vries trans., frontispiece.
81. Also noted by Saparamadu, 'Introduction'.
82. Knox, *t'Eyland Ceylon*, de Vries trans., frontispiece and 162–163.
83. Robert Knox, *Relation Ou Voyage De L'isle De Ceylon, Dans Les Indes Orientales* (Pieter Schoonebeek, and Jan Branen trans?) (Paul Marret: Amsterdam, 1693).
84. I consulted the copy of Sieur de l'Isle's map in the Maps Department, National Archives of India. It is reproduced in João Riberio, *Histoire de L'Isle de Ceylan*, trans. by L'Abbé le Grand (Amsterdam: Chez J.l. de Lorme, 1701), and elsewhere. The de l'Isle family (also written Delisle or de Lisle) were influential French cartographers during the seventeenth and eighteenth centuries (Josef W. Konvitz, *Cartography in France, 1660–1848: Science, Engineering, and Statecraft*. Chicago: University of Chicago Press, 1987, xvii, 70–73).

85. Madras to the Court dated 8 December 1664 (cf. Ferguson, *Captain Robert Knox*, 12) on the royal correspondence. A section from the Dagh-Register of 30 January 1664 summarised by Ferguson, *Captain Robert Knox*, 13, describes Edward Winter's negotiations. See also London to Fort St George, 7 December 1669 (cf. Ferguson, *Captain Robert Knox*, 20), and London to Fort St George, 15 December 1676 (c.f. Ferguson, *Captain Robert Knox*, 25). In 1689, the factory at Fort St George made a further attempt, writing to 'Raja Devora of Tutecaree' (from Tamil Tūttukkudi, *Hobson-Jobson*, 946) to request his intercession with Rajasingha II. (BL IOR G/19/6, 26 August 1689, transcribed in *Records of Fort St George, Diary and Consultation Book of 1689* (Madras: Government Press, 1916), 72).
86. *Autobiography* 63–64 (596–597) – a copy of a letter from Knox to his fellow captive in Ceylon, John Morgan and William Vassall. On these men, see E. Reimers, 'Raja Singh and His British Captives', *Journal of the Ceylon Branch of the Royal Asiatic Society*, 30 (1925), 13–36.
87. Londa L. Schiebinger, *Plants and Empire: Colonial Bioprospecting in the Atlantic World* (Cambridge, MA: Harvard University Press, 2004), 92–93, gives some further examples.
88. BL IOR O.C. 8188, Condore, 8 July 1703, reproduced in Yule, *Diary of William Hedges*, II, cccxxix–cccxxxvii. Again, this Henry Smith may be the same man who claimed authorship of the *Dictionary English and Malayo, Malayo and English* (Chapter 3).
89. Hooke Folio, fol. 469, 16 December 1680.
90. Paul Hermann, *Paradisus Batavus* (Leiden: Elzevier, 1698) is mentioned in Royal Society, JB, X, fol. 82, 24 August 1698.
91. Royal Society, JB, X, fol. 217, 7 May 1701. Sloane presents the expressed oil of ricinus (castor oil) given to him by Hermann and said to be useful in the treatment of French pox (syphilis).
92. Royal Society, JB, fol. 241, 25 Feb 1702.
93. Paulusz, i, 400. For the making of the copper plate and comparison between Knox' and Hermann's vocabularies, Thomas Hyde to Thomas Bowrey, 7 February 1700, IOPP, MSS Eur. E. 192. 2 (a), Item 4, fol. 1 and Item 5, fol. 1, respectively. The vocabulary of Sinhala in this volume (MSS Eur 192 (b) J. 766, item 8) is a copy with translations of the Latin into English of Hermann's *Vocabularium Selanense seu Insulae Qeylon in India Orientali*, Or. Reg. 16 B. XX. See also Don Martino de Zilva Wickremasinghe, *Catalogue of the Sinhalese Manuscripts in the British Museum* (London: Longmans & Co, 1900). The plate is among those published in Hyde, *Syntagma dissertationum*.
94. IOPP MSS Eur. E. 192. 2 (a), Item 8, fol. 1: Thomas Hyde to Thomas Bowrey, Oxon, 7 February 1700 (1701).
95. D. Strachan, 'An Account of the Taking and Taming of Elephants in Zeylan, by Mr. Strachan, a Physician, Who Lived 17 Years There', 'Observations on the Planting and Culture of Tobacco in Zeylan, by Mr Strachan', 'Observations Made in the Island of Ceilan, by Mr Strachan on the Way of Catching Fowl and Deer, of Serpents, of the Antbear and of Cinnamon', and 'Some Observations on Coral, Large Oysters, Rubies, the Growing of a Sort of Ficus Indica, the Gods of the Ceylanese, Etc. Made in Ceilan, by Mr Strachan', all in *Philosophical Transactions*, 23 (1702/1703).

96. Cf. Paulusz, *Relation*, I, ii–xvi, Frank, *Crusoe*, 5–7.
97. Emma Spary 'Of Nutmegs and Botanists: The Colonial Cultivation of Botanical Identity', in Schiebinger and Swan, *Colonial Botany*, 187–203.
98. F. Guerra, 'Drugs from the Indies and the Political Economy of the Sixteenth Century', *Analecta Médico-Histórica*, 1 (1966), 29–54.
99. C. Mukerji 'Dominion, Demonstration and Domination: Religious Doctrine, Territorial Politics and French Plant Collection', 19–32; M. T. Bravo 'Mission Garden, Natural History and Global Expansion, 1720–1820', 49–65; and L. Schiebinger, 'Prospecting for Drugs: European Naturalists in the West Indies', 119–132, all in Schiebinger and Swan, *Colonial Botany*; Schiebinger, *Plants and Empire*, 73–74, discuss the European concern over the drain of money to the East and West Indies as the new cures became increasingly popular. See also Chapter 4.
100. Schiebinger, *Plants and Empire*; Grove, *Green Imperialism*.
101. See, for example, the preface of Draakenstein et al., *Hortus Malabaricus* for the acknowledgements of the contribution of Itty Achuden's work and the book *Manhaningattnam*. However, as Manilal also notes, no manuscript in Malayalam or Sanskrit that corresponds with the description given has so far been discovered. K. S. Manilal, *Botany and History of Hortus Malabaricus* (Rotterdam: Balkema, 1980), 1–5.
102. This does not, of course, imply that the information of Knox' informants could not be ultimately traced back to Sinhalese Buddhist variations on ayurvedic medicine with its extensive pharmacopoeia or to the knowledge of the village-level medical professionals (*vedarāla*).
103. Knox refers in the interleaved version to the dedication of food to demons, noting that he himself became ill after eating a cock dedicated to these demons (*Relation* (interleaved) fol. 184, inserted after iii.11.144). See Gananath Obeyesekere, 'The Ritual Drama of the Sanni Demons: Collective Representations of Disease in Ceylon', *Comparative Studies in Society and History*, 11 (1969), 174–216, for discussion of the Kōla Sanniya ritual and B. Kapferer, *A Celebration of Demons: Exorcism and the Art of Healing in Sri Lanka*, 2nd edn. (Oxford: Smithsonian Institution Press, 1991). And for a critique of these approaches, D. Scott, *Formations of Ritual* (Minneapolis: University of Minnesota Press, 1994).
104. Knox, *Relation*, i.5.19.
105. Knox, *Relation*, i.5.20. Paulusz, *Relation*, II, 75, fn. 1, identifies this as Kāmaranga – Lat.: *Averrhoa carambola* (family Oxalidaceae). The identification was confirmed by Siril Wijesundara, the Director of the National Botanic Gardens of Sri Lanka.
106. Knox, *Relation*, i.5.19 (see Paulusz, II, 74, fn. 3, for identification of the plants for which Knox gives the approximations of the Sinhala names).
107. A transcription of this journal is Appendix 2 in Winterbottom, *Company Culture*.
108. Knox, *Relation*, I, IV, 14. Paulusz identifies the 'jack' with *Artocarpus integrifolia*, from the Portuguese 'jaca', 'warracha' with Sinhala 'varāka', and 'vella' with 'vāla' (Paulusz, *Relation*, II, 50, fn. 2). According to Siril Wijesundara (personal communication), all in fact come from the same jack tree, *Artocarpus heterophyllus*.

109. Identified by Siril Wijesundara (personal communication) as *Ipomoea pes-caprae* (Linn.) Roth. (family Convolvulaceae).
110. Identified by Siril Wijesundara (personal communication) as castor *Ricinus communis* (family Euphorbiaceae), one of the most poisonous plants in the world.
111. BL MS Lansdowne 1197.
112. Knox, *Relation* (interleaved), folios inserted after i.4.18 (Paulusz, II, 61–62).
113. Knox, *Relation* (interleaved), folios inserted after i.4.18 (Paulusz, II, 63–64). The specific plantation, which Knox gives as 'Jehew Halls', could not be identified.
114. Grove, *Green Imperialism.*
115. The EIC's relationship with the colonies in the West Indies was ambiguous: see Chapter 6.
116. BL MS Lansdowne 1197.
117. 'Instructions to Captain Knox', London, 4 April 1684, IOR E/3/90 fols. 182–183. In his autobiography, Knox notes that he reached the latitude of Tristan d'Acuna but was unable to land there due to bad weather and lack of time.
118. 'Instructions to Captain Knox', London, 4 April 1684, IOR E/3/90 fols. 182–183.
119. For an account of the French Ambassador de la Hays' survey of possible sites for the *Compagnie des Indes* in the Indian Ocean, du Bois, *Les voyages fait par le Sievr D B* (Paris, 1674). See also Abraham du Quesne, *A Voyage to the East Indies in the Years 1690 and 1691* (London: Printed for Daniel Dring, 1696), for the brief settlement by French Huguenot refugees of the island of Rodriguez, near Mauritius. A reference in Hooke's diary (BL MS Sloane 4024, entry for 24 October 1689) to Diego Rodriguez after a conversation with Knox also describes it explicitly in terms of a potential settlement.
120. The Dutch and Portuguese fought battles over the island of St Helena from 1625; the Dutch formally claimed it in 1633, and recaptured it from the English, who had claimed it in 1659, between January and May 1673. See S. B. Royle, *The Company's Island: St Helena and the Colonial Endeavour* (London and New York: I. B. Tauris, 2007), 153–155. For the suggestion in 1715 of Governor Isaac Pyke of St Helena that the English claim Mauritius, recently abandoned by the Dutch, and transplant the population of St Helena there see Philip Gosse, *St Helena 1502–1938* (London: Cassell and Co., 1938), 137–138.
121. See, for example, W. Cockburn, *An Account of the Nature, Causes, Symptoms and Cure of the Distempers That Are Incident to Seafaring People* (London: printed for Hugh Newman, 1696), for the classification of diseases according to meridian.
122. On Hooke's instruments in general, J. A. Bennett, 'Robert Hooke as Mechanic and Natural Philosopher', *Notes and Records of the Royal Society of London*, 35:1 (1980), 33–48. For a reconstruction of the longitude clock, Michael Wright, 'Robert Hooke's Longitude Timekeeper', in Michael Hunter and Simon Schaffer (eds.) *Robert Hooke: New Studies* (Boydell Press: Woodbridge, 1989). On the portable camera obscura, there remains little clear information.
123. Cañizares-Esguerra, *Nature, Empire, and Nation*; Guerra, 'Drugs from the Indies'.

124. BL IOR B/40 (Court Minutes 1690–1695) fol. 15, 12 September 1690.
125. IOR E/3/90, fols. 87v–98v, London to St Helena, 1 August 1683, fol. 95v. Also noted in Royle, *The Company's Island*, 27. See also E/3/90 fol. 178v, London to St Helena, 5 April 1684, which qualifies the discussion of latitude, noting that the mountainous parts of the island are likely to be colder and more suited to European plants while the hotter parts could support crops transplanted from the West and East Indies.
126. IOR E/3/90, fols. 182–183, 4 April 1684.
127. Royle, *The Company's Island*, 79–80. E/3/29, OC 3144 fols. 152–162, Sir George Oxenden and Council in Surat to EIC in London, 1 January 1666.
128. IOR E/3/90, fols. 182–183, 4 April 1684.
129. IOR E/3/90, fol. 175, London to St Helena, 5 April 1684/1685. For discussion of slavery on St Helena, see Royle, *The Company's Island*, 84–103, and Chapter 6.
130. Wickremesekera, *Kandy at War*, 50–53.
131. Chaudhuri, *Asia before Europe*, 238, notes that it is hard to assign a date to the original development of quick-ripening rice (which allows for a double crop in a year), but that it seems to have been introduced from the ancient kingdom of Champa to China in around 1012 and spread from there across South Asia.
132. Knox, *Relation*, I, III, 8. See Paulusz, II, 34, note 1, for an identification of the types of rice.
133. Knox, *Relation*, i:3, 9–12.
134. Knox, *Relation* (interleaved), folios inserted between 9 and 12 (37–46).
135. It is not named in the original, Paulusz II, 37, fn. 1, speculates that it may refer to Suvanda-āl. In the interleaved copy, Knox names two types: one with a red husk called 'Kere-All-wea' and the other with a white husk known as 'Meputt-all-wea', interleaves inserted after 9 (37).
136. IOR E/3/90, fol. 170: London to Surat, 7 April 1684.
137. IOR E/3/90, fol. 170: London to Surat, 7 April 1684.
138. IOR E/3/90; fol. 175. London to St Helena, 5 April 1684. The EIC displays more concern for the conservation of timber here than Grove, *Green Imperialism*, generally credits them with in the early period, noting that 'we will never permit any to be made on your Island lest it should consume our wood'.
139. IOR E/3/90, fol. 170: London to Surat, 7 April 1684, fol. 175. Ashmole and Ashmole (74 and 440) note that the fossil evidence from St Helena shows that palms were once important; they appear to have been extinct for some time. The few surviving date palms were planted in the sixteenth century by the Portuguese: J. C. Meliss, *St Helena: A Physical, Historical and Topographical History of the Island* (London, 1875). R. O. Roberts, The plants of St Helena Island (unpublished MS notes compiled during the 1970s and kept in the St Helena library), mentions four types of palm but not the *Cocos nucifera*. Gosse, *St Helena*, 169, mentions that coconut trees sent from Bombay were planted by Isaac Pyke in 1733–1734.
140. *Autobiography*, 100 (641).
141. London to St Helena, 5 April 1684, E/3/90, fol. 175.
142. SHA, *Consultations*, II, fol. 52 (48), 11 April 1684.
143. SHA, *Letter Book*, I, 23 January, 1679; see also Royle, *The Company's Island*, 87.

144. Knox does appear in St Helena records, where he was made a temporary member of council (SHA *Consultations*, ii, fol. 164, 7 May 1685).
145. Birch, *History of the Royal Society*, ii, 226.
146. Ashmole and Ashmole, *St Helena and Ascension Island*, 65–67.
147. Joseph Banks to William Devaynes 27 December 1788, BL IOPP MS D993. Transcribed in N. Chambers (ed.) *The Letters of Sir Joseph Banks, A Selection, 1768–1820* (London, Imperial College Press, 2000). Letter 35, 114–119.
148. British Library, copy at shelfmark 983 h 12. At the back of the book, Banks has added cross-references to the *Hortus Malabaricus* and a work of Paul Hermann and other names for some of the plants Knox lists. I am grateful to Katherine Frank for bringing this volume to my attention.
149. Frank, *Crusoe*, 5–7, 231, 267.
150. Boxer, 'Ceylon through Puritan Eyes'.
151. *Autobiography*, v (517).
152. *Autobiography*, 6 (532).
153. *Autobiography*, 116 (655).
154. Frank, *Crusoe*, 184–185.
155. Knox, *Relation*, iii:4, 78.
156. Frank, *Crusoe*, 203–209.
157. *Autobiography*, 15 (542–543).
158. Raj, *Relocating Modern Science*, 252.
159. Knox told the Dutch that he was able to escape because knowing the language and wearing Sinhalese clothing they were 'not regarded as Europeans', or were seen as Portuguese, who were sufficiently integrated not to wish to leave.
160. Knox, *Relation,* (interleaved) IV, VIII, 152, (396), and *Autobiography*, 63–64 (596–597), for Knox' adoption of a female child of one of his shipmates, Lucea, whom he taught to read and write English and to whom he bequeathed his land and possessions and later sent his picture. Frank, *Crusoe*, 105–109, and others have argued that Lucea was Knox' child by birth. For the marriage and settlement in Kandy of many of his fellow captives, Knox, *Relation*, iv:7, 146–147.
161. BL Lansdowne MS 1197. This appears to be an unconscious deviation from his normal 'the Chingulais have…' but perhaps all the more revealing as such. Attang-cola is a reference to the intoxicant *Datura stramonium*. (M. C. M. Iqbal, personal communication; see also Iqbal, Raheem, and Tennakone, 'Fascinating Facets', 17, for Knox and Hooke's discussion of *Datura*. Fn. 26 notes Hooke's reference to the use of *Datura* in Ceylon.)
162. See *Autobiography*, 9–23 (537–551). Knox' description of Madagascar is another element of his later work that warrants more attention.
163. SHA, *Consultations*, fol. 175, 8 June 1685.

6 Transportation and Transplantation: Slave Knowledge and Company Plantations

1. John Emerson, 'Chardin, Sir John', *Encyclopaedia Iranica*, V, Fasc. 4, 369–377. (Originally published: 15 December 1991; online version http:

//www.iranicaonline.org/articles/chardin-sir-john; last updated 13 October 2011, last accessed 18 September 2014.) For the contract, Baladouni and Makepeace, *Armenian Merchants*, document 112.

2. A recent exception is Richard B. Allen, *European Slave Trading in the Indian Ocean, 1500–1850* (Athens: Ohio University Press, 2014). I am very grateful to Richard Allan for sharing his work in advance of publication. Other relevant works include Indrani Chatterjee, *Gender, Slavery and Law in Colonial India* (New York, 1999); Anthony Reid (ed.) *Slavery, Bondage and Dependency in South East Asia* (St. Lucia, QLD, 1983); Basanta Kumar Basu, 'Notes on Slavery in India during the Early Days of John Company', *Muslim Review*, 4:4 (1930), 21–34; Marina Carter, 'Slavery and Unfree Labour in the Indian Ocean', *History Compass*, 4:5 (2006), 800–813; and Frenise A. Logan, 'The British East India Company and African Slavery in Benkulen, Sumatra, 1687–1792', *Journal of Negro History*, 41:4 (1956), 339–348.
3. Judith Carney and Richard Nicholas Rosomoff, *In the Shadow of Slavery: Africa's Botanical Legacy in the Atlantic World* (Berkeley: University California Press, 2009).
4. Chatterjee, *Gender, Slavery, and Law*; Gwyn Campbell (ed.) *The Structure of Slavery in Indian Ocean Africa and Asia* (London: Frank Cass, 2004); Gwyn Campbell, Suzanne Miers, and Joseph C. Miller, *Women and Slavery*, 2 vols., I: *Africa, the Indian Ocean World, and the Medieval North Atlantic* (Athens: Ohio University Press, 2007); Emma Christopher, Cassandra Pybus and Marcus Rediker (eds.) *Many Middle Passages: Forced Migration and the Making of the Modern World* (Berkeley: University of California Press, 2007).
5. S. Arasaratnam, 'Slave Trade in the Indian Ocean in the Seventeenth Century', in K. S. Mathew (ed.) *Mariners, Merchants and Oceans: Studies in Maritime History* (New Delhi: Manohar, 1995), 195–208; Carter, 'Slavery and Unfree Labour'.
6. Robert Harms, Bernard K. Freamon and David Blight (eds.) *Indian Ocean Slavery in the Age of Abolition* (Yale University Press, 2013), 183–199.
7. Pedro Machado, 'A Forgotten Corner of the Indian Ocean: Gujarati Merchants, Portuguese India and the Mozambique Slave Trade, c. 1730–1830', in Campbell (ed.) *Structure of Slavery*, 16–33; Rudy Bauss, 'The Portuguese Slave Trade from Mozambique to Portuguese India and Macau and Comments on Timor, 1750–1850: New Evidence from the Archives,' *Camoes Center Quarterly*, VI–VII, No. 1–2 (1997); C. R. Boxer, *Portuguese Society in the Tropics: The Municipal Councils of Goa, Macao, Bahia, and Luanda, 1501–1800* (Madison: University of Wisconsin Press, 1965); and Jeanette Pinto, *Slavery in Portuguese India, 1510–1842* (Bombay: Himalaya Publishing House, 1992). Goa Central Archives (Panaji), item 860, 'Alforria dada aos escravos, 1682–1759', a list of domestic slaves, gives castes and/or places of origin. Most were acquired within India through purchase, transfer or capture in conflict. More African slaves appear in the list after around 1740, perhaps as a result of trade with the French.
8. Vink, 'The World's Oldest Trade'; Ward, *Networks of Empire*.
9. Edward A. Alpers, 'The French Slave Trade in East Africa (1721–1810)', *Cahier d'Études Africaines*, 10:1 (1970), 80–124; Holden Furber, *Rival Empires of Trade in the Orient, 1600–1800* (Minneapolis: University of Minnesota Press, 1976); Teotonio R. de Souza, 'French Slave-Trading in Portuguese

Goa (1772–1791),' in de Souza (ed.) *Essays in Goan History* (New Delhi: Concept Pub. Co., 1989), 119–131; Richard B. Allen, 'Carrying away the Unfortunate: The Exportation of Slaves from India during the Late Eighteenth Century,' in Jacques Weber (ed.) *Le monde créole: peuplement, sociétés et condition humaine, XVIIe–XXe siècles* (Les Indes savants: Paris, 2005), 289–302.

10. Gillian Tindall, *City of Gold: The Biography of Bombay* (London, 1992), also notes this map and identifies some slave graves in Mumbai.
11. 'A Plan of Fort St George and the Bounds of Madraspatnam', surveyed and drawn by FL Conradi, 1755. Eight sheets. National Archives of India (NAI), Cartography Room, 912.548 M 267, Acc. No. 3565, 1755. 'Copied by W. Woodrow from the original in the British Museum.'
12. IOR, G/19/4, 68, 19 November 1686, Fort St George (FSG). The slaves were sold in Aceh and Quedah 'not having Rice to maintain them here, the famine still increasing'.
13. Arasaratnam, 'Slave Trade in the Indian Ocean', 205.
14. TNSA, *Public Dispatches from Court*, I, fols. 161–169, London to Surat, 15 March 1677. Chatterjee, *Gender, Slavery and Law*, notes a similar problem of terminology in Asian languages.
15. Shanti Sadiq Ali, *The African Dispersal in the Deccan: From Medieval to Modern Times* (Hyderabad, 1996), 4, argues that the word 'Siddi', when applied to slaves, normally refers to those from the southern region of Arabia, and 'Habshi' refers to those from the Red Sea region.
16. St Helena Archives (SHA), *Consultations*, XVII, ff. 205–206, 24 January 1716.
17. Gwyn Campbell, et al., 'Introduction: Slavery, Forced Labour, and Resistance in Indian Ocean Africa and Asia', in Edward A. Alpers, et al. (eds.) *Slavery and Resistance in Africa and Asia* (London, 2005), 1–20, notes that 'received distinctions between "slave" and "free" are not particularly helpful tools of analysis in most regions of [Indian Ocean Africa] and Asia'.
18. Gosse, *St Helena*, 149–150.
19. Arasaratnam, 'Slave Trade in the Indian Ocean', 206. He also notes a 1678 proclamation by Maratha leader Shivaji prohibiting Europeans from transporting slaves.
20. IOR, G/19/5, fol. 43, 14 May 1688.
21. IOR, G/19/5, fol. 43, 14 May 1688. The 'choultry' was a place for the transaction of public business. Yule, *Hobson-Jobson*, 211–212.
22. TNSA, *Public Consultations*, XX, 73–74, 14 November 1693.
23. As Vink, 'World's Oldest Trade', notes, jurists including Grotius recognised the following as legitimate routes into slavery: voluntary sale in order to escape famine and starvation, capture in a 'just war', judicial slavery, and birth into slavery.
24. Margaret Makepeace, 'English Traders on the Guinea Coast, 1657–1668: An Analysis of the East India Company Archive', *History in Africa*, 16 (1989), 237–284, 239.
25. Foster, *English Factories in India, 1661–1664*, 282.
26. *Records of FSG*, Public Despatches from Court, I, 93, fols. 97–100, 107 and 108, 15 December 1676.
27. Vink, 'World's Oldest Trade'.

28. Furber, 'Rival Empires', in Arasaratnam (ed.) *Maritime India*, 79, uses an example from the Dutch factory in Cochin in 1716 said to be typical of European factories. The Governor owned 26 slaves, the second in command had 16, 'all European' households had a total of 117 and 'half-European' households had 252.
29. SHA, *Letter Book*, vol. 2, fol. 88, London to St Helena by Princess Amelia, 21 March 1717; E/3/100, fol. 146, London to St Helena, 9 March 1719, giving the number of Company slaves as 203.
30. Ward, *Networks of Empire*, 3.
31. Mughal records of slavery in Allahabad in the NAI are described in R. K. Perti (ed.) *Descriptive List of Acquired Documents (1356–1790 AD)* (Delhi: National Archives of India, 1912), 3. The first is dated 15 March 1722. All refer to women or girls. See also Alpers, et al., *Slavery and Resistance*, 3, and Chatterjee, *Gender, Slavery and Law*.
32. For example, E/3/92, fol. 60r London to Bencoulen, 3 October 1690 and E/3/98, fol. 128v, London to Bencoulen, 22 January 1713. Vink, 'World's Oldest Trade', notes that there were more male than female slaves in VOC settlements. Allen, *European Slave Trading*, 59, estimates an average ratio of men to women of 2:1 in the settlements of all the European trading companies.
33. For a description of the housing of the Bencoulen slaves, IOR, G/35/3, 15 July 1695.
34. IOR, E/3/90, fol. 158, 5 March 1685; and SHA, Letter Book, II, St Helena to London, 23 February 1722.
35. Allen, *European Slave Trading*, 44.
36. For public sales of EIC slaves, see SHA, Consultations, III, fol. 286, 5 August 1686; for an order to transport slaves from Madras to St Helena to sell to the free inhabitants, see Records of FSG, *Public Consultations*, 15 February 1688, 30.
37. For estimates of the total numbers of slaves acquired and transferred between factories between 1622 and 1772, Allen, *European Slave Trading*, tables 6 and 8.
38. TNSA, *Consultations*, III, fols. 135–136, 30 December 1679.
39. Allen, *European Slave Trading*, 49.
40. Foster, 'An English Settlement in Madagascar', p. 240.
41. Walter Hammond, *Madagascar, the Richest and Most Fruitful Island in the World* (London, 1643).
42. BL Add Ms. 14,037, 'A Booke of Consultations Belonging to the Plantation of Madagascar al[ia]s the Island of St. Lawrence', 12 May 1645.
43. BL, Add Ms. 14,037, fol. 810, 12 May 1645. A consultation dated 14 June 1645 reported that the expedition had been frustrated by the French and Dutch who were settled in similar strength at Antongill. The ship returned to Augustine's Bay, 'the design for slaves etc. being utterly void'. See also Aniruddha Ray, *The Merchant and the State: The French in India, 1666–1739* (2 vols., New Delhi, 2004); and Mike Pearson Parker, 'Close Encounters of the Worst Kind: Malagasy Resistance and Colonial Disasters in Southern Madagascar,' *World Archaeology*, 28:3 (1997), 393–417.
44. James C. Armstrong, 'Madagascar and the Slave Trade in the Seventeenth Century', *Omaly sy Anio*, 17–20 (1983–1984), 214–218.

45. Examples include a letter dated 26 July 1630 off Madagascar from William Rastell to the Council at Surat, which reads 'at my arrival at St. Lawrence...wee most prepostrously found these ships the Charles and Jonas...', NAI, Foreign Department Records, V, fol. 7.
46. Philip Stern, 'Politics and Ideology in the Early East India Company-State: The Case of St. Helena, 1673–1709', *Journal of Imperial and Commonwealth History*, 35:1 (2007), 1–23.
47. The list of items is taken from Bodleian Library, Ms. Rawl, A 334, 61.
48. Drury, *Madagascar*, describes a Madagascan ruler surrounded by his weapons. See also Marcus Rediker, 'Postscript: Gun-Slave Cycle', in Christopher, et al. (eds.) *Many Middle Passages*, 235–237.
49. 'Guinea' or 'Gana', also called 'Ghanah, Fuinoa and Geneoa', derived from the name 'Jenné' given to the main city described by both Ibn Battuta and Leo Africanus. Gomes Eannes de Azurara, *The Chronicle of the Discovery and Conquest of Guinea*, trans. Charles Raymond Beazley and Edgar Prestage (London: Hakluyt Society, 1899), 'Introduction', xlix; and Margaret Makepeace, 'English Traders on the Guinea Coast'.
50. Stephen Royle, *The Company's Island*, 85–96; and Makepeace, 'English Traders on the Guinea Coast'.
51. MS Bod Rawl A. 170, fol. 267.
52. Surendranath Sen (ed.) *Indian Travels of Thevenot and Careri* (National Archives of India: New Delhi, 1949), 188.
53. Allen, *European Slave Trading*, 46; E/3/91, fol. 256, 28 Jan 1697/1698. Instructions to Capt. William Freke, Commander of the Anne.
54. Guildhall Library (GL), Ms. 3041(ii). The description of Madagascar contained in Bowrey's papers (although not the projected plantation) is discussed in Bialuschewski, 'Thomas Bowrey's Madagascar Manuscript of 1708'.
55. LMA, Ms. 3041/3, iv (bound unnumbered papers), 'On "Mannigaro" on the west side of the island (15: 40 S° Latt)'.
56. Virginia B. Platt, 'The East India Company and the Madagascar Slave Trade', *William and Mary Quarterly*, 3rd ser., 26:4 (1969), 548–577.
57. One reference is in the Maharastra State Archives (MSA), Secretariat Outward Letter Book, fol. 165, regarding instructions to returning ships: 'we will alter from the method we pursued the last year, which we found inconvenient by the sickness and death of many of our seamen at Barbados, where there hath been a very raging plague and distemper'. For two trips from Madagascar to Barbados in 1686 and 1688 'by order', Robert Knox, *Autobiography* (Bodleian Library, Ms. Rawl Q., c. 15, fol. 561).
58. James Delbourgo, personal communication. St Helena, the Cape and Madagascar provided potential meeting points for EIC ships with sailors from the Americas.
59. Platt, 'The East India Company and the Madagascar Slave Trade'.
60. BL, IOR, G/19/4, fol. 33, 21 June 1686.
61. BL, Add Ms. 34,123, Copybook of Henry Vansittart, Governor of Bombay, 1756. Trade with Aceh declined after a political crisis in 1699; see Anthony Reid, 'The Seventeenth-Century Crisis in Southeast Asia', *Modern Asian Studies*, 24:4 (1990), 645.

62. A letter from London noted at Fort St George expressed '[d]isapproval of slave selling at Fort St. David.' TNSA, *Public Despatches from Court*, XII, 6 March 1701. This letter is not preserved in the IOR records. Transfers of slaves from Fort St David to the Anjengo factory are recorded in *Records of FSG, Public Consultations*, XXVI, fols. 81–84, 15 November 1697; *Letters from FSG*, VII, fols. 83–84, 28 December 1697; *Letters from FSG*, VII, fols. 76–77, 18 December 1697; and Letters from FSG, VII, fols. 80–82, 27 December 1697.
63. TNSA, *Letters from FSG*, X, fols. 104–105. (Check original for date.)
64. Kerry Ward, 'Slavery in Southeast Asia, 1420–1804', in David Eltis and Stanley L. Engerman (eds.) *The Cambridge World History of Slavery, III: AD 1420–AD 1804*, 4 vols. (Cambridge: Cambridge University Press, 2011), 163–185.
65. IOR, G/35/2, fol. 22, 10 December 1686; G/35/3, fols. 64–65, 14 August 1695.
66. Ward, 'Slavery in Southeast Asia'; Allen, *European Slave Trading*, 48.
67. Gosse, *St Helena*.
68. Gosse, *St Helena*, 422–423.
69. Grove, *Green Imperialism*, 42–47, discusses English, Dutch and French Huguenot conceptualisations of the tropical island in this period.
70. [Francis Godwin], *The Man in the Moone: Or a Discourse of a Voyage Thither by Domingo Gonsales the Speedy Messenger* (London: printed by John Norton, for Ioshua Kirton and Thomas Warren, 1638). For a modern edition, Francis Godwin, *The Man in the Moone*, ed. William Poole (Peterborough: Broadview Editions, 2009).
71. Dampier, *A New Voyage around the World*.
72. Leiden University Library, Bodel Nieuwenhui collection, COLLBN 002-12-037.
73. IOR E/3/91, fol. 95, London to Surat and Bombay, 15 October 1686.
74. IOR G/35/1, fol. 10, York Fort to William Gyfford at Fort St George, 3 October 1685.
75. Anon, Royal Society, *Register Book*, vol. 9, fol. 71: 'A Memoriall to live under the Equinoctial and between the Tropicks; the Common Distemper incident to the Inhabitants of those Climates with ye Manner of Cure by the Natives and likewise by other European practitioners in Physick' (no author or date, placed between documents dated 1701 and 1703).
76. Bod. MS Rawl Q. c. 15., fols. 39–40 (Robert Knox, *An Historical Relation of Ceylon*, ed. by J. H. O. Paulusz, ii, 568–569).
77. William Foster, 'The Acquisition of St. Helena', *English Historical Review*, 34 (1919), 285. Royle, *The Company's Island*, 170–171, gives a list of the produce recommended for St Helena from 1659 to 1698.
78. Royle, *The Company's Island*, 28–29. Both were briefly profitable, but flax, like many other introduced species, soon became considered detrimental to the survival of indigenous and endemic plants.
79. Yams were introduced at around the time of the original settlement and became an important foodstuff; cinchona was introduced in 1867–1868 but did not thrive. See Philip Ashmole and Myrtle Ashmole, *St Helena and Ascension Island: A Natural History* (Oswestry: A. Nelson, 2000), 156–157.

80. Philip and Myrtle Ashmole, *St Helena and Ascension Islands* (Shropshire: Anthony Nelson, 2000).
81. For the sale of tobacco in Bencoulen, see Chapter 2. The introduction of tobacco was attempted in St Helena, and SHA, *Consultations*, V, fol. 62, 19 April 1697, reports 'green tobacco growing on the Company's common', but the crop never took off. A walk on Lot's Wife in 2007 with botanists from Kew Gardens revealed that a few wild plants still survive. For sugar canes in Bencoulen, Collet MS D1153/2, fols. 62–64, 1 July 1713, in which he claims 'this place produces the best Sugar Canes in India'.
82. SHA, *Letters Sent, 1706–1714*, fols. 30–32, 7 March 1710; IOR G/35/7, fol. 112, Collet to London, 10 September 1713, mentions raising coffee plants, and a letter of 10 February 1718 notes that a few coffee trees planted in Collet's time show promise. Coffee is still grown in very small quantities on St Helena.
83. Stern, 'Politics and Ideology in the Early East India Company-State', 4.
84. SHA, *Consultations*, vol. 6, fol. 20.
85. Royle, *The Company's Island*, 45–46 and table 3.1, appendix 3.
86. Royle, *The Company's Island*, 48–49.
87. John Ovington, *A Voyage to Surat in the Year 1689* (Delhi: Asian Educational Services, 1994), 59–60.
88. SHA, *Consultations*, XII, fols. 200–204, 24 January 1716. 'Guinea' includes parts of modern Nigeria and Cameroon as well as Equatorial Guinea.
89. Royle, *The Company's Island*, 52–54.
90. IOR G/35/3, fols. 53–54, 5 August 1685.
91. Allen, *European Slave Trading*, 55–56.
92. Gosse, *St Helena*, 96.
93. IOR G/35/7, fol. 4, York Fort, Bencoulen to London, 12 February 1704.
94. IOR E/3/91, fol. 179, London to St Helena, 3 August 1687.
95. 'Ovington, *A Voyage to Surat*, p. 63.'
96. London to Madras, 27 November 1668, cf. Yule *Diary of William Hedges*, II, cccliv. 'Gentues' refers to Hindus and 'Aracans' to people from modern Burma.
97. IOR E/3/90, fols. 89–98, 1 August 1683.
98. Nehemiah Grew, *The Anatomy of Plants, with an Idea of a Philosophical History, and Several Other Lectures Read before the Royal Society of London* (London: Printed by W. Rawlins, for the author, 1682); Frank Nicholls, 'An Account of the Veins and Arteries of Leaves', *Philosophical Transactions*, 36 (1729–1730), 371–372.
99. Jean Bodin, *Method for the Easy Comprehension of History* (1566), Beatrice Reynolds (ed.) (New York: Octagon Books, 1966).
100. Harrison, *Climates and Constitutions*; Ivan Hannaford, *Race: The History of an Idea in the West* (Washington, DC: Woodrow Wilson Center Press, 1996).
101. SHA, *Letters Sent 1717–1720*, fols. 137–143, Isaac Pyke to London, 3 November 1718.
102. For discussion of the Iberian precedents, Cañizares-Esguerra, *Nature, Empire, and Nature.*
103. Knox, *Autobiography*, fols. 39–40 (pp. 568–569).

104. Grove, *Green Imperialism*; Drayton, *Nature's Government*, 235–238, for an important qualification of Grove's arguments about the radical nature of early environmentalism.
105. SHA, *Consultations*, vol. 3, fol. 129, Court of Justice, Session House, 07 October 1689.
106. SHA, *Letters Received 1717–1725*, fol. 42, London to St Helena by the *Princess Amelia*, 21 March 1717.
107. Morgan Godwyn [Godwin], *The Negro's & Indians Advocate, Suing for Their Admission into the Church* (London: printed for the author, by J. D., 1680), Dedication to William, Archbishop of Canterbury.
108. SHA, *Letters Received 1673–1683*, fol. 9, 19 December 1673.
109. Robert Boyle, *Papers* (online edition by Birkbirk available at http://www.bbk.ac.uk/Boyle/boyle_papers/boylepapers_index.htm, accessed 17 May 2010), IV, fol. 144. 'Clause in the Company's: letter written to St: Hellena Dated 9th: Decr: 1670. Sent on Ship Unicorne'. No letter with this date appears to be preserved on St Helena or in the British Library, but the instruction is repeated in SHA, *Letters Received 1673–1683*, fol. 5, 19 December 1673, London.
110. IOR E/3/87 [n.f.] p. 419, London to Bantam,18 January 1670.
111. Register of Baptisms, St Mary's Church, I (now in the Fort Museum, Chennai). Each slave was provided with godmothers and/or godfathers. In a few cases, the names imply that these were also slaves or former slaves.
112. *Records of FSG, Diary and Consultation Book*, 1678–1679, 88–89, 4 April 1678, and 88–89, 4 April 1678.
113. IOR G/35/2, fol. 28 (49r), London to Priaman, 21 October 1686: the Priaman factors are urged to train Madagascan slaves as 'ships' carpenters, smiths, & [in] other handicrafts, as the Dutch do to their great advantage at Batavia.
114. SHA, *Letters from England, 1717–1725*, 21 March 1717, fol. 36.
115. Godwin, *The Negro's and Indian's Advocate*, 13, notes that slaves were commonly taught to read and write, were appointed as overseers to other slaves, and 'show so much discretion in managing business'.
116. IOR E/3/90, 05 March 1685, fol. 158, London to Bengal.
117. IOR G/35/2, fols. 98–99r, London to Bencoulen, 31 August 1687.
118. IOR G/35/9, fol. 212, Fort Malborough to Court, 30 November 1748.
119. IOR G/35/3, fols. 38–41, Consultation, 18 July 1695.
120. SHA, *Consultations*, XVI, fols. 205–206, 24 January 1716.
121. SHA, *Letters Sent 1717–1720*, fols. 50–51, St Helena to London, 6 January 1718. Godwin, *The Negro's and Indians Advocate*, 101, mentions slaves being baptised and trained in England before being sent elsewhere.
122. See also Allen, *European Slave Trading*, 56–57.
123. IOR E/3/91, fols. 1–4, 21 October 1685.
124. IOR G/35/2, fol. 133v, Sir Josiah Child to Pryaman (diverted to Bencoulen), sent 26 June 1686, received 17 September 1688. 'Portugall' and 'Mesier' refer to mixed-race people with some Portuguese heritage.
125. IOR G/35/2, fol. 137, Hamon Gibbon in Indrapore to London, 20 September 1688. Whether the factory ever actually received this many is doubtful; the same letter reports that the previous ship had brought only 18 slaves because of conflicts in Madagascar.

126. IOR G/35/2, fol. 85, Elihu Yale at Madras to Benjamin Bloome at Bencoulen, 8 September 1687.
127. IOR G/35/2, fol. 95, Commission from London to Capt. John Harding, 3 August 1687.
128. FSG, *Public Despatches from Court*, XI, 101–119.
129. FSG, *Letters to Fort St George*, V, 107–122 and 127–128, Bencoulen to Madras, 2 November 1693. Seven more 'coffrey' soldiers are reported in IOR G/35/3, fol. 216, 18 January 1696.
130. Yule, *Diary of William Hedges*, ii, ccclvi, 6 March 1694.
131. IOR G/35/2, fol. 24 (47v), London to Priaman, 21 October 1685.
132. Gosse, *St Helena*, 171.
133. Ali, *African Dispersal*; Joseph E. Harris, *The African Presence in Asia: Consequences of the East African Slave Trade* (Evanston: Northwestern University Press, 1971), 78.
134. Ali, *African Dispersal*, 79–80.
135. Ali, *African Dispersal*, 87–88.
136. D. R. Seth, 'The Life and Times of Malik Ambar', *Islamic Culture*, 31:1 (1957), 142–155; Edward A. Alpers, 'Sailing into the Past: The African Experience in India', *SAMAR*, 13 (2000), http://www.samarmagazine.org/archive/articles/20 (last accessed 19 September 2014).
137. Vink, 'World's Oldest Trade'; Shihan S. Jayasuriya, 'The African Diaspora in Sri Lanka', Shihan S. Jayasuriya and Richard Pankhurst (eds.) *The African Diaspora in the Indian Ocean* (Trenton: Africa World Press, 2003), 251–288: 255.
138. R. R. S. Chauhan, *Africans in India: From Slavery to Royalty* (New Delhi: Asian Publication Services, 1995).
139. BL, IOR, OC, No. 2238, 30 December 1651.
140. MSA, Secretariat Outwards, VI, fols. 155–156, n.d. [September 1698].
141. MSA, Secretariat Outwards, VII, fols. 208–211. John Leckie at Bombay Castle, 24 May 1695, described an attempt to entice Portuguese slaves to join the English in return for freedom and wages.
142. Yule (ed.) *Hobson-Jobson*, 507–508, and Amitav Ghosh, 'Of Fanás and Forecastles: The Indian Ocean and Some Lost Languages of the Age of Sail', *Economic and Political Weekly*, 21 July 2008, show that 'lascar' was freely applied to people of several different races.
143. TNSA, *Consultations 1686–1688*, 14–17, 21–31 January 1688.
144. IOR, G/35/2, fol. 144, Thos. Lench (onboard *Royal James*) to Bencoulen, 13 November 1688.
145. IOR, G/35/2, fols. 144–144v, William Akid to Bencoulen, 13 November 1688.
146. Janet J. Ewald, 'Crossers of the Sea: Slaves and Migrants in the Western Indian Ocean, c. 1800–1900', *American Historical Review*, 55:1 (2000), 76.
147. MSA, *Surat Inwards*, 39, fols. 26–27.
148. Drury, Madagascar, relates that in Mauritius, the Dutch received them kindly and gave them the news that 'Capt. Boon, a Pyrate (English) had been at the island two months earlier and had taken prisoner several Lascars'.
149. IOR, G/19/5, fol. 1, 20 February 1688.
150. SHA, *Consultations*, III, fol. 34, 18 June 1688.

151. TNSA, *Letters from FSG*, IV, fols. 102–108, FSG to John Goldsborough (Bengal), 10 November 1693.
152. IOR, G/35/2, fol. 68v, Hamon Gibbon (Indrapoore) to Bencoulen, 19 July 1687.
153. IOR G/19/5, fol. 21v, FSG, Consultation, 22 March 1688.
154. IOR E/3/91, fol. 35, London to FSG, 16 March 1685.
155. Shihan S. Jayasuriya, 'Identifying Africans in Asia: What's in a Name?', *African and Asian Studies*, 5 (2006), 275–303.
156. MSA, *Secretariat Outwards*, 1694–1696, V, fols. 211–212. See also Christopher, *Slaves Ship Sailors*.
157. Bod. MS Rawl. C. 450: the logbook of the *Nathaniel* on a voyage to the East Indies and back via Cape of Good Hope, Mocha, etc. 1714–1716.
158. Farrington et al., *English Factory in Taiwan*, 662 and 707–708.
159. Court to Fort St. David, 6 March 1694/1695. Cf. Yule (ed.) *Diary of William Hedges*, ccclvi. 'Mosambeek' is Mozambique and 'Mumbas' is Mombasa (Kenya), the site of the notorious Portuguese Fort Jesus.
160. BL Collet Ms., D 1153/2, letter to Daniel Dolins, 15 July 1717.
161. IOR G/35/3, fols. 38–41, 18 July 1695. Perpetuanos were a light and glossy twilled stuff of wool; Yule (ed.) *Hobson-Jobson*, 699.
162. SHA, *Letters Sent 1717–1718*, 171, 3 November 1718.
163. TNSA, *Letters from FSG*, VII, fols. 39–49, 26 October 1697, re: 'employment of Coffries in the pepper plantations' in Bencoulen.
164. IOR E/3/90, fols. 89–98, 1 August 1682.
165. Judith A. Carney, *Black Rice: The African Origins of Rice Cultivation in the Americas* (Cambridge, MA: Harvard University Press, 2001); Carney and Rosomoff, *In the Shadow of Slavery*.
166. Dampier, *New Voyage*, 107.
167. Gosse, *St Helena*, 280.
168. SHA, *Letters Sent, 1716–1717*, 57–57a, 1 July 1717. See also Gosse, *St Helena*, 134–135, for Pyke's efforts to cultivate yams on a larger scale.
169. As Carney, *Black Rice*, 49–55, notes, in West Africa, the cultivation of crops and transplantation was done by women.
170. Susan Scott Parrish, 'Diasporic Sources of African Enlightenment' in Delbourgo and Dew (eds.) *Science and Empire in the Atlantic World*, 281–310.
171. Godwin, *Man in the Moone*.
172. Court to Fort St George, 20 February 1663. Cf. Yule (ed.) *Diary of William Hedges*, II, cccliv.
173. IOR E/3/91, fol. 177v, 3 August 1687.
174. Pratik Chakrabarti, 'Neither of Meate nor Drinke, But What the Doctor Alloweth: Medicine amidst War and Commerce in Eighteenth-Century Madras', *Bulletin of the History of Medicine*, 80:1 (2006), 4–5.
175. SHA, *Godwin's Abstracts*, 'Laws and Orders Constituted for the Negroe Slaves', fols. 71–76.
176. SHA, *Letters Sent, 1707–1714*, XXIV, fol. 24, Captain ApRice to St. Helena Council, 4 August 1709.
177. Royle, *The Company's* Island, 88. SHA, *Letter Book*, II, fol. 142, London to St Helena, 31 May 1721. Vink, 'The World's Oldest Trade', notes that female slaves followed similar professions in the VOC. See also Chatterjee, *Gender and Slavery*; Lionel Caplan, 'Power and Status in South Asian Slavery',

in *Asian and African Systems of Slavery*, ed. by James L. Watson (Oxford: Blackwell, 1980), 169–194.

178. IOR G/19/1, fols. 16–17, 5 May 1655.
179. SHA, *Consultations*, IV, fol. 237, 2 December 1695.
180. SHA, *Consultations*, IV, fol. 259, 16 December 1695.
181. IOR, G/35/3, fols. 80–81, 22 August 1695.
182. IOR G/35/4, [not foliated], 10 October 1699.
183. SHA, *Consultations*, IV, fol. 340.
184. SHA, *Consultations*, XV, fol. 286–288, 10 June 1718. Similar acts of marine maroonage are discussed in Richard B. Allen, 'A Serious and Alarming Daily Evil: Marronage and Its Legacy in Mauritius and the Colonial Plantation World', in Alpers, et al. (eds.) *Slavery and Resistance*, 20–37.
185. SHA, *Consultations*, IX, fols. 7–13, 10 January 1707.
186. SHA, *Consultations*, IV, fol. 238, 2 December 1695; IOR, G/35/5 [not foliated], letter of 14 April 1701.
187. SHA, *Consultations*, I, fol. 36, 27 January 1679.
188. SHA, *Letters Sent to England 1717–1720*, fols. 137–143, 3 November 1718.
189. IOR G/35/4 [not foliated], 20 October 1699.
190. IOR G/35/2, fol. 22, 10 December 1686, for an agreement made with a local ruler, Raja Kalippa, that he would receive customs on slaves bought and sold; G/35/2, fol. 295, 9 June 1692, for payment for the capture and return of a slave.
191. See, for example, John Ovington, *A Voyage to Surat in the Year 1689* (London, 1929; reprint, Asian Educational Services: New Delhi, 1996), 54.
192. SHA, *Consultations*, I, 67, 3 November 1679.
193. SHA, *Consultations*, I, 72, 6 November 1679.
194. SHA, *Consultations*, II, 389, 24 November 1687, and III, fol. 140, 23 December 1689.
195. Vincent Brown, 'Spiritual Terror and Sacred Authority in Jamaican Slave Society', *Slavery and Abolition*, 24 (2003), 24–53, 36, notes the use of glass, blood and grave dirt as well as other materials in 'Obeah' rituals in Jamaica, derived ultimately from West African practice.
196. SHA, *Consultations*, V, fol. 156, 13 November 1698 (which concerns a claim for compensation for the slave in question; the *Consultations* for 1693 omit the trial).
197. SHA, *Consultations*, XIV, 24 January 1716, fols. 205–206.
198. Cf. Gosse, *St Helena*, 173.
199. 'House of Lords Journal Volume 24: February 1736, 21–28', *Journal of the House of Lords volume 24: 1732–1737* (1767–1830), 595–598. URL: http://www.british-history.ac.uk/report.aspx?compid=114044&strquery=witch craft 1736 act (accessed: 20 September 2014).
200. SHA, *Letters Sent 1714–1715*, fols. 5–5a, 29 July 1714, St Helena to Bencoulen, sent on the *Rochester*. Note that suicide was considered a crime: in this period, white people on the island who were found to have 'murdered themselves' were buried at crossroads with stakes through their hearts (e.g. SHA, *Consultations*, IV, fol. 273, 2 March 1696).
201. Brown, 'Spiritual Terror and Sacred Authority'.
202. Isaac Pyke, 'The Method of Making the Best Mortar at Madras', *Philosophical Transactions*, 37 (1731/1732), 231–235, notes on other experiments made

by Pyke are in his contributions to Senate House MS 56 and his journal in the British Library (Mss Eur D 5).

203. SHA, *Letters Sent 1714–1715*, fols. 24–24a, 8 December 1714.
204. SHA, *Godwin's Abstracts*, 1708, 'Laws and Orders Constituted for the Negroe Slaves', fols. 71–76.
205. BL Collet MS D. 1153/2, fol. 44, in a letter to Nathaniel Hodges (2 November 1712) mentions freeing slaves as a personal good deed. For examples of slaves freed in wills in Madras, see Yule, *Diary of William Hedges*, p. cccliv.
206. IOR G/35/3, fols. 53–54, 5 August 1695.
207. Royle, *The Company's Island*, 100.
208. Royle, *The Company's Island*, 176–177.
209. SHA, *Consultations*, XVII, fols. 142–145, 3 November 1723.
210. SHA, *Letters Received 1717–1725*, fol. 217, London to St Helena, 12 February 1724.
211. SHA, *Consultations*, III, fol. 175, 12 March 1690; V, fol. 57, 5 April 1697 (Marcy Oliver is lashed for bearing an illegitimate child); V, fol. 242, 4 July 1699 (Martha Oliver is sent to Bencoulen); IX, fol. 111, 27 January 1708.
212. IOR G/35/4 [no folio numbers], Consultation, 28 November 1699.
213. SHA, *Consultations*, VI, fol. 116v, 18 November 1701.
214. SHA, *Consultations*, IV, fol. 57, 5 April 1697.
215. SHA, *Consultations*, VI, fol. 106, 16 October 1701. It is not clear whether Elizabeth Lansdowne was free or enslaved.
216. SHA, *Consultations*, XVI, fols. 138–139, 6 November 1716.
217. IOR G/35/4, fol. 197, 6 January 1696.
218. Collet MS D1153/2, fols. 71–72, Collet to Rev. Giles Dent, 1 March 1713.
219. Transcribed in Antony Reid, *Witness to Sumatra: A Traveller's Anthology* (Oxford: Oxford University Press, 1995), 128.
220. Gosse, *St Helena*, uses this term.
221. Allen, *European Slave Trading in the Indian Ocean*, 85.
222. Raj, *Relocating Modern Science*.
223. Heyne, *Statistical Tracts*, 388–389, notes the work of convict prisoners and debt bondsmen in Bencoulen. See also Clare Anderson, 'Convict Passages in the Indian Ocean, c. 1790–1860' in Christopher, Pybus and Rediker (eds.) *Many Middle Passages*, 129–149, 129. Settlements followed in the Andaman Islands, Penang, Malacca, and Singapore. Matthew P Guterl, 'After Slavery: Asian Labour, the American South, and the Age of Emancipation', *Journal of World History*, 14 (2003), 209–241, discusses similarities between the conditions of African slave labourers in Cuba and the Chinese 'coolies' who replaced them. Ward, *Networks of Empire*, also compares slavery and other forms of coerced labour.
224. Gosse, *St Helena*, 245–246.
225. Gosse, *St Helena*, 310–312.

Conclusion

1. For models, reconstructions of the observatories and a bibliography, Barry Perlus, 'Jantar Mantar: The Astronomical Observatories of Jai Singh', http://www.jantarmantar.org (updated 23 May 2014, accessed 16 December 2014). The order in which the observatories were constructed is uncertain, but Jai

Singh's *Zij-i Muhammad Shahi* implies that the structure at Delhi was built first and the others were built later to check the measurements. Historians vary in their estimates of the date of the Delhi observatory, but it is likely to be between 1710 and 1724, while the Jaipur observatory, the largest, was complete in 1734. Andreas Volwahsen, *Cosmic Architecture in India: The Astronomical Monuments of Maharaja Jai Singh II* (Munich: Prestel-Verlag, 2001), 65.

2. Virendra Nath Sharma, 'Sawai Jai Singh's Hindu Astronomers', *Indian Journal of History of Science*, 23 (1993).
3. Ulugh Beg (Muhammad Tūrghāy, 1393–1449), *Zidj-i-djedīd-i sulṭānī,* 1437. For a recent translation, Ulugh Beg and E. B. Knobel, *Ulugh Beg's Catalogue of Stars: Rev. from All Persian Manuscripts Existing in Great Britain, with a Vocabulary of Persian and Arabic Words* (Washington, DC: The Carnegie Institution of Washington, 1917).
4. Volwahsen, *Cosmic Architecture in India.*
5. John Greaves, *Quibus accesserunt, Insigniorum aliquot Stellarum Longitudines, et Latitudines, Ex Astronomicis Observationibus Ulug Beigi, Tamerlani Magni Nepotis* (Oxoniae, 1648), *Binae Tabulae Geographicae, una Nassir Eddini Persae, altera Vlug Beigi Tatari* (Oxoniae, 1648), *Epochæ Celebriores, Astronomis, Historicis, Chronologis, Chataiorvm, Syro-Græcorvm Arabvm, Persarvm Chorasmiorvm, Usitatæ* (London: Jacobi Flesher, Cornelium Bee, 1650).
6. Royal Society, *Minutes of Council*, I, May 1664, fols. 4–5; Royal Society, *Early Letters*, I, fol. 4, John Wallis, dated at Oxford, to Henry Oldenburg, 6 April 1664, on Hyde's edition of Ulug Begh.
7. Uluġ Beg, Thomas Hyde, *Jadāvil-i mavāẕi'-i s̱avābit dar ṭūl va 'arẕ ... sive Tabvlæ long. ac lat. stellarum fixarvm* (Oxford: Henry Hall, 1665). Shen's annotations to the tables derived from Uluġ Beg are in Reg. 16 B. XVIII (now classified as a British Library printed book 10055.ee.32), fol. 28.
8. Sharma, 'Sawai Jai Singh's Hindu Astronomers', 137.
9. David Turnbull, *Masons, Tricksters and Cartographers: Comparative Studies in the Sociology of Scientific and Indigenous Knowledge* (London: Routledge, 2000), 'Introduction'.
10. Drayton, *Nature's Government*, 271–272.
11. Sanjay Subrahmanyam, 'Between a Rock and a Hard Place' in *The Brokered World*, 429–443; Greenblatt (ed.) *Cultural Mobility.*
12. Raj, *Relocating Modern Science*, 226.
13. Nicholas Hudson, 'From "Nation" to "Race": The Origin of Racial Classification in Eighteenth Century Thought', *Eighteenth-Century Studies*, 29.3 (1996), 265.
14. Robert N. Proctor, 'Agnotology: A Missing Term to Describe the Cultural Production of Ignorance (and Its Study)', in Robert N. Proctor and London Schiebinger (eds.) *Agnotology: The Making and Unmaking of Ignorance* (Stanford University Press: Stanford, 2008), 1–33.
15. See Koen Vermeir and Dániel Margócsy, 'States of Secrecy: An Introduction' *British Journal for the History of Science*, 45 (2) (2012), 153–164 and other papers in this special issue.
16. Simon Schaffer, 'The Asiatic Enlightenments of British Astronomy', in Schaffer, Roberts, Raj and Delbourgo (eds.) *The Brokered World*, 49–107.

17. A reassessment of the purpose of the observatories in terms of cosmic conceptions of kingship is given by Volwahsen, *Cosmic Architecture in India*.
18. Ronald B. Inden, *Imagining India* (Oxford: Basil Blackwell, 1990).
19. Robert Barker, 'An Account of the Bramins Observatory at Benares', *Philosophical Transactions* 67 (1777), 601. See also Robert Barker and John Lloyd Williams, 'Further Particulars Respecting the Observatory at Benares', *Philosophical Transactions*, 83 (1793), 45–49.
20. Raj, *Relocating Modern Science*, 233.
21. Robert Liss, 'Frontier Tales: Tokugawa Japan in Translation', in Schaffer, Roberts, Raj and Delbourgo (eds.) *The Brokered World*, 1–47.
22. Schottenhammer (ed.) *Trade and Transfer*.
23. Newbury, *Patrons, Clients, and Empire*, 12.
24. Jos J. L. Gommans, *Mughal Warfare: Indian Frontiers and Highroads to Empire, 1500–1700* (London: Routledge, 2003), 203–206.
25. Edney, *Mapping an Empire*.
26. Menon, 'British Naturalists in Eighteenth Century India'.
27. Noltie, *Robert Wight and the Botanical Drawings of Rungiah & Govindoo*.
28. Steadman-Jones, *Colonialism and Grammatical Representation*.
29. Newbury, *Patrons, Clients, and Empire*.
30. Ranajit Guha, *Dominance without Hegemony: History and Power in Colonial India* (Cambridge, MA: Harvard University Press, 1997), 23–55.
31. Bruno Latour, *We Have Never Been Modern*; *On the Modern Cult of the Factish Gods* (Durham, NC: Duke University Press, 2010).
32. Prakash, *Another Reason*, 13.
33. Mohamad Tavakoli-Targhi, 'Early Persianate Modernity', in Pollock (ed.) *Forms of Knowledge in Early Modern Asia*, 257–287.
34. Pollock (ed.) 'Introduction', in *Forms of Knowledge in Early Modern Asia*, 2.
35. Prasannan Parthasarathi, *Why Europe Grew Rich and Asia Did Not* (Cambridge: Cambridge University Press).
36. Parthasarathi, *Why Europe Grew Rich and Asia Did Not*; Tavakoli-Targhi, 'Early Persianate Modernity'.
37. Kapil Raj, 'The Historical Anatomy of a Contact Zone', *Indian Economic and Social History Review*, 48.1 (2011), 55–82.
38. Schaffer, Roberts, Raj and Delbourgo, *The Brokered World*. See also my review of this work, 'Hybrid Knowledge', *History Workshop Journal*, 71:1 (2011), 267–273.
39. Tavakoli-Targhi, 'Early Persianate Modernity', 269.
40. Subrahmanyam, 'Conclusion', in *The Brokered World*', 440.
41. Durba Ghosh, 'Decoding the Nameless: Gender, Subjectivity, and Historical Methodologies in Reading the Archives of Colonial India', in Kathleen Wilson (ed.) *A New Imperial History: Culture, Identity and Modernity in Britain and the Empire* (Cambridge: Cambridge University Press, 2004).
42. Particularly Amitav Ghosh, *Sea of Poppies* (London: John Murrey, 2008) and *River of Smoke* (London: John Murray, 2011). See also Anderson, *Subaltern Lives*.

Bibliography

Archival sources

Holland

Leiden University Library: Bodel Nieuwenhui map collection

India

Goa State Archives: Item 860, Alforria dada aos escravos, 1682–1759 and Bengala 2760

Maharashtra State Archives (MSA): Surat Diaries; Outwards or order books; Inwards or order books; Secretariat Records (Bombay)

National Archives of India (NAI): Foreign Department; Home Department; Misc.; Maps

Panaji Central Library: Roberto Nobili materials

St Mary's Church, Chennai: Baptism and marriage records

Tamil Nadu State Archives (TNSA): General and territorial ledgers of Fort St George; Mayor's Court records; Public Sundries

Sri Lanka

State Archives of Sri Lanka: Translated records of the Dutch administration

St Helena

St Helena Archives (SHA): Diary and consultation books; Letter books (LB)

United Kingdom

Bodleian Library (Bod): Manuscript collections: Ashmole; Hyde (Or); Marshall (Or); Rawlinson (Rawl); Smith; Sherard; Tanner

British Library (BL): 1) Asia Pacific and Africa Collection: India Office Records (IOR): Court minutes (B); Factory records ('G'); Correspondence ('E'); Private Papers (IOPP). 2) Department of Western Manuscripts: Sloane manuscripts (MS Sloane); Additional manuscripts (Add MS); Harley manuscripts

Cambridge University Library: Correspondence of Robert Knox and John Strype (Add 1–9)

Royal Society (RS): Letter books (LB); Journal books (JB); Council minutes; Hooke folio (HF); Boyle papers

Primary texts

(Pre-c. 1800)

Anon., 'A Narrative Concerning the Success of Pendulum-Watches at Sea for the Longitudes', *Philosophical Transactions (1665–1678)*, 1(1665), 13–15.

———, *A True and Perfect Description of the Strange and Wonderful Elephant Sent from the East-Indies and Brought to London on Tuesday the Third of August, 1675: With a Discourse of the Nature and Qualities of Elephants in General.* ([London]: printed for J. Conniers, 1675).

Baldaeus, Philippus, *Naauwkeurige Beschryvinge Van Malabar En Choromandel, Der Zelver Aangrenzende Ryken, En Het Machtige Eyland Ceylon* (Amsterdam: Johannes Janssonius van Waesberge, 1672).

Barker, Robert, 'An Account of the Bramins Observatory at Benares', *Philosophical Transactions of the Royal Society of London*, 67(1777), 601.

Barker, Robert and John Lloyd Williams, 'Further Particulars Respecting the Observatory at Benares…', *Philosophical Transactions of the Royal Society of London*, 83(1793), 45–49.

Beg, Ulugh, Muḥammad ibn Muẓammad al-Tizīnī and Hyde, Thomas, *Jadāvil-i mavāẕiʿ-i ṣavābit Dar Ṭūl Va ʿarẕ … Sive Tabvlæ Long. Ac Lat. Stellarum Fixarvm* (Oxonii: H. Hall., 1665).

Beg, Ulugh and John Greaves, *Epochæ Celebriores, Astronomis, Historicis, Chronologis, Chataiorvm, Syro-Græcorvm Arabvm, Persarvm Chorasmiorvm, Usitatæ* (Londini: Jacobi Flesher, 1650).

Birch, Thomas, *The History of the Royal Society of London for Improving of Natural Knowledge from Its First Rise* (London: 1757).

Bois, Sieur du, *Les Voyages Fait Par Le Sievr D B* (Paris: 1674).

Bowrey, Thomas, *A Dictionary, English and Malayo, Malayo and English* (London: printed by S. Bridge for the author, 1701).

Boyle, Robert, *General Heads for the Natural History of a Country Great or Small Drawn Out for the Use of Travellers and Navigators* (London: printed for John Taylor…and S. Hedford…, 1692).

Camelli, George, 'A Description and Figure of the True Amomum, or Tugus. Sent from the Reverend Father George Camelli, at the Phillipine Isles, to Mr. John Ray and Mr. James Petiver, Fellows of the Royal Society', *Philosophical Transactions (1683–1775)*, 21(1699), 2–4.

Child, Josiah, *A New Discourse of Trade Wherein Is Recommended Several Weighty Points Relating to Companies of Merchants* (London: John Everingham, 1693).

Churchill, Awnsham, John Churchill and Ann Cary, *A Collection of Voyages and Travels: Some Now First Printed from Original Manuscripts, Others Now First Published in English* (London: printed for the authors, 1732).

Cockburn, W., *An Account of the Nature, Causes, Symptoms and Cure of the Distempers That Are Incident to Sea-Faring People* (London: Hugh Newman, 1696).

Cudworth, Ralph, *True Intellectual System of the Universe…* (London: R Royston, 1678).

Dampier, William, A Supplement of the Voyage round the World: Voyages and Descriptions, Vol. II, in 3 parts (London: J. Knapton, 1700).

Donne, John, *A Sermon upon the VIII. Verse of the I. Chapter of Acts: Preached to the Honourable Company of the Virginian Plantation* (London: Jones, 1622).

East India Company, *A True and Large Discourse of the Voyage of the Whole Fleete of Ships Set Forth the 20. of Aprill 1601…* (London: Thomas Thorpe, 1603).

Godwin, Francis, *The Man in the Moone, or a Discourse of a Voyage Thither by Domingo Gonsales the Speedy Messenger* (London: printed by John Norton, and are to be sold by Ioshua Kirton, and Thomas Warren, 1638).

Godwyn, Morgan, *The Negro's & Indians Advocate Suing for Their Admission into the Church* (London: printed for the author, by J. D., 1680).

Grew, Nehemiah, *Musaeum Regalis Societatis, or, a Catalogue & Description of the Natural and Artificial Rarities Belonging to the Royal Society and Preserved at Gresham Colledge* (London: printed by W. Rawlins, for the author, 1681).

———, *The Anatomy of Plants: With an Idea of a Philosophical History of Plants, and Several Other Lectures, Read before the Royal Society* (London: printed by W. Rawlins for the author, 1682).

Gueynier, F., 'Vocabulaer, Ofte Woorden-Boeck, Naer Ordre Van Den Alphabet in't Duytsch Ende Maleys' (Batavia: Eede, 1677).

Haex, David, *Dictionarium Malaico-Latinum Et Latino-Malaicum* (Typis Sac. Congr. de Propag. Fide: Roma, 1631).

Hamilton, Alexander, *A New Account of the East Indies, Being the Observations and Remarks of Capt. Alexander Hamilton, Who Spent His Time There from the Year 1688 to 1723* (Edinburgh: 1727).

Hamond, Walter, *Madagascar, the Richest and Most Fruitfull Island in the World* (London: printed for Nicolas Bourne, and are to be sold at his shop, at the south entrance of the Royall Exchange, 1643).

Hasel, Jan, Justus Heurnius, and Albert Ruyl trans., Thomas Hyde, and Thomas Marshall ed. *Jang Ampat Evangelia ... Daan Berboatan Derri Jang Apostoli Bersacti, Bersalin Dallam Bassa Malayo* (H. Hall: Oxford, 1677).

Hermann, Paul, *Paradisus Batavus* (Leiden: Elzevier, 1698).

Houtman, Frederick d., *Spraecke Ende Woord-Boeck, in De Maleysche Ende Madagaskarsche Talen* (Amstelredam: 1603).

Hunter, William, 'Some Accounts of the Astronomical Labours of Jai Singh', *Asiatick Researches*, 5(1799), 177–213.

[Hyde, Thomas], *An Account of the Famous Prince Giolo, Son of the King of Gilolo, Now in England: With an Account of His Life, Parentage, and His Strange and Wonderful Adventures, the Manner of His Being Brought for England* (London: printed and sold by R. Taylor, 1692).

———, *Historia Religionis Veterum Persarum* (Oxonii: e Theatro Sheldoniano, 1700).

Hyde, Thomas and Gregory Sharpe, *Syntagma Dissertationum* (Oxonium: Clarendon, 1767).

James II, H. R. H., *By the King. A Proclamation for the Recalling All His Majesties Subjects from the Service of Foreign Princes in East India: [17 July 1686].* (London: printed by C. Bill, H. Hills and T. Newcomb, 1686).

Jones, William, *The Works of Sir William Jones*, 13 vols (London: Stockdale, 1807).

Kaempfer, Engelbert, *The History of Japan Together with a Description of the Kingdom of Siam, 1690–1692*, trans. John Gaspar Scheuchzer (London: 1727).

Knox, Robert, *An Historical Relation of the Island Ceylon, in the East-Indies...* (London: printed by Richard Chiswell..., 1681).

———, *'t Eyland Ceylon in Sijn Binnenste, of, 't Koningrijck Candy...*, trans. S. de Vries (Utrecht: by Wilhelm Broedelet..., 1692).

———, *Relation Ou Voyage De l'Isle De Ceylan Dans Les Indes Orientales*, trans. Pieter Schoonebeek and Jan van Branen (A Amsterdam: chez Paul Marret, 1693).

Lewis, William, *An Experimental History of the Materia Medica; or, of the Natural and Artificial Substances Made Use of in Medicine ...* (London: J. Johnson, 1791).

Locke, John, *An Essay Concerning Humane Understanding* (London: printed by Eliz. Holt for Thomas Basset..., 1690).

Lockyer, Charles, *An Account of the Trade in India...: To Which Is Added an Account of the Management of the Dutch in Their Affairs in India* (London: 1711).

Lord, Henry, *A Display of Two Forraigne Sects in the East Indies* (imprinted at London: T. and R. Cotes for Francis Constable, 1630)

Marini, Giovanni F. d., *Relation Nouvelle Et Curieuse Des Royaumes De Tunquin Et De Lao...* (Paris: Chez Gervais Clouzier..., 1666)

Marsden, William, *A Dictionary of the Malayan Language in Two Parts, Malayan and English, and English and Malayan... by William Marsden* (London: printed by Cox and Baylis, 1812a).

———, *A Dictionary of the Malayan Language: To Which Is Prefixed a Grammar, with an Introduction and Praxis* (Longman: London, 1812b).

Marshall, John, 'A Letter from the East Indies, Giving an Account of the Religion, Rites, Notions, Manners of the Heathen Priests Commonly Called Bramines Communicated by the Reverend Mr Abraham De La Pryme', *Philosophical Transactions (1683–1775)*, 22 (1700–1701), 729–738.

Martin, Pierre, 'Lettres Du Père Pierre Martin...*Lettres Edifiantes*, 10 (1730–1732), 22–81.

More, Henry, *A Collection of Several Philosophical Writings of Dr Henry More...* (London: printed by James Flesher for William Morden, 1662).

———, *Divine Dialogues Containing Disquisitions Concerning the Attributes and Providence of God. In Three Volumes. By Henry More, D.D.* (Glasgow: Robert Foulis, 1743).

Newton, Isaac, *The Chronology of Ancient Kingdoms Amended...* (London: printed for J. Tonson in the Strand, and J. Osborn and T. Longman in Pater-noster Row, 1728).

Nicholls, Frank, 'An Account of the Veins and Arteries of Leaves, by Frank Nicholls, M. D. Prael. Anat. Oxon. F. R. S.', *Philosophical Transactions (1683–1775)*, 36(1729), 371–372.

Nobili, Robert, *Cahecismo Manduerense* (Travancorensi, 1674).

Oldenburg, Henry, 'An Account of Some Books; 1. The History of the Royal Society of London for the Advancement of Experimental Philosophy, by Tho. Spratt', *Philosophical Transactions (1665–1678)*, 2 (1665), 502.

Petiver, James, *Opera*, 3 vols (London: Smith and Bateman, 1695).

———, 'Some Attempts Made to Prove That Herbs of the Same Make or Class for the Generallity, Have the Like Vertue and Tendency to Work the Same Effects...', *Philosophical Transactions (1683–1775)*, 21 (1699), 289–294.

———, *Gazophylacii Naturae & Artis Decas Prima-[Quinta]:...Descriptionibus Brevibus & Iconibus Illustrantur* (Londini: Christ. Bateman, 1702).

———, 'Some Farther Account of Divers Rare Plants, Lately Observed in Several Curious Gardens about London, and Particularly in the Company of Apothecaries Physick-Garden at Chelsey. By Mr. James Petiver, F. R. S.', *Philosophical Transactions (1683–1775)*, 27 (1712), 416–426.

Petiver, James and Samuel Brown, 'An Account of Mr Sam. Brown His Second Book of East India Plants...', *Philosophical Transactions (1683–1775)*, 22 (1700), 699–721.

———, 'An Account of Mr Sam. Brown, His Third Book of East India Plants, with Their Names, Vertues, Description, &c. By James Petiver, Apothecary, and

Fellow of the Royal Society. To Which Are Added Some Animals Sent Him from Those Parts', *Philosophical Transactions (1683–1775)*, 22 (1700), 843–862.

———, 'A Description of Some Shells Found on the Molucca Islands: As Also an Account of Mr Sam. Brown, His Fourth Book of East India Plants ...', *Philosophical Transactions (1683–1775)*, 22 (1700), 927–946.

———, 'An Account of Mr Sam. Brown His Fifth Book of East India Plants', *Philosophical Transactions (1683–1775)*, 22 (1700), 1007–1029.

———, 'An Account of Mr. Sam. Brown His Sixth Book of East India Plants, with Their Names, Vertues, Description, Etc. By James Petiver, Apothecary, and Fellow of the Royal Society. To These Are Added Some Animals, Etc. Which the Reverend Father George Joseph Camel, Very Lately Sent Him from the Philippine Isles', *Philosophical Transactions (1683–1775)*, 23 (1702), 1055–1068.

———, 'Mr Sam. Brown His Seventh Book of East India Plants, with an Account of Their Names, Vertues, Description, Etc. By James Petiver, Apothecary, and Fellow of the Royal Society. These Plants Were Gathered between the 15th and 20th of June, A. D. 1696. In the Ways between Fort St George and Trippetee, Which Is about 70 Miles Off', *Philosophical Transactions (1683–1775)*, 23 (1702), 1251–1566.

———, 'An Account of Part of a Collection of Curious Plants and Drugs, Lately Given to the Royal Society by the East India Company', *Philosophical Transactions (1683–1775)*, 22 (1700), 579–594.

Pitt, Moses, *The English Atlas* ... (Oxford: 1680).

Psalmanazar, George, *An Historical and Geographical Description of Formosa, an Island Subject to the Emperor of Japan* ... (London: D. Brown, 1704).

Pyke, Isaac and Edmund Halley, 'The Method of Making the Best Mortar at Madrass in East India ...', *Philosophical Transactions (1683–1775)*, 37 (1732), 231–235.

Quesne, Abraham D. and Jacques-Joseph Le Maire, *A New Voyage to the East-Indies in the Years 1690 and 1691 Being a Full Description of the Isles of Maldives, Cicos, Andamants, and the Isle of Ascention* ... (London: printed for Daniel Dring ..., 1696).

Ray, John, *Historia Plantarum*, 3 vols. Samuel Smith and Benjamin Walford, 1693–1704).

———, *Synopsis methodica avium et piscium: opus posthumous* (London: William Innys, 1713).

Ribeiro, João, *Histoire De l'Isle De Ceylan*, trans. Joachim Le Grand (A Amsterdam: chez J. L. de Lorme, 1701).

———, *The Historical Tragedy of the Island of Ceilao*, trans. P. E. Pieris (Colombo: 1930).

Saar, J. J., *Ost Indianisge* (Nuremburg: 1662).

Spalding, Augustine, *Dialogues in the English and Malaiane Languages* ... (London: Felix Kyngston for William Welby, 1614).

Sprat, Thomas, *The History of the Royal-Society of London, for the Improving of Natural Knowledge* (London: J. Martyn and others, 1667).

Strachan, 'An Account of the Taking and Taming of Elephants in Zeylan, by Mr. Strachan, a Physician, Who Lived 17 Years There', *Philosophical Transactions (1683–1775)*, 23 (1703), 1051–1054.

———, 'Observations on the Planting and Culture of Tobacco in Zeylan, by Mr Strachan', *Philosophical Transactions (1683–1775)*, 23 (1703), 1164–1166.

———, 'Some Observations on Coral, Large Oysters, Rubies, the Growing of a Sort of Ficus Indica, the Gods of the Ceylanese, Etc. Made in Ceilan, by Mr Strachan', *Philosophical Transactions (1683–1775)*, 23 (1703), 1248–1250.

Tavernier, Jean Baptiste, *Les Six Voyages De J. B. Tavernier... En Turquie, En Perse Et Aux Indes* (Paris: 1676).

———, *The Six Voyages of John Bapista Tavernier* (London: Daniel Cox, William Gobbid and Moses Pitt, 1677).

Tavernier, Jean-Baptiste and Edmund Everard, *A Collection of Several Relations and Treatises Singular and Curious of John Baptista Tavernier, Baron of Aubonne Not Printed among His First Six Voyages...* (London: printed by A. Godbid and J. Playford for Moses Pitt..., 1680).

University of Oxford, *Pietas Universitatis Oxoniensis in Obitum Augustissimae & Desideratissimae Reginae Mariae* (Oxonii: E Theatro Sheldoniano, 1695).

van Rheede tot Drakestein, Hendrik, et. al. *Hortus Indicus Malabaricus*, 12 vols (1678–1703) (Amstelodamum: 1703).

Werndly, George H., *Maleische Spraakkunst* (Amsterdam: Bewindhebberen, 1736).

Wilkins, John, *An Essay towards a Real Character, and a Philosophical Language* (London, 1668).

Wiltens, C. and S. Danckaerts, *Vocabularium Ofte Woort-Boek Naer Order Vanden Alphabet in 't Duytsch-Malaysch-Duytsch* (Gravenhage: 1623).

Secondary sources and later editions

Adams, Julia, *The Familial State: Ruling Families and Merchant Capitalism in Early Modern Europe* (Ithaca: Cornell University Press, 2005).

Adelaar, K. A., 'Malay: A Short History', *Oriente Moderno*, 19 (80).2 (2000), 225–242.

Ahmed, S. J., 'Mimicry and Counter-Discourse on the Palimpsest of Nagori: The Play of Saint Anthony and His Double, Sādhu Āntoni', *Asian Theatre Journal*, 27.1 (2010), 40–75.

Alam, Muzaffar and Sanjay Subrahmanyam, *Indo-Persian Travels in the Age of Discoveries, 1400–1800* (Cambridge and New York: Cambridge University Press, 2007).

———, *Writing the Mughal World: Studies on Culture and Politics* (New York: Columbia University Press, 2012).

Alavi, Seema, *Islam and Healing: Loss and Recovery of an Indo-Muslim Medical Tradition, 1600–1900* (New York: Palgrave Macmillan, 2008).

Aljunied, Syed M. K., 'British Discourses and Malay Identity in Colonial Singapore', *Indonesia and the Malay World*, 37.107 (2009), 1–21.

Allen, Richard Blair, *European Slave Trading in the Indian Ocean, 1500–1850* (Athens, OH: Ohio University Press, 2014).

Alpers, Edward A., 'The French Slave Trade in East Africa (1721–1810)', *Cahiers d'Études Africaines*, 10 (1970), 80–124.

Alpers, Edward A., Gwyn Campbell, and Michael Salman, *Slavery and Resistance in Africa and Asia* (London and New York: Routledge, 2005).

Amaladass, Anand and Ines G. Županov, *Intercultural Encounter and the Jesuit Mission in South Asia (16th–18th Centuries)* (Bangalore: Asian Trading Corporation, 2014).

Anderson, Clare, *Subaltern Lives: Biographies of Colonialism in the Indian Ocean World, 1790–1920* (Cambridge: Cambridge University Press, 2012).

Anderson, John, *English Intercourse with Siam in the Seventeenth Century* (London: K. Paul, Trench, Trübner, & Co, 1890).

Andrade, Tonio, 'Pirates, Pelts, and Promises: The Sino-Dutch Colony of Seventeenth-Century Taiwan and the Aboriginal Village of Favorolang', *Journal of Asian Studies*, 64.2 (2005), 295–321.

Appadurai, Arjun, ed., *The Social Life of Things: Commodities in Cultural Perspective* (Cambridge: Cambridge University Press, 1986).

Arasaratnam, Sinnappah, *Maritime India in the Seventeenth Century* (Delhi and New York: Oxford University Press, 1994).

———, *Maritime Trade, Society and European Influence in South Asia, 1600–1800* (Aldershot and Brookfield, VT: Variorum, 1995).

Armitage, David and M. J. Braddick, *The British Atlantic World, 1500–1800* (Basingstoke and New York: Palgrave Macmillan, 2002).

Armstrong, James C., 'Madagascar and the Slave Trade in the Seventeenth Century', *Omaly Sy Anio* (1983), 17–18.

Arnold, David, *Colonizing the Body State Medicine and Epidemic Disease in Nineteenth-Century India* (Berkeley: University of California Press, 1993).

———, *Warm Climates and Western Medicine: The Emergence of Tropical Medicine, 1500–1900* (Amsterdam and Atlanta, GA: Rodopi, 1996).

Asmole, Philip and Mrytle Ashmole, *St Helena and Ascension Islands* (Shropshire: Anthony Nelson, 2000).

Avicenna (Ibn Sina), *The Canon of Medicine (Al-Qanun Fi'l-Tibb)*, trans. Laleh Bakhtiar (Chicago, IL: Great Books of the Islamic World, distributed by KAZI Publications, 1999).

Bacon, Francis, *The Essays or Counsel, Civill and Moral (1625)* (Oxford: Clarendon Press, 1985).

Baladouni, Vahé and Margaret Makepeace, *Armenian Merchants of the Seventeenth and Early Eighteenth Centuries: English East India Company Sources* (Philadelphia: American Philosophical Society, 1998).

Baldwin, Anna P. and Sarah Hutton, *Platonism and the English Imagination* (Cambridge and New York: Cambridge University Press, 1994).

Barnes, Geraldine, 'Curiosity, Wonder, and William Dampier's Painted Prince', *Journal for Early Modern Cultural Studies*, 6.1 (2006), 31–50.

Basalla, George, 'The Spread of Western Science. A Three-Stage Model Describes the Introduction of Modern Science into Any Non-European Nation', *Science (New York, N.Y.)*, 156.3775 (1967), 611–622.

Bastin, John, *The British in West Sumatra (1685–1825)* (Kuala Lumpur: University of Malaya Press, 1965).

Basu, Basanta K., 'Notes on Slavery during the Early Days of John Company', *Muslim Review*, 4.4 (1930), 21–34.

Bauss, Rudy, *The Portuguese Slave Trade from Mozambique to Portuguese India and Macau and Comments on Timor, 1750–1850: New Evidence from the Archives* (New York: Camões Center for the Study of the Portuguese-Speaking World, 1997).

Bayly, C. A. *Imperial Meridian: The British Empire and the World, 1780–1830* (London and New York: Longman, 1989).

———, 'Knowing the Country: Empire and Information in India', *Modern Asian Studies*, 27.1 (1993), 3–43.

———, *Empire and Information: Intelligence Gathering and Social Communication in India, 1780–1870* (Cambridge and New York: Cambridge University Press, 1996).

Beg, Ulugh and E. B. Knobel, *Ulugh Beg's Catalogue of Stars, Rev. from all Persian Manuscripts Existing in Great Britain, with a Vocabulary of Persian and Arabic Words* (Washington, DC: The Carnegie Institution of Washington, 1917).

Ben-Zaken, Avner, *Reading Hayy Ibn-Yaqzan: A Cross-Cultural History of Autodidacticism* (Baltimore: Johns Hopkins University Press, 2011).

Bhabha, Homi K., *The Location of Culture* (London and New York: Routledge, 1994).

Bhattacharya, Bhaswati, 'Armenian European Relationship in India, 1500–1800: No Armenian Foundation for European Empire?', *Journal of the Economic and Social History of the Orient*, 48.2 (2005), 277–322.

Bialuschewski, Arne, 'Thomas Bowrey's Madagascar Manuscript of 1708', *History in Africa*, 34.1 (2007), 31–42.

———, 'A True Account of the Design, and Advantages of the South-Sea Trade: Profits, Propaganda, and the Peace Preliminaries of 1711', *Huntington Library Quarterly*, 73.2 (2010), 273–285.

Bleichmar, Daniela, *Visible Empire: Botanical Expeditions and Visual Culture in the Hispanic Enlightenment* (Chicago: The University of Chicago Press, 2012).

Blussé, Leonard, *Strange Company: Chinese Settlers, Mestizo Women, and the Dutch in VOC Batavia* (Dordrecht and Riverton: Foris Publications, 1986).

Bodart-Bailey, Beatrice M. and Derek Massarella *The Furthest Goal: Engelbert Kaempfer's Encounter with Tokugawa Japan* (Sandgate: Japan Library, 1995).

Borschberg, Peter, *Iberians in the Singapore-Melaka Area and Adjacent Regions (16th to 18th Century)* (Wiesbaden and Lisboa: Harrassowitz; Fundação Oriente, 2004).

———, *The Singapore and Melaka Straits: Violence, Security, and Diplomacy in the 17th Century* (Singapore: NUS Press, 2010).

Bowe, Patrick, 'The Indian Gardening Tradition and the Sajjan Niwas Bagh, Udaipur', *Garden History*, 27.2 (1999), 189–205.

Bowen, H. V., *Elites, Enterprise, and the Making of the British Overseas Empire, 1688–1775* (Basingstoke: Macmillan, 1996).

Bowen, H. V., Margarette Lincoln, and Nigel Rigby, *The Worlds of the East India Company* (Rochester, NY: D. S. Brewer, 2002).

Bowrey, Thomas, *A Geographical Account of the Countries round the Bay of Bengal, 1669 to 1679*, ed. Richard Carnac Temple (Liechtenstein: Kraus Reprint, 1967).

Boxer, C. R., 'Ceylon through Puritan Eyes: Robert Knox in the Kingdom of Kandy 1660–1679', *History Today*, 4 (1954), 660–667.

———, *Portuguese Society in the Tropics: The Municipal Councils of Goa, Macao, Bahia, and Luanda, 1510–1800* (Madison: University of Wisconsin Press, 1965).

Boyle, Richard, *Knox's Words: A Study of the Words of Sri Lankan Origin or Association First Used in English Literature by Robert Knox and Recorded in the Oxford English Dictionary* (Colombo: Visidunu Publication, 2004).

Breen, B., 'No Man Is an Island: Early Modern Globalization, Knowledge Networks, and George Psalmanazar's Formosa', *Journal of Early Modern History*, 17.4 (2013), 391–417.

Brockway, Lucile, *Science and Colonial Expansion: The Role of the British Royal Botanic Gardens* (New York: Academic Press, 1979).
Brook, Timothy, *Mr. Selden's Map of China: Decoding the Secrets of a Vanished Cartographer* (London: Profile Books, 2013).
Brookes, John, *Gardens of Paradise: The History and Design of the Great Islamic Gardens* (New York: New Amsterdam, 1987).
Brown, Cheever Mackenzie, *The Triumph of the Goddess: The Canonical Models and Theological Visions of the Devī-Bhāgavata Purāṇa* (Delhi: Sri Satguru Publications, 1992).
Brown, Katherine Butler, 'Did Aurangzeb Ban Music? Questions for the historiography of his Reign,' *Modern Asian Studies*, 41.1 (2007), 77–120.
Brown, Vincent, 'Spiritual Terror and Sacred Authority in Jamaican Slave Society', *Slavery and Abolition*, 24.1 (2003), 24–53.
———, *The Reaper's Garden: Death and Power in the World of Atlantic Slavery* (Cambridge, MA: Harvard University Press, 2008).
Bruce, John, *Annals of the Honorable East-India Company, from Their Establishment by the Charter of Queen Elizabeth, 1600, to the Union of the London and English East-India Companies, 1707–8* (London: Black, Parry, and Kingsbury, 1810).
Buch, W. J. M., 'La Compagnie Des Indes Néerlandaises Et l'Indochina', *Bulletin De l'Ecole Français d'Extrême-Orient*, 36 (1936), 97–196.
———, 'II. La Compagnie Des Indes Néerlandaises Et l'Indochina', *Bulletin De l'Ecole Française d'Extrême-Orient*, 37 (1937), 121–237.
Burgess, Ebenezer, 'Translation of the Sûrya-Siddhânta, a Text-Book of Hindu Astronomy; with Notes, and an Appendix, Containing Additional Notes and Tables, Calculations of Eclipses, a Stellar Map, and Indexes', *Journal of the American Oriental Society*, 6.6 (1860), 141–498.
Campbell, Gwyn, *The Structure of Slavery in Indian Ocean Africa and Asia* (London and Portland, OR: Frank Cass, 2004).
Campbell, Gwyn, Suzanne Miers, and Joseph Calder Miller, *Women and Slavery* (Athens, OH: Ohio University Press, 2007).
Cañizares-Esguerra, Jorge, *Nature, Empire, and Nation: Explorations of the History of Science in the Iberian World* (Stanford, CA: Stanford University Press, 2006).
Canny, Nicholas P. and Alaine M. Low, *The Origins of Empire British Overseas Enterprise to the Close of the Seventeenth Century* (Oxford: Oxford University Press, 2001).
Carey, Daniel, 'Compiling Nature's History: Travellers and Travel Narratives in the Early Royal Society', *Annals of Science*, 54.3 (1997), 269–292.
———, 'Anthropology's Inheritance: Renaissance Travel, Romanticism and the Discourse of Identity', *History and Anthropology*, 14.2 (2003), 107–126.
———, *Asian Travel in the Renaissance* (Malden, MA: Blackwell Publishing, 2004).
Carney, Judith A., *Black Rice: The African Origins of Rice Cultivation in the Americas* (Cambridge, MA: Harvard University Press, 2001).
Carney, Judith A. and Richard Nicholas Rosomoff, *In the Shadow of Slavery: Africa's Botanical Legacy in the Atlantic World* (Berkeley, CA: University of California Press, 2011).
Carter, Marina, 'Slavery and Unfree Labour in the Indian Ocean', *History Compass*, 4.5 (2006), 800–813.
Chakrabarti, Kunal, *Religious Process: The Purāṇas and the Making of a Regional Tradition* (New Delhi: Oxford University Press, 2001).

Chakrabarti, Pratik, 'Empire and Alternatives: Swietenia Febrifuga and the Cinchona Substitutes', *Medical History*, 54.1 (2010a), 75–94.

———, *Materials and Medicine: Trade, Conquest, and Therapeutics in the Eighteenth Century* (Manchester: Manchester University Press, 2010b).

Chambers, Neil ed. *The Letters of Sir Joseph Banks: A Selection, 1768–1820* (London: River Edge, 2000).

Chandra, Satish, 'Jizyah and the State in India during the 17th Century', *Journal of the Economic and Social History of the Orient*, 12.3 (1969), 322–340.

Chatterjee, Anjali, *Bengal in the Reign of Aurangzib, 1658–1707* (Calcutta: Progressive Publishers, 1967).

Chatterjee, Indrani, *Gender, Slavery, and Law in Colonial India* (New Delhi: Oxford University Press, 1999).

Chaudhuri, K. N. *The English East India Company* (London: Cass, 1965).

———, *The Trading World of Asia and the English East India Company, 1660–1760* (Cambridge: Cambridge University Press, 1978).

———, *Trade and Civilisation in the Indian Ocean: An Economic History from the Rise of Islam to 1750* (Cambridge: Cambridge University Press, 1985).

———, *Asia before Europe: Economy and Civilisation of the Indian Ocean from the Rise of Islam to 1750* (Cambridge: Cambridge University Press, 1990).

Chauhan, R. R. S., *Africans in India: From Slavery to Royalty* (New Delhi: Asian Publication Services, 1995).

Child, Josiah, *Selected Works, 1668–1697* (London: Gregg, 1968).

Chomsky, Noam, *Cartesian Linguistics: A Chapter in the History of Rationalist Thought* (New York: Harper & Row, 1966).

Christopher, Emma, *Slave Ship Sailors and Their Captive Cargoes, 1730–1807* (Cambridge: Cambridge University Press, 2006).

Christopher, Emma, Cassandra Pybus, and Marcus Rediker, *Many Middle Passages Forced Migration and the Making of the Modern World* (Berkeley: University of California Press, 2007).

Cohn, Bernard S., *Colonialism and Its Forms of Knowledge: The British in India* (Princeton, NJ: Princeton University Press, 1996).

Collins, James T., 'Language Death in Maluku: The Impact of the VOC', *Bijdragen Tot De Taal, Land- En Volkenkunde*, 159.2–3 (2003), 247–289.

Considine, John, *Ashgate Critical Essays on Early English Lexicographers, Volume 4* (Farnham: Ashgate, 2012).

———, *Academy Dictionaries 1600–1800* (Cambridge: Cambridge University Press, 2014).

Cook, Alan, *Edmond Halley: Charting the Heavens and the Seas* (Oxford: Clarendon Press, 1998).

Cook, Harold J., *Matters of Exchange: Commerce, Medicine, and Science in the Dutch Golden Age* (New Haven, CT: Yale University Press, 2007).

Cooke, Nola, Tana Li, and James Anderson, *The Tongking Gulf through History* (Philadelphia: University of Pennsylvania Press, 2011).

Cooper, Alix, *Inventing the Indigenous: Local Knowledge and Natural History in Early Modern Europe* (Cambridge: Cambridge University Press, 2007).

Cowan, William, *Papers of the Tenth Algonquian Conference* (Ottawa: Carleton University, 1979).

Crawford, D. G., *A History of the Indian Medical Service 1600–1913* (London: W. Thacker, 1914).

Crawfurd, John M., 'On the History and Migration of Cultivated Narcotic Plants in Reference to Ethnology', *Transactions of the Ethnological Society of London*, 7 (1868), 78–91.
Dampier, William, *A New Voyage around the World* (London: printed for James Knapton, 1984).
Dandy, J. E., *The Sloane Herbarium: An Annotated List of the Horti Sicci Composing It* (London: British Museum, 1958).
Đặng Nam, *Tranh Dân Gian Việt Nam = Vietnamese Folk Pictures = Imagerie Folklorique Vietnamienne* (Hà Nội: Văn hóa dân tộc, 1995).
Dange, Sindhu S., *The Bhāgavata Purāṇa: Mytho-Social Study* (Delhi: Ajanta Publications, 1984).
da Orta, Garcia, *Colloquies on the Simples and Drugs of India*, trans. Clements Robert Markham (London: Sotheran, 1913).
Das, Harihar, Susobhan Chandra Sarkar, and L. S. Amery, *The Norris Embassy to Aurangzib (1699–1702)* (Calcutta: Firma K. L. Mukhopadhyay, 1959).
Das Gupta, Ashin and M. N. Pearson, *India and the Indian Ocean, 1500–1800* (Calcutta and New York: Oxford University Press, 1987).
Davis, Natalie Z., *Trickster Travels: A Sixteenth-Century Muslim between Worlds* (London: Faber, 2007).
de Beer, E. S., ed., *The Correspondence of John Locke* (Oxford: Clarendon Press, 1976).
Delbourgo, James, 'Exceeding the Age in Everything', *Spontaneous Generations: A Journal for the History and Philosophy of Science*, 3 (2009), 41–54.
Delbourgo, James and Nicholas Dew, *Science and Empire in the Atlantic World* (New York: Routledge, 2008).
Dellios, Paulette, 'A Lexical Odyssey from the Malay World', *Studia Universitatis Petru Maior – Philologia*, 4 (2005), 141–144.
Denoon, Donald and Gavan McCormack *Multicultural Japan: Palaeolithic to Postmodern* (New York: Cambridge University Press, 1996).
de Rhodes, Alexandre, *Histoire Du Royaume Du Tonkin* (Paris: Éditions Kimé, 1999).
Descartes, René, Stanley Tweyman, Elizabeth Sanderson Haldane, and G. R. T. Ross, *Meditations on First Philosophy in Focus* (London and New York: Routledge, 1993).
de Silva Jayasuriya, Shihan, 'Identifying Africans in Asia: What's in a Name?', *African and Asian Studies*, 5.3/4 (2006), 275–303.
De Souza and R. Teotonio, *Essays in Goan History* (New Delhi: Concept Pub. Co., 1989).
Devi Chand, Joshi, *The Sāmaveda* (New Delhi: Munshiram Manoharlal Publishers, 1981).
de Vries, Jan, 'The Economic Crisis of the Seventeenth Century after Fifty Years', *Journal of Interdisciplinary History*, 40.2 (2009), 151–194.
Dewaele, Jean-Marc, Alex Housen, and Li Wei, *Bilingualism: Beyond Basic Principles* (Clevedon and Buffalo: Multilingual Matters, 2003).
de Zilva Wickremasinghe, Martino, *Catalogue of the Sinhalese Manuscripts in the British Museum* (London: British Museum, 1900).
Dimock, Edward C., *The Place of the Hidden Moon: Erotic Mysticism in the Vaiṣṣava-Sahajiyā Cult of Bengal* (Chicago: University of Chicago Press, 1966).
Doniger, Wendy, ed. and trans. *The Laws of Manu* (London and New York: Penguin, 1991).

Doniger, Wendy, *Purāna Perennis Reciprocity and Transformation in Hindu and Jaina Texts* (Albany, NY: State University of New York Press, 1993).

———, *The Hindus: An Alternative History* (New York: Penguin Press, 2009).

Drayton, Richard H., *Nature's Government: Science, Imperial Britain, and the 'Improvement' of the World* (New Haven: Yale University Press, 2000).

Dror, Olga and Keith Taylor, *Views of Seventeenth-Century Vietnam: Christoforo Borri on Cochinchina & Samuel Baron on Tonkin* (Ithaca, NY: Southeast Asia Program Publications, Southeast Asia Program, Cornell University, 2006).

Drury, Robert, *The Pleasant and Surprising Adventures of Robert Drury, during His Fifteen Years' Captivity on the Island of Madagascar* (London: Whittaker, Treacher, and Arnot, 1831).

Eaton, Richard M., *The Rise of Islam and the Bengal Frontier, 1204–1760* (Berkeley: University of California Press, 1993).

Echols, John M., 'Presidential Address. Dictionaries and Dictionary Making: Malay and Indonesian', *Journal of Asian Studies*, 38.1 (1978), 11–24.

Edelmann, Jonathan B., *Hindu Theology and Biology: The Bhāgavata Purāṇa and Contemporary Theory* (Oxford and New York: Oxford University Press, 2012).

Edney, Matthew H., *Mapping an Empire: The Geographical Construction of British India, 1765–1843* (Chicago and London: University of Chicago Press, 1997).

Elsner, Jas and Joan-Pau Rubiés, *Voyages and Visions: Towards a Cultural History of Travel* (London: Reaktion Books, 1999).

Ernst, Waltraud and Projit B. Mukharji, 'From History of Colonial Medicine to Plural Medicine in a Global Perspective: Recent Works on History of Medicine in Colonial/Postcolonial Contexts', *NTM Journal of the History of Science, Technology and Medicine*, 17.4 (2009), 447–458.

Errington, James J., *Linguistics in a Colonial World: A Story of Language, Meaning, and Power* (Malden, MA, and Oxford: Blackwell Publishing, 2008).

Errington, Joseph, 'Colonial Linguistics', *Annual Review of Anthropology*, 30 (2001), 19–39.

Farooqui, R., 'Medicine in the Deccan', *Bulletin of the Indian Institute of History of Medicine (Hyderabad)*, 16 (1986), 31–52.

Farrington, Anthony, Chang Hsiu-jung, Huang Fu-san, and Wu Mi-tsa, *The English Factory in Taiwan, 1670–1685* (Taipei: National Taiwan University, 1995).

Farrington, Anthony and Thīrawat Na Pō̜mphet, *The English Factory in Siam, 1612–1685* (London: British Library, 2007).

Feener, R. Michael, Patrick T. Daly, and Anthony Reid, *Mapping the Acehnese Past* (Leiden: KITLV Press, 2011).

Ferguson, Donald W., 'Robert Knox's Sinhalese Vocabulary', *Journal of the Ceylon Branch of the Royal Asiatic Society*, 14 (1896a), 155–200.

Ferguson, Donald W., *Captain Robert Knox: Contributions towards a Biography* (Croydon: 1896b).

Findlen, Paula, *Athanasius Kircher: The Last Man Who Knew Everything* (New York: Routledge, 2004).

———, *Early Modern Things: Objects and Their Histories, 1500–1800* (Abingdon and New York: Routledge, 2013).

Flamsteed, John, Margaret Flamsteed, and James Hodgson, *Atlas Coelestis, by the Late Reverend Mr John Flamsteed… [Edited by Margaret Flamsteed and James Hodgson.]* (London: 1729).

Flores, Jorge Manuel Costa Da Silva, *Re-Exploring the Links: History and Constructed Histories between Portugal and Sri Lanka* (Wiesbaden: Harrassowitz, 2007).

Forest, Alain, *Falcon: L'Imposteur De Siam: Commerce, Politique Et Religion Dans La Thaïlande Du XVIIe siècle* (Paris: Les Indes savantes, 2010).

Foster, William, 'An English Settlement in Madagascar in 1645–6', *English Historical Review*, 27.106 (1912), 239–250.

———, 'The Acquisition of St. Helena', *English Historical Review*, 34.135 (1919), 281–289.

———, *The English Factories in India, 1655–1660* (Oxford: Clarendon Press, 1921).

———, *The East India House, Its History and Associations* (London: John Lane, 1924).

———, *John Company* (London: John Lane, 1926).

Frank, Katherine, *Crusoe: Daniel Defoe, Robert Knox, and the Creation of a Myth* (New York: Pegasus Books, 2012).

Fryer, John and William Crooke, *A New Account of East India and Persia: Being Nine Years' Travels, 1672–1681* (London: Hakluyt Society, 1909).

Furber, Holden, *Rival Empires of Trade in the Orient, 1600–1800* (Minneapolis: University of Minnesota Press, 1976).

———, *Private Fortunes and Company Profits in the India Trade in the 18th Century* (Brookfield, VT: Variorum, 1997).

Gallop, Annabel Teh and Bernard Arps, *Golden Letters: Writing Traditions of Indonesia = Surat Emas: Budaya Tulis Di Indonesia* (London and Jakarta: British Library; Yayasan Lontar, 1991).

Games, Alison, *The Web of Empire: English Cosmopolitans in an Age of Expansion, 1560–1660* (New York and Oxford: Oxford University Press, 2010).

Gascoigne, John, *Science in the Service of Empire: Joseph Banks, the British State and the Uses of Science in the Age of Revolution* (Cambridge and New York: Cambridge University Press, 1998).

———, 'The Royal Society, Natural History and the Peoples of the "New World(s)", 1660–1800', *British Journal for the History of Science*, 42.155 (2009), 539–562.

Ghosh, Amitav, 'Of Fanas and Forecastles: The Indian Ocean and Some Lost Languages of the Age of Sail', *Economic and Political Weekly*, 43.25 (2008), 56.

———, *Sea of Poppies* (London: J. Murray, 2009).

———, *River of Smoke* (New York: Farrar, Straus and Giroux, 2011).

Gilroy, Paul, *The Black Atlantic: Modernity and Double Consciousness* (London and New York: Verso, 1993).

Godwin, Francis, *The Man in the Moone*, ed. William Poole (Peterborough, ON: Broadview Press, 2009).

Gole, Susan ed. *Maps of Mughal India: Drawn by Colonel Jean-Baptiste-Joseph Gentil, Agent for the French Government to the Court of Shuja-Ud-Daula at Faizabad, in 1770* (London and New York: Kegan Paul International, 1988).

Golinski, Jan, *Making Natural Knowledge Constructivism and the History of Science* (Chicago: University of Chicago Press, 2005).

Gommans, Jos J. L., *Mughal Warfare Indian Frontiers and Highroads to Empire, 1500–1700* (London and New York: Routledge, 2003).

Goody, Jack, *Renaissances: The One or the Many?* (Cambridge and New York: Cambridge University Press, 2010).

Goonetileke, H. A. I., 'Robert Knox in the Kandyan Kingdom, 1660–1679. A Bio-Bibliographical Commentary', *Sri Lankan Journal of the Humanities*, 1(1975), 81–151.

Gosse, Philip, *St. Helena, 1502–1938* (London: Cassell and Co., 1938).

Greenblatt, Stephen, *Renaissance Self-Fashioning: From more to Shakespeare* (Chicago: University of Chicago Press, 1980).

———, *Marvelous Possessions: The Wonder of the New World* (Chicago: University of Chicago Press, 1991).

———, *Cultural Mobility: A Manifesto* (Cambridge and New York: Cambridge University Press, 2010).

Greentree, Richard and Edward Williams Byron Nicholson, *Catalogue of Malay Manuscripts and Manuscripts Relating to the Malay Language in the Bodleian Library* (Oxford: Clarendon Press, 1910).

Grigg, David, 'The Worlds of Tea and Coffee: Patterns of Consumption', *Geojournal*, 57.4 (2002), 283–294.

Groeneboer, Kees, 'The Dutch Language in Maluku under the VOC', *Cakalele*, 5 (1994), 1–10.

Grove, Richard, *Green Imperialism: Colonial Expansion, Tropical Island Edens, and the Origins of Environmentalism, 1600–1860* (Cambridge and New York: Cambridge University Press, 1995).

Gruzinski, Serge, *The Mestizo Mind: The Intellectual Dynamics of Colonization and Globalization* (New York: Routledge, 2002).

Guerra, F., 'Drugs from the Indies and the Political Economy of the Sixteenth Century', *Analecta Médico-Histórica*, 1 (1966), 29–54.

Guha, Ranajit, 'Not at Home in Empire', *Critical Inquiry*, 23.3 (1997), 482–493.

———, *Dominance without Hegemony: History and Power in Colonial India* (Cambridge, MA: Harvard University Press, 1997).

Guha, Sumit, 'Serving the Barbarian to Preserve the Dharma', *Indian Economic and Social History Review*, 47.4 (2010), 497–525.

Gunn, Geoffrey C., *First Globalization: The Eurasian Exchange, 1500–1800* (Lanham, MD: Rowman & Littlefield, 2003).

Gupta, Ashin D., *Merchants of Maritime India, 1500–1800* (Aldershot and Brookfield, VT: Variorum; Ashgate, 1994).

Gupta, R. K. and S. R. Bakshi, *Rajasthan through the Ages* (New Delhi: Sarup & Sons, 2008).

Gupta, Ravi M. and Kenneth Russell Valpey, *The Bhāgavata Purāṇa: Sacred Text and Living Tradition* (New York: Columbia University Press, 2013).

Guterl, Matthew P., 'After Slavery: Asian Labor, the American South, and the Age of Emancipation', *Journal of World History*, 14.2 (2003), 209–241.

Haakonssen, Knud, *The Cambridge History of Eighteenth-Century Philosophy* (Cambridge and New York: Cambridge University Press, 2006).

Habib, Irfan, *An Atlas of the Mughal Empire: Political and Economic Maps with Detailed Notes, Bibliography, and Index* (Delhi and New York: Centre of Advanced Study in History, Aligarh Muslim University; Oxford University Press, 1982).

Hadfield, Andrew, *Literature, Travel, and Colonial Writing in the English Renaissance, 1545–1625* (Oxford and New York: Clarendon Press; Oxford University Press, 1998).

Hakluyt, Richard, *The Principall Nauigations, Voiages, and Discoueries of the English Nation* (imprinted at London: by George Bishop and Ralph Newberie . . . , 1589).

Hall, D. G. E., 'From Mergui to Singapore, 1686–1819: A Neglected Chapter in the Naval History of the Indian Ocean', *Journal of the Siam Society*, 5 (1953), 1–18.

Hall, Marie B., *Promoting Experimental Learning: Experiment and the Royal Society 1660–1727* (Cambridge and New York: Cambridge University Press, 1991).

———, *The Library and Archives of the Royal Society, 1660–1990* (London: Royal Society, 1992).

———, *Henry Oldenburg: Shaping the Royal Society* (Oxford and New York: Oxford University Press, 2002).

Halliday, Fred, ' "Orientalism" and Its Critics', *British Journal of Middle Eastern Studies*, 20.2 (1993), 145–163.

Haneda, Masashi, *Asian Port Cities, 1600–1800: Local and Foreign Cultural Interactions* (Singapore and Kyoto: NUS Press; in association with Kyoto University Press, 2009).

Hannaford, Ivan, *Race: The History of an Idea in the West* (Washington, DC, and Baltimore, MD: Woodrow Wilson Center Press; order from the Johns Hopkins University Press, 1996).

Harris, Joseph E., *The African Presence in Asia: Consequences of the East African Slave Trade* (Evanston: Northwestern University Press, 1971).

Harrison, Mark, *Climates and Constitutions: Health, Race, Environment and British Imperialism in India* (New Delhi: Oxford University Press, 2002).

Henderson, Felicity, 'Unpublished Material from the Memorandum Book of Robert Hooke, Guildhall Library MS 1758', *Notes and Records of the Royal Society of London*, 61.2 (2007), 129.

Henry, John, 'A Cambridge Platonist's Materialism: Henry More and the Concept of Soul', *Journal of the Warburg and Courtauld Institutes*, 49 (1986), 172–195.

Hickey, Raymond, *Legacies of Colonial English: Studies in Transported Dialects* (Cambridge and New York: Cambridge University Press, 2004).

Hodgen, Margaret T., *Early Anthropology in the Sixteenth and Seventeenth Centuries* (Philadelphia: University of Pennsylvania Press, 1964).

Hodges, Ian, 'Western Science in Siam: A Tale of Two Kings', *Osiris*, 13(1998), 80–95.

Home, R. W., 'The Royal Society and the Empire: The Colonial and Commonwealth Fellowship Part 1. 1731–1847', *Notes and Records*, 56.3 (2002), 307–332.

Hostetler, Laura, 'Qing Connections to the Early Modern World: Ethnography and Cartography in Eighteenth-Century China', *Modern Asian Studies*, 34.3 (2000), 623–662.

Hudson, D. Dennis, *Krishna's Mandala: Bhagavata Religion and Beyond* (New Delhi: Oxford University Press, 2010).

Hudson, Nicholas, 'From "Nation" to "Race": The Origin of Racial Classification in Eighteenth- Century Thought', *Eighteenth-Century Studies*, 29.3 (1996), 265.

Hunter, Michael, *The Correspondence of Robert Boyle, 1636–1691* (London and Brookfield, VT: Pickering & Chatto, 2001).

Hunter, Michael C. W., *Establishing the New Science: The Experience of the Early Royal Society* (Woodbridge, UK, and Wolfeboro, NH: Boydell Press, 1989).

Hutchinson, E. W., *Adventurers in Siam in the Seventeenth Century* (London: Royal Asiatic Society, 1940).

Hymavathi, P., *History of Āyurvēda in Āndhradēśa, A.D. 14th C.–17th C.* (Warangal: Bhargava Publishers, 1993).

Iliffe, Rob, 'The Masculine Birth of Time: Temporal Frameworks of Early Modern Natural Philosophy', *British Journal for the History of Science*, 33 (2000), 427–453.

Iliffe, Robert, 'Newton: The Making of a Politician', *Newton Project* (2013), 4 March 2014.

Inden, Ronald B., *Imagining India* (Oxford and Cambridge, MA: Basil Blackwell, 1990).

Irving, David R. M., 'Lully in Siam: Music and Diplomacy in French-Siamese Cultural Exchanges, 1680–1690', *Early Music*, (2012), 393–420.

Israel, Jonathan I., *Radical Enlightenment: Philosophy and the Making of Modernity, 1650–1750* (Oxford and New York: Oxford University Press, 2001).

Jackson, Anna and Amin Jaffer, *Encounters: The Meeting of Asia and Europe, 1500–1800* (London and New York: V & A; distributed in North America by Harry N. Abrams, 2004).

Jameson, J. Franklin, *Privateering and Piracy in the Colonial Period: Illustrative Documents* (New York: Macmillan Co., 1923).

Janisch, Hudson Ralph, *Extracts from the St. Helena Records* (St Helena: B. Grant, 1885).

Jardine, Lisa, *On a Grander Scale: The Outstanding Career of Sir Christopher Wren* (London: HarperCollins, 2002).

———, *The Curious Life of Robert Hooke: The Man Who Measured London* (New York: HarperCollins, 2004).

Jarow, Rick, *Tales for the Dying: The Death Narrative of the Bhāgavata-Purāṇa* (Albany: State University of New York Press, 2003).

Jayasuriya, Shihan de Silva and Jean-Pierre Angenot, *Uncovering the History of Africans in Asia* (Leiden and Boston: Brill, 2008).

Jayasuriya, Shihan de Silva and Richard Pankhurst, *The African Diaspora in the Indian Ocean* (Trenton, NJ: Africa World Press, 2003).

Jayawardena, Kumari, *Erasure of the Euro-Asian: Recovering Early Radicalism and Feminism in South Asia* (Colombo: Social Scientists' Association, 2007).

Jayawickrama, Sarojini, *Writing That Conquers: Re-Reading Knox's an Historical Relation of the Island Ceylon* (Colombo: Social Scientists' Association, 2004).

Jenner, Mark S. R. and Patrick Wallis, *Medicine and the Market in England and Its Colonies, c. 1450–c. 1850* (Basingstoke: Palgrave Macmillan, 2007).

Jernudd, B. H. and Michael J. Shapiro, *The Politics of Language Purism* (Berlin and New York: Mouton de Gruyter, 1989).

Jones, Russell, 'The First Indonesian Mission to London', *Indonesia Circle*, 28(1982), 9–19.

———, 'Malay Studies and the British: I: An Outline History of the Early XXth Century', *Archipel*, 28 (1984), 117–148.

Kamferer, B., *A Celebration of Demons: Exorcism and the Art of Healing in Sri Lanka* (Oxford: Smithsonian Institution Press, 1991).

Keay, John, *The Honourable Company: A History of the English East India Company* (New York: Macmillan, 1994).

Keevak, Michael, *The Pretended Asian: George Psalmanazar's Eighteenth-Century Formosan Hoax* (Detroit: Wayne State University Press, 2004).

Khan, Iqtidar A., 'The Middle Classes in the Mughal Empire', *Social Scientist*, 5.1 (1976), 28–49.

Khan, Shafa'at A., *The East India Trade in the 17th Century in its Political and Economic Aspects* (London: Oxford University Press, 1923).

——, *John Marshall in India: Notes and Observations in Bengal, 1668–1672* (London: Oxford University Press; H. Milford, 1927).

——, *Sources for the History of British India in the Seventeenth Century* (London and Totowa, NJ: Curzon Press; Rowman and Littlefield, 1975).

Kircher, Athanasius, *China Illustrata* (Muskogee, OK: Indian University Press, Bacone College, 1987).

Kleinen, J., *Lion and Dragon: Four Centuries of Dutch-Vietnamese Relations* (Amsterdam: Boom, 2008).

Knox, Robert, *An Historical Relation of Ceylon Together with Somewhat Concerning Severall Remarkeable Passages of My Life That Hath Hapned since My Deliverance Out of My Captivity* (Glasgow: James MacLehose and Sons, 1911).

——, *An Historical Relation of the Island Ceylon/Robert Knox* (Dehiwala: Tisara Prakasakayo, 1989).

Knox, Robert and Sumana D. Saparamadu, *An Historical Relation of Ceylon ... with an Introduction by S. D. Saparamadu ...* (Dehiwala: Ceylon Historical Journal, 1958).

Konvitz, Josef W., *Cartography in France, 1660–1848: Science, Engineering, and Statecraft* (Chicago: University of Chicago Press, 1987).

Kruijtzer, Gijs, *Xenophobia in Seventeenth-Century India* (Leiden: Leiden University Press, 2009).

Kumar, Deepak, *Science and the Raj: A Study of British India* (New Delhi and New York: Oxford University Press, 2006).

Lach, Donald F. and Edwin J. Van Kley, *Asia in the Making of Europe,* 2 Vols. in 5 Books (Chicago: University of Chicago Press, 1965–1970).

Lacouperie, Terrien d., *Formosa Notes on Mss., Languages and Races ...* (Hertford: S. Austin & Sons, 1887).

Lambert, David and Alan Lester, *Colonial Lives across the British Empire: Imperial Careering in the Long Nineteenth Century* (Cambridge and New York: Cambridge University Press, 2006).

Laslett, Peter, 'The English Revolution and Locke's "Two Treatises of Government" ', *Cambridge Historical Journal*, 12.1 (1956), 40–55.

Latif, Abdul, *The Muslim Mystic Movement in Bengal, 1301–1550* (Calcutta: K. P. Bagchi, 1993).

Latour, Bruno, *Science in Action: How to Follow Scientists and Engineers through Society* (Cambridge, MA: Harvard University Press, 1987).

——, *We Have Never Been Modern* (Cambridge, MA: Harvard University Press, 1993).

——, *On the Modern Cult of the Factish Gods* (Durham, NC, and London: Duke University Press, 2010).

Lehmann, E. A., *It Began at Tranquebar* (Madras: 1956).

Lepschy, Giulio, *History of Linguistics* (London: Longman, 1998).

Leur, J. C. v., *Indonesian Trade and Society: Essays in Asian Social and Economic History* (The Hague: W. Van Hoeve, 1955).

Lieberman, Victor B., 'Transcending East-West Dichotomies: State and Culture Formation in Six Ostensibly Disparate Areas', *Modern Asian Studies*, 31.3 (1997), 463–546.

——, *Strange Parallels: Southeast Asia in Global Context, c 800–1830. Volume 1* (New York: Cambridge University Press, 2003).

Lincoln, Bruce, 'Isaac Newton and Oriental Jones on Myth, Ancient History, and the Relative Prestige of Peoples', *History of Religions*, 42.1 (2002), 1–18.

Linebaugh, Peter and Marcus Rediker, *The Many-Headed Hydra: Sailors, Slaves, Commoners, and the Hidden History of the Revolutionary Atlantic* (Boston: Beacon Press, 2000).

Lineham, W., 'The Earliest Word-Lists and Dictionaries of the Malay Language', *Journal of the Royal Asiatic Society Malayan Branch*, 22 (1949), 183–187.

Livingstone, David N. and Charles W. J. Withers, *Geography and Enlightenment* (Chicago: University of Chicago Press, 1999).

Logan, Frenise A., 'The British East India Company and African Slavery in Benkulen, Sumatra, 1687–1792', *Journal of Negro History*, 41.4 (1956), 339–348.

Louis, William Roger, Nicholas P. Canny, Alaine M. Low, P. J. Marshall, A. N. Porter, Judith M. Brown, and Robin W. Winks eds., *The Oxford History of the British Empire* (Oxford: Oxford University Press, 1998).

Love, Henry D., *Vestiges of Old Madras, 1640–1800* (London: published for the Government of India, 1913).

Love, Ronald S., 'Monarchs, Merchants, and Missionaries in Early Modern Asia: The Missions Etrangeres in Siam, 1662–1684', *International History Review*, 21.1 (1999), 1–27.

Lux, David S. and Harold J. Cook, 'Closed Circles or Open Networks?: Communicating at a Distance during the Scientific Revolution', *History of Science*, 36.2 (1998), 179–211.

M, Baldwin, 'The Snakestone Experiments. An Early Modern Medical Debate', *Isis: An International Review Devoted to the History of Science and Its Cultural Influences*, 86.3 (1995), 394–418.

Machin, Roger and Shimizu Hirokazu, *Experiment and Return: Documents Concerning the Japan Voyage of the English East India Company, 1671–3* (Kyoto: Richard Cocks Society, 1978).

MacLean, Gerald M., *The Rise of Oriental Travel: English Visitors to the Ottoman Empire, 1580–1720* (New York: Palgrave Macmillan, 2004).

Madras Record Office, *Diary and Consultation Book, 1672–[1756]* (Madras: printed by the Superintendent, Government Press, 1910–1946).

———, Letters from Fort St George 1679-1765 (Madras: Government Press, 1915–1941).

———, *Letters to Fort St. George 1681–1765* (Madras: Superintendent, Government Press, 1916–1946).

Madras Tercentenary Celebration Committee, *The Madras Tercentenary Commemoration Volume* (London and Madras: Indian Branch, Oxford University Press, 1939).

Magalotti, Lorenzo, *Travels of Cosmo the Third, Grand Duke of Tuscany, through England during the Reign of King Charles the Second (1669)* (London: printed by J. Mawman, 1821).

Makepeace, Margaret, 'English Traders on the Guinea Coast, 1657–1668: An Analysis of the East India Company Archive', *History in Africa*, 16 (1989), 237–284.

Mancall, Peter C., *Bringing the World to Early Modern Europe: Travel Accounts and Their Audiences* (Leiden and Boston: Brill, 2007).

Manilal, K. S., *Botany and History of Hortus Malabaricus* (Rotterdam: Balkema, 1980).

———, *Van Rheede's Hortus Malabaribus = Malabar Garden: English Edition* (Thiruvanathapuram, India: University of Kerala, 2003).

Marcinkowski, M. I., 'The Iranian-Siamese Connection: An Iranian Community in the Thai Kingdom of Ayutthaya', *Iranian Studies*, 35 (2002), 23–46.

Marsden, William, *A Dictionary and Grammar of the Malayan Language* (Singapore and New York: Oxford University Press, 1984).

Marshall, P. J. and Glyndwr Williams, *The Great Map of Mankind: Perceptions of New Worlds in the Age of Enlightenment* (Cambridge, MA: Harvard University Press, 1982).

Massarella, Derek, *A World Elsewhere: Europe's Encounter with Japan in the Sixteenth and Seventeenth Centuries* (New Haven: Yale University Press, 1990).

———, 'Chinese, Tartars and "Thea" or a Tale of Two Companies: The English East India Company and Taiwan in the Late Seventeenth Century', *Journal of the Royal Asiatic Society*, 3.3 (1993), 393–426.

Mathew, Kuzhippalli Skaria, *Mariners, Merchants, and Oceans: Studies in Maritime History* (New Delhi: Manohar, 1995).

Mcclellan, J. E. and F. Regourd, 'The Colonial Machine: French Science and Colonization in the *Ancien Régime*', *Osiris*, 15 (2001), 31–50.

McPherson, Kenneth, *The Indian Ocean: A History of People and the Sea* (Delhi and New York: Oxford University Press, 1993).

Mee, R., 'An Old Malay Dictionary', *Journal of the Malayan Branch, Royal Asiatic Society*, 7 (1929), 316–326.

Meer, Sjoerd de and Frits Loomeijer, *De Schat Van Corpus Christi: VOC-Kaarten Boven Water* (Zutphen: De Walburg Pers, 2007).

Melliss, John C., *St. Helena: A Physical, Historical, and Topographical Description of the Island, Including Its Geology, Fauna, Flora and Meteorology* (London: L. Reeve & Co, 1875).

Mentz, Søren, *The English Gentleman Merchant at Work: Madras and the City of London 1660–1740* (Copenhagen: Museum Tusculanum Press, University of Copenhagen, 2005).

More, J. B. P., *Bagavadam Ou Bhagavata Purana Ouvrage Religieux Et Philosophique Indien Traduit Par Maridas Poullé De Pondichéry En 1769* (Paris: Irish, 2004).

Moreland, William H., *Relations of Golconda in the Early Seventeenth Century* (London: Hakluyt Soc., 1931).

Muhḥammad Rabīʿibn Muhḥammad Ibraāhīm, *The Ship of Sulaimān* (New York: Columbia University Press, 1972).

Mukherjee, Rila, 'The Story of Kasimbazar: Silk Merchants and Commerce in Eighteenth-Century India', *Review (Fernand Braudel Center)*, 17.4 (1994), 499–554.

Nah, Alice M. '(Re)Mapping Indigenous Race/Place in Postcolonial Peninsular Malaysia', *Geografiska Annaler, Series B, Human Geography*, 88 (2006), 285–297.

Nair, Savithri P., 'Native Collecting and Natural Knowledge (1798–1832): Raja Serfoji II of Tanjore as a Centre of Calculation', *Journal of the Royal Asiatic Society*, 15.3 (2005), 279–302.

———, *Raja Serfoji II: Science, Medicine and Enlightenment in Tanjore* (London and New York: Routledge, 2012).

Narayana Rao, Velcheru, David Dean Shulman, and Sanjay Subrahmanyam, *Symbols of Substance, Court and State in Nāyaka Period Tamilnadu* (Delhi: Oxford University Press, 1992).

Neild-Basu, Susan, 'The *Dubashes* of Madras', *Modern Asian Studies*, 18.1 (1984), 1–31.

Neill, Stephen, *A History of Christianity in India, 1707–1858* (Cambridge: Cambridge University Press, 1985).

Newbury, C. W., *Patrons, Clients, and Empire: Chieftaincy and Over-Rule in Asia, Africa, and the Pacific* (Oxford: Oxford University Press, 2003).

Newton, Isaac, *The Correspondence of Isaac Newton, II: 1676–1687*, ed. H. W. Turnball (Cambridge: Cambridge University Press, 1960).

Nightingale, C. H., 'Before Race Mattered: Geographies of the Color Line in Early Colonial Madras and New York', *American Historical Review*, 113.1 (2008), 48–71.

Noltie, Henry J., *Robert Wight and the Botanical Drawings of Rungiah & Govindoo* (Edinburgh: Royal Botanic Garden, 2007).

North, Michael, *Artistic and Cultural Exchanges between Europe and Asia, 1400–1900: Rethinking Markets, Workshops and Collections* (Farnham and Burlington, VT: Ashgate, 2010).

Novetzke, Christian L., 'The Brahmin Double: The Brahminical Construction of Anti-Brahminism and Anti-Caste Sentiment in the Religious Cultures of Precolonial Maharashtra', *South Asian History and Culture*, 2.2 (2011), 232–252.

Obeyesekere, Gananath, 'The Ritual Drama of the Sanni Demons: Collective Representations of Disease in Ceylon', *Comparative Studies in Society and History* 11.2 (1969), 174–216.

Ogborn, Miles, *Indian Ink Script and Print in the Making of the English East India Company* (Chicago: University of Chicago Press, 2007).

O'Hanlon, Rosalind, 'Performance in a World of Paper: Puranic Histories and Social Communication in Early Modern India', *Past and Present*, 219.1 (2013), 87–126.

O'Hanlon, Rosalind and Christopher Minkowski, 'What Makes People Who They Are? Pandit Networks and the Problem of Livelihoods in Early Modern Western India', *Indian Economic and Social History Review*, 45.3 (2008), 381–416.

O'Hanlon, Rosalind and David A. Washbrook, *Religious Cultures in Early Modern India: New Perspectives* (New Delhi: Routledge, 2011).

Oliver, William, 'A Letter from Dr William Oliver, Physician and Fellow of the Royal Society, to Mr James Petiver, F. R. S. Concerning the Jesuits Bark', *Philosophical Transactions (1683–1775)*, 24 (1705), 1596.

Ovington, J., *A Voyage to Surat: In the Year 1689* (New Delhi: Asian Educational Services, 1994).

Pagden, Anthony, *European Encounters with the New World: From Renaissance to Romanticism* (New Haven: Yale University Press, 1993).

Papac, R. J., 'Origins of Cancer Therapy', *Yale Journal of Biology and Medicine*, 74.6 (2001).

Parker, Geoffrey, *Global Crisis: War, Climate Change and Catastrophe in the Seventeenth Century* (New Haven: Yale University Press, 2013).

Parker, Geoffrey and Lesley M. Smith, *The General Crisis of the Seventeenth Century* (London; New York: Routledge, 1997).

Parkinson, John and Thomas Cotes, *Theatrum Botanicum* (London: printed by Tho. Cotes, 1640).

Parthasarathi, Prasannan, *Why Europe Grew Rich and Asia Did Not: Global Economic Divergence, 1600–1850* (Cambridge and New York: Cambridge University Press, 2011).

Pearson, M. N., *The World of the Indian Ocean, 1500–1800: Studies in Economic, Social, and Cultural History* (Burlington, VT: Ashgate, 2005).

Pearson, Mike P., 'Close Encounters of the Worst Kind: Malagasy Resistance and Colonial Disasters in Southern Madagascar', *World Archaeology*, 28.3 (1997), 393.

Perpustakaan Negara Malaysia, *Tradisi Penulisan Manuskrip Melayu* (Kuala Lumpur: Perpustakaan Negara Malaysia, 1997).

Perti, R. K., *Descriptive List of Acquired Documents (1356–1790 AD). Volume III* (New Delhi: National Archives of India, 1992).

Peterson, Willard J., *The Cambridge History of China. Vol. 9., Pt. 1, the Ch'ing Empire to 1800* (Cambridge: Cambridge University Press, 2002).

Pickstone, John V., *Ways of Knowing: A New History of Science, Technology, and Medicine* (Chicago: University of Chicago Press, 2001).

Pinto, Jeanette, *Slavery in Portuguese India, 1510–1842* (Bombay: Himalaya Publishing House, 1992).

Platt, Virginia B., 'The East India Company and the Madagascar Slave Trade', *William and Mary Quarterly: A Magazine of Early American History*, 26.4 (1969), 548–577.

Pollock, Sheldon I., *Forms of Knowledge in Early Modern Asia: Explorations in the Intellectual History of India and Tibet, 1500–1800* (Durham, NC: Duke University Press, 2011).

Prakash, Gyan, *Another Reason: Science and the Imagination of Modern India* (Princeton, NJ: Princeton University Press, 1999).

Prakash, Om, 'The European Trading Companies and the Merchants of Bengal 1650–1725', *Indian Economic and Social History Review*, 1.3 (1964), 37–63.

———, *The New Cambridge History of India. II.5, European Commercial Enterprise in Pre-Colonial India* (Cambridge: CUP, 2008).

Pratt, Mary Louise, *Imperial Eyes: Travel Writing and Transculturation* (London and New York: Routledge, 1992).

Pratt, Peter and Montague Paske-Smith, *History of Japan Compiled from the Records of the English East India Company at the Instance of the Court of Directors* (Rep: Curzon Press; London: Frobe, 1972).

Proctor, Robert and Londa L. Schiebinger, *Agnotology: The Making and Unmaking of Ignorance* (Stanford, CA: Stanford University Press, 2008).

Raaflaub, Kurt A. and Richard J. A. Talbert, *Geography and Ethnography: Perspectives of the World in Pre-Modern Societies* (Oxford: Blackwell, 2010).

Raffles Museum and Library, *Catalogue of the Raffles Library, Singapore, 1900* (Singapore: American Mission Press, 1905).

Raj, Kapil, 'Colonial Encounters and the Forging of New Knowledge and National Identities: Great Britain and India, 1760–1850', *Osiris*, 15 (2000), 119–134.

———, *Relocating Modern Science: Circulation and the Construction of Knowledge in South Asia and Europe, 1650–1900* (Basingstoke: Palgrave Macmillan, 2007).

———, 'The Historical Anatomy of a Contact Zone', *Indian Economic and Social History Review*, 48.1 (2011), 55–82.

Raj, Selva J. and Corinne G. Dempsey, *Sacred Play Ritual Levity and Humor in South Asian Religions* (Albany: State University of New York Press, 2010).

Ram, H. Y. M., 'On the English Edition of Van Rheede's *Hortus Malabaricus* by K. S. Manilal (2003)', *Current Science*, 89.10 (2005), 1672–1680.

Rama Rao B., 'Medico Historical Information from Non-Medical Sources', *Bulletin of the Indian Institute of History of Medicine (Hyderabad)*, 16 (1986), 1–9.

Raman, Bhavani, *Document Raj: Scribes and Writing under Early Colonial Rule in Madras, 1771–1860* (Chicago: The University of Chicago Press, 2012).

Rao, Rama, 'Medical Allusions in the Inscriptions of AP', *Bulletin of the Indian Institute for the History of Medicine*, 5(1975), 198–206.

Rao, Velcheru N., 'Coconut and Honey: Sanskrit and Telugu in Medieval Andhra', *Social Scientist*, 23.10/12 (1995), 24–40.

Ray, Aniruddha, *The Merchant and the State: The French in India, 1666–1739* (New Delhi: Munshiram Manoharlal Publishers, 2004).

Reddy, D. V. S., 'John Fryer – A Traveller of XVII Century', *Bulletin of the Department of the History of Medicine*, 2 (1964).

Reid, Antony, 'Merchant Princes and Magic Mediators', *Indonesia and the Malay World*, 36.105 (2008), 253–267.

———, 'The Structure of Cities in Southeast Asia, Fifteenth to Seventeenth Centuries', *Journal of Southeast Asian Studies*, 11.2 (1980), 235–250.

———, 'From Betel-Chewing to Tobacco-Smoking in Indonesia', *Journal of Asian Studies*, 44.3 (1985), 529–547.

———, Southeast Asia in the Age of Commerce, 2 vols., II: Expansion and Crisis (New Haven: Yale University Press, 1993).

———, 'The Seventeenth-Century Crisis in Southeast Asia', *Modern Asian Studies*, (1990).

Reid, Anthony and Jennifer Brewster, *Slavery, Bondage, and Dependency in Southeast Asia* (New York: St Martin's Press, 1983).

Reimers, E., 'Raja Singh and His British Captives', *Journal of the Ceylon Branch of the Royal Asiatic Society*, 30(1925), 13–36.

Reyes, R. A., 'Botany and Zoology in the Late Seventeenth-Century Philippines: The Work of Georg Josef Camel SJ (1661–1706)', *Archives of Natural History*, 36.2 (2009), 262–276.

Richards, John F., 'The Hyderabad Karnatik, 1687–1707', *Modern Asian Studies*, 9.2 (1975a), 241–260.

———, *Mughal Administration in Golconda* (Oxford: Clarendon Press, 1975b).

———, 'The Seventeenth-Century Crisis in South Asia', *Modern Asian Studies*, 24.04 (1990).

———, *The Mughal Empire* (Cambridge and New York: Cambridge University Press, 1993).

———, 'Early Modern India and World History', *Journal of World History*, 8.2 (1997), 197–209.

Riddle, John M., 'Ancient and Medieval Chemotherapy for Cancer', *Isis*, 76.283 (1985).

Robins, Nick, *The Corporation That Changed the World: How the East India Company Shaped the Modern Multinational* (London and Ann Arbor, MI: Pluto Press, 2006).

Robinson, Henry W. and Walter Adams, *The Diary of Robert Hooke, 1672–1680* (London: Wykeham Publications, 1968).

Rocher, Ludo, *Ezourvedam: A French Veda of the Eighteenth Century* (Amsterdam and Philadelphia: J. Benjamins, 1984).

Rostenberg, Leona, *The Library of Robert Hooke: The Scientific Book Trade of Restoration England* (Santa Monica, CA: Modoc Press, 1989).

Roxburgh, William, *Plants of the Coast of Coromandel, Selected from Drawings and Descriptions Presented to the Hon. Court of Directors of the East India Company* (London: printed by W. Bulmer and Co. for G. Nicol, bookseller, 1795).

Royle, Stephen A., *The Company's Island: St Helena, Company Colonies and the Colonial Endeavour* (London and New York: I. B. Tauris, 2007).

Rubiés, Joan Pau, 'Hugo Grotius's Dissertation on the Origin of the American Peoples and the use of Comparative Methods', *Journal of the History of Ideas*, 52.2 (1991), 221–244.

———, 'Instructions for Travellers: Teaching the Eye to See', *History and Anthropology*, 9.2–3 (1996), 139–190.

———, *Travel and Ethnology in the Renaissance: South India through European Eyes, 1250–1625* (Cambridge and New York: Cambridge University Press, 2000).

Sadiq Ali, Shanti, *The African Dispersal in the Deccan: From Medieval to Modern Times* (New Delhi: Orient Longman, 1996).

Safier, Neil, *Measuring the New World Enlightenment Science and South America* (Chicago: University of Chicago Press, 2008).

Said, Edward W., *Culture and Imperialism* (New York: Knopf; distributed by Random House, 1993).

Sainsbury, Ethel B. and William Foster, *A Calendar of the Court Minutes, Etc. of the East India Company, 1635–1639* (Oxford: Clarendon Press, 1907).

Sainsbury, Ethel B. and William Foster, *A Calendar of the Court Minutes, Etc., of the East India Company, 1668–1670* (Oxford: Clarendon Press, 1929).

Sainsbury, William Noel, *Calendar of State Papers: Colonial Series* (Burlington, ON: Tanner Ritchie Pub, 2005).

Salmon, Claudine and Trọng Hiêp Tạ, 'Les Récits De Voyage Chinois Comme Source Pour l'Étude Du Viêtnam (Xe–XXe Siècle)', *Bulletin De l'École Française d'Extrême-Orient*, 83 (1996), 67–87.

Salmon, Vivian, *The Study of Language in 17th-Century England* (Amsterdam and Philadelphia: J. Benjamins Pub. Co., 1988).

Sarkar, Jadunath, *Shivaji and His Times* (London and New York: Longmans, Green and Co., 1920).

Schiebinger, Londa, *Plants and Empire: Colonial Bioprospecting in the Atlantic World* (Cambridge, MA: Harvard University Press, 2004).

Schiebinger, Londa and Claudia Swan, *Colonial Botany: Science, Commerce, and Politics in the Early Modern World* (Philadelphia: University of Pennsylvania Press, 2005).

Schottenhammer, Angela, *Trade and Transfer across the East Asian 'Mediterranean'* (Wiesbaden: Harrassowitz, 2005).

Schweig, Graham M., *Dance of Divine Love: The Rāsa Līlā of Krishna from the Bhāgavata Purāṇa, India's Classic Sacred Love Story* (Princeton, NJ: Princeton University Press, 2005).

Scott, David, *Formations of Ritual: Colonial and Anthropological Discourses on the Sinhala Yaktovil* (Minneapolis: University of Minnesota Press, 1994).

Screech, Timon, 'Pictures (the Most Part Bawdy): The Anglo-Japanese Painting Trade in the Early 1600s', *Art Bulletin*, 87.1 (2005), 50–72.

Seeley, John R., *The Expansion of England* (Chicago: University of Chicago Press, 1971).

Sen, Amrita, 'Early Liaisons: East India Company, Native Wives and Inscription in the Seventeenth Century ', *South Asian Review*, 33.2 (2012), 101–116.

Sen, Amrita and Jyotsna G. Singh, 'Classifying the Natives in Early Modern Ethnographies: Henry Lord's *A Display of Two Foreign Sects in the East Indies* (1630)', *Journeys*, 14.2 (2013), 69–84.

Sen, Sudipta, *Empire of Free Trade: The East India Company and Making of the Colonial Marketplace* (Philadelphia: University of Pennsylvania Press, 1998).

Sen, Surendranath, *Indian Travels of Thevenot and Careri* (New Delhi: 1949).

Seuren, Pieter A. M., *Western Linguistics: An Historical Introduction* (Oxford and Malden, MA: Blackwell Publishers, 1998).

Shapin, Steven, *A Social History of Truth: Civility and Science in Seventeenth-Century England* (Chicago: University of Chicago Press, 1994).

Sharma, Virendra N., 'Sawai Jai Singh's Hindu Astronomers', *Indian Journal of History of Science*, 23 (1993), 131–155.

Shaw, William A., *Calendar of Treasury Books, VII: 1681–1685* (London: His Majesty's Stationery Office, 1916).

Shellabear, W. G., *Some Old Malay Manuscripts* (Singapore: American Mission Press, 1901).

———, 'Hikayat Seri Rama: Introduction to the Text of the Ms. in the Bodleian Library at Oxford', *Journal of the Straits Branch, Royal Asiatic Society*, 70 (1917).

Silva, K. M. D., *A History of Sri Lanka* (London: Vijitha Yapa Publications, 2005).

Singh, Jyotsna G., *Colonial Narratives/Cultural Dialogues: Discoveries of India in the Language of Colonialism* (London and New York: Routledge, 1996).

Sivasundaram, Sujit, *Islanded: Britain, Sri Lanka, and the Bounds of an Indian Ocean Colony* (Chicago: The University of Chicago Press, 2013).

———, 'Sciences and the Global: On Methods, Questions, and Theory', *Isis: An International Review Devoted to the History of Science and Its Cultural Influences*, 101.1 (2010), 146–158.

———, 'Tales of the Land: British Geography and Kandyan Resistance in Sri Lanka, c. 1803–1850', *Modern Asian Studies*, 41.5 (2007), 925–965.

Smith, Brian K., 'Eaters, Food, and Social Hierarchy in Ancient India: A Dietary Guide to a Revolution of Values', *Journal of the American Academy of Religion*, 58.2 (1990), 177–205.

Smith, Pamela H. and Paula Findlen, *Merchants & Marvels: Commerce, Science and Art in Early Modern Europe* (New York: Routledge, 2002).

Smithies, Michael, 'Tachard's Last Appearance in Ayutthaya, 1699', *Journal of the Royal Asiatic Society*, 12 (2002), 67–78.

Smithies, Michael, *Witnesses to a Revolution: Siam, 1688: Twelve Key Texts Describing the Events and Consequences of the Phetracha Coup d'État and the Withdrawal of French Forces from the Country* (Bangkok: Siam Society, 2004).

Souza, G. B., 'Opium and the Company: Maritime Trade and Imperial Finances on Java, 1684–1796', *Modern Asian Studies*, 43.1 (2009), 113–133.

Srivastava, Ashok K., *Mughal Painting: An Interplay of Indigenous and Foreign Traditions* (New Delhi: Munshiram Manoharlal Publishers, 2000).

Stagl, Justin, *A History of Curiosity: The Theory of Travel, 1550–1800* (Australia: Harwood Academic Publishers, 1995).

Steadman-Jones, Richard, *Colonialism and Grammatical Representation: John Gilchrist and the Analysis of the 'Hindustani' Language in the Late Eighteenth and Early Nineteenth Centuries* (Oxford and Boston: Blackwell Publishing, 2007).

Steensgaard, Niels, *The Asian Trade Revolution of the Seventeenth Century: The East India Companies and the Decline of the Caravan Trade* (Chicago: University of Chicago Press, 1974).

Steingass, Francis Joseph and Francis Johnson, *A Comprehensive Persian-English Dictionary*... (Beirut: Librarie du Liban, 1970).

Stephens, Thomas, *Father Thomas Stephens' Kristapurāṇa: Purāṇa 1 & II*, ed. Nelson M. Falcao (Bengaluru and Vasai, India: Kristu Jyoti Publications; distributors: Jeevan Darshan Kendra, 2012).

Stern, Philip J., 'Politics and Ideology in the Early East India Company-State: The Case of St Helena, 1673–1709', *Journal of Imperial and Commonwealth History*, 35.1 (2007), 1–23.

———, 'History and Historiography of the English East India Company: Past, Present, and Future!', *History Compass*, 7.4 (2009), 1146–1180.

———, *The Company-State: Corporate Sovereignty and the Early Modern Foundations of the British Empire in India* (New York: Oxford University Press, 2011).

Stoler, Ann L., *Carnal Knowledge and Imperial Power: Race and the Intimate in Colonial Rule* (Berkeley: University of California Press, 2002).

Stolte, Carolien, *Philip Angel's Deex-Autaers: Vaiṣṇava Mythology from Manuscript to Book Market in the Context of the Dutch East India Company, c. 1600–1672*, Dutch Sources on South Asia, c. 1600–1825, 5 vols (New Delhi: Manohar, 2012).

Strathern, A., 'Sri Lanka in the Long Early Modern Period: Its Place in a Comparative Theory of Second Millennium Eurasian History', *Modern Asian Studies*, 43.4 (2009), 815–869.

Subrahmanyam, Sanjay, 'Connected Histories: Notes towards a Reconfiguration of Early Modern Eurasia', *Modern Asian Studies*, 31.3 (1997), 735–762.

———, *From the Tagus to the Ganges* (New Delhi: Oxford University Press, 2005a).

———, *Mughals and Franks: Explorations in Connected History* (New Delhi and Oxford: Oxford University Press, 2005b).

———, *Three Ways To Be Alien: Travails and Encounters in the Early Modern World* (Waltham, MA: Brandeis University Press, 2011).

Sweet, James H., *Domingos Álvares, African Healing, and the Intellectual History of the Atlantic World* (Chapel Hill: University of North Carolina Press, 2011).

Tachot, Louise B. and Serge Gruzinski, *Passeurs Culturels: Mécanismes De Métissage* (Paris and Marne-la-Vallée, France: Ed. de la Maison des sciences de l'homme; Presses universitaires de Marne-La-Vallée, 2002).

Tagliacozzo, Eric and Wen-Chin Chang, *Chinese Circulations: Capital, Commodities, and Networks in Southeast Asia* (Durham, NC: Duke University Press, 2011).

Taylor, E. G. R., 'Robert Hooke and the Cartographical Projects of the Late Seventeenth Century (1666–1696)', *Geographical Journal*, 90.6 (1937), 529–540.

Taylor, Nora A., *Painters in Hanoi: An Ethnography of Vietnamese Art* (Honolulu: University of Hawaii Press, 2004).

Teltscher, Kate, *India Inscribed: European and British Writing on India, 1600–1800* (Delhi: Oxford University Press, 1995).

———, ' "Maidenly and Well Nigh Effeminate": Constructions of Hindu Masculinity and Religion in Seventeenth-Century English Texts', *Postcolonial Studies: Culture, Politics, Economy*, 3.2 (2000), 159–170.

Temple, Richard Carnac, *The Diaries of Streynsham Master, 1675–1680, and Other Contemporary Papers Relating Thereto* (London: J. Murray, 1911).

———, *The Papers of Thomas Bowrey 1669–1713* (London: Hakluyt Society, 1927).

———, *New Light on the Mysterious Tragedy of the Worcester, 1704–1705; an Episode in the Long Struggle between England and Scotland for Union as Great Britain, 1603–1707* (London: E. Benn Ltd, 1930).

Thiong'o, Ngũgĩ., *Decolonising the Mind: The Politics of Language in African Literature* (London; Portsmouth, NH: J. Currey; Heinemann, 1986).

Tindall, Gillian, *City of Gold: The Biography of Bombay* (London: Temple Smith, 1982).

Tooley, Marian J., 'Bodin and the Mediaeval Theory of Climate', *Speculum, A Journal of Mediaeval Studies*, 28.1 (1953), 64–83.

Trocki, Carl A., 'Opium and the Beginnings of Chinese Capitalism in Southeast Asia', *Journal of Southeast Asian Studies*, 33.2 (2002), 297–314.

Tsing, Anna L., *Friction: An Ethnography of Global Connection* (Princeton, NJ: Princeton University Press, 2005).

Turnbull, David, *Masons, Tricksters and Cartographers Comparative Studies in the Sociology of Scientific and Indigenous Knowledge* (London and New York: Routledge, 2000).

Veen, Ernst v. and Leonard Blussé, *Rivalry and Conflict: European Traders and Asian Trading Networks in the 16th and 17th Centuries* (Leiden: CNWS Publications, 2005).

Veevers, D., ' "The Company as Their Lords and the Deputy as a Great Rajah": Imperial Expansion and the English East India Company on the West Coast of Sumatra, 1685–1730', *Journal of Imperial and Commonwealth History*, 41.5 (2013), 687–709.

Vermeir, Koen and Dániel Margócsy. 'States of Secrecy: An Introduction', *The British Journal for the History of Science*, 45.2 (2012), 153–164.

Vink, Markus, 'The World's Oldest Trade: Dutch Slavery and Slave Trade in the Indian Ocean in the Seventeenth Century', *Journal of World History*, 14.2 (2003), 131–177.

Vissière, Arnold, 'Ngan-Nan Ki Yeou: Relation d'Un Voyage Au Tonkin', *Bulletin De Georgraphie Historique Et Descriptive*, 4(1889), 70–86.

Volwahsen, Andreas, *Cosmic Architecture in India: The Astronomical Monuments of Maharaja Jai Singh II* (Munich and London: Prestel, 2001).

Vos, F. H. d., 'Extracts Relating to Ceylon from the Dag-Register, Batavia, AD. 1678–1680', *Journal of the Ceylon Branch of the Royal Asiatic Society*, 16 (1918), 181–198.

Wainwright, Mary D. and Noel Matthews, *A Guide to Western Manuscripts and Documents in the British Isles Relating to South and South East Asia* (London and New York: Oxford University Press, 1965).

Wanley, Humphrey and P. L. Heyworth, *Letters of Humfrey Wanley: Palaeographer, Anglo-Saxonist, Librarian, 1672–1726: With an Appendix of Documents* (Oxford: Oxford University Press, 1989).

Ward, Kerry, *Networks of Empire: Forced Migration in the Dutch East India Company* (Cambridge and New York: Cambridge University Press, 2009).

Watson, Ian B., *Foundation for Empire: English Private Trade in India, 1659–1760* (New Delhi: Vikas, 1980).

Watson, James L., *Asian and African Systems of Slavery* (Berkeley: University of California Press, 1980).

Weber, Jacques, *Le Monde Créole: Peuplement, Sociétés Et Condition Humaine, XVIIe–XXe Siècles* (Paris: Les Indes savants, 2005).

Wells, Wilfrid H., *Perspective in Early Chinese Painting* (London: E. Goldston, Ltd, 1935).

Wescoat, James L. and Joachim Wolschke-Bulmahn, *Mughal Gardens: Sources, Places, Representations, and Prospects* (Washington, DC: Dumbarton Oaks Research Library and Collection, 1996).

Wheeler, James T., *India under British Rule from the Foundation of the East India Company* (London: Macmillan and Co., 1886).

White, H., 'Knox in Its Literary Aspect', *Journal of the Ceylon Branch of the Royal Asiatic Society*, 13 (1893), 23–35.

Wickremesekera, Channa, *Kandy at War: Indigenous Military Resistance to European Expansion in Sri Lanka 1594–1818* (Colombo: Vijitha Yapa Publications, 2004).

Wild, Antony, *The East India Company Trade and Conquest from 1600* (London: HarperCollins, 2000).

Wilson, C. R., *Old Fort William in Bengal: A Selection of Official Documents Dealing with Its History* (London: J. Murray, 1906).

Wilson, H. H., 'Notes on the Sabhá Parva of the Mahábhárata, Illustrative of Some Ancient Usages and Articles of Traffic of the Hindus', *Journal of the Royal Asiatic Society of Great Britain and Ireland*, 7.1 (1843), 137–144.

Wilson, Kathleen, *A New Imperial History: Culture, Identity, and Modernity in Britain and the Empire, 1660–1840* (Cambridge and New York: Cambridge University Press, 2004).

Winterbottom, Anna, 'Hybrid Knowledge', *History Workshop Journal*, 71.1 (2011), 267–273.

———, 'Of the China Root: A Case Study of the Early Modern Circulation of *Materia Medica*', *Social History of Medicine*, 28.1 (2015), 22–44.

Withers, Charles W., *Encyclopaedism, Modernism and the Classification of Geographical Knowledge*, Transactions of the Institute of British Geographers, 21 (1996), 275–298.

Wright, C. E., *Fontes Harleiani: A Study of the Sources of the Harleian Collection of Manuscripts Preserved in the Department of Manuscripts in the British Museum* (London: British Museum, 1972).

Wujastyk, Dagmar and Frederick M. Smith, *Modern and Global Ayurveda: Pluralism and Paradigms* (Albany: State University of New York Press, 2008).

Yothers, Brian, 'Global Captivities: Robert Knox's An Historical Relation of Ceylon and the New England Captivity Narrative Tradition', *Journeys* 8.1 (2007), 41–55.

Young, Robert, *Colonial Desire: Hybridity in Theory, Culture, and Race* (London and New York: Routledge, 1995).

Yu, In-son, *Law and Society in Seventeenth and Eighteenth Century Vietnam* (Seoul: Asiatic Research Center, Korea University, 1990).

Yule, Henry ed., *The Diary of William Hedges, Esq* (London: printed for the Hakluyt Society, 1887).

Yule, Henry, A. C. Burnell, and William Crooke, *Hobson-Jobson: A Glossary of Colloquial Anglo-Indian Words and Phrases, and of Kindred Terms, Etymological, Historical, Geographical and Discursive* (London: J. Murray, 1903).

Zeng, Minhao, 'Subaltern Cosmopolitanism: Concept and Approaches', *SORE the Sociological Review*, 62.1 (2014), 137–148.

Županov, Ines G., 'Conversion Historiography in South Asia: Alternative Indian Christian Counter-Histories in Eighteenth Century Goa', *Medieval History Journal*, 12.2 (2009), 303–325.

———, *Disputed Mission: Jesuit Experiments and Brahmanical Knowledge in Seventeenth-Century India* (New Delhi and New York: Oxford University Press, 1999).

———, *Missionary Tropics: The Catholic Frontier in India, 16th–17th Centuries* (Ann Arbor: University of Michigan Press, 2005).

Zysk, K. G., 'The Evolution of Anatomical Knowledge in Ancient India, with Special Reference to Cross-Cultural Influences', *American Oriental Society*, 106.4 (1986), 687–705.

Index

CPSIA information can be obtained
at www.ICGtesting.com
Printed in the USA
LVHW080800271118
598240LV00014B/468/P

9 781137 380197